Python Machine Learning By Example

Example

Fourth Edition

Unlock machine learning best practices with real-world use cases

Yuxi (Hayden) Liu

Packt and this book are not officially connected with Python. This book is an effort from the Python community of experts to help more developers.

Python Machine Learning By Example

Fourth Edition

Senior Publishing Product Manager: Bhavesh Amin

Acquisition Editor – Peer Reviews: Tejas Mhasvekar

Project Editor: Janice Gonsalves

Content Development Editor: Rebecca Youé

Copy Editor: Safis Editing

Technical Editor: Karan Sonawane

Proofreader: Safis Editing

Indexer: Rekha Nair

Presentation Designer: Ajay Patule

Developer Relations Marketing Executive: Monika Sangwan

First published: May 2017

Second edition: Feb 2019

Third edition: Oct 2020

Fourth edition: July 2024

Production reference: 2300525

Published by Packt Publishing Ltd.

Grosvenor House

11 St Paul's Square

Birmingham

B3 1RB, UK.

ISBN: 978-1-83508-562-2

www.packt.com

Contributors

About the author

Yuxi (Hayden) Liu was a Machine Learning Software Engineer at Google. With a wealth of experience from his tenure as a machine learning scientist, he has applied his machine learning expertise across data-driven domains, computational advertising, cybersecurity, and information retrieval.

He is the author of a series of influential machine learning books and an education enthusiast. His debut book, also the first edition of *Python Machine Learning by Example*, ranked the #1 bestseller in Amazon and has been translated into many different languages.

About the reviewers

Aayush Mudgal is an expert in deep learning, an investor and advisor to multiple ML startups, and also a 40 Under 40 Data Scientist. Currently he is a Staff Machine Learning Engineer and is responsible for developing personalization models used by millions of users around the world. His expertise is in large-scale recommendation systems, personalization, and ad marketplaces.

Before entering the industry, Aayush conducted research on intelligent tutoring systems, developing data-driven feedback to aid students in learning computer programming. He holds a Masters in Computer Science from Columbia University and a Bachelor of Technology in Computer Science from Indian Institute of Technology Kanpur. He has served as a program committee member for many top-tier conferences including Neurips, ICML, ICLR, Recsys, and WWW and has also been a regular hackathon judge.

Juantomás Garcia is Chief Envisioning Officer at SNGULAR. Since joining SNGULAR in 2018, Juantomás has leveraged his extensive experience to harness the potential of new technologies and implement them across the company's solutions and services.

Juantomás is a Google Developer Expert for Cloud and Machine Learning, co-author of the open software book *La Pastilla Roja* and creator of "AbadIA", the artificial intelligence platform built to solve the popular Spanish game *La Abadía del Crimen*. He's an expert on free software technologies and has been a speaker at more than 300 international industry events.

He studied IT engineering at the Universidad Politécnica de Madrid and plays an active role as a tech contributor and mentor to various academic organizations and startups. He regularly organizes Machine Learning Spain and GDG cloud Madrid meetups, is a mentor at Google Launchpad for entrepreneurs and Women Techmakers, and is also technical director of the Smart Global Forest Universidad de Valladolid and SNGULAR chair. He is also an advisor on the Google Developers Advisory Board.

With all my love to my wife, Elisa, and my sons, Nico and Oli.

Roman Tezikov is a Machine Learning Engineer and enthusiast with over seven years of experience in Computer vision, NLP, LLM, and MLOps. As a co-creator of the ML-REPA community, he has organized workshops and meetups focused on ML reproducibility and pipeline automation. His recent interests lie in AI for fitness at Zing Coach.

Roman also served as a technical reviewer for Sebastian Raschka's book, *Machine Learning with PyTorch and Scikit-Learn*, ensuring the comprehensive coverage and quality of content on topics from machine learning fundamentals to deep learning innovations.

Big thanks to my wife, Anna, for her support and encouragement. I would also like to thank Packt for the opportunity to be a part of this journey.

Join our book's Discord space

Join our community's Discord space for discussions with the authors and other readers:

https://packt.link/yuxi

Table of Contents

Preface

The fourth edition of *Python Machine Learning by Example* is a comprehensive guide for beginners, and experienced **Machine Learning** (**ML**) practitioners who want to learn more advanced techniques like multimodal modeling. This edition emphasizes best practices, providing invaluable insights for ML engineers, data scientists, and analysts.

Explore advanced techniques, including two new chapters on NLP transformers with BERT and GPT and multimodal computer vision models with PyTorch and Hugging Face. You'll learn key modeling techniques using practical examples, such as predicting stock prices and creating an image search engine.

This book navigates through complex challenges, bridging the gap between theoretical understanding and practical application. Elevate your ML expertise, tackle intricate problems, and unlock the potential of advanced techniques in machine learning with this authoritative guide.

Who this book is for

If you're a machine learning enthusiast, data analyst, or data engineer who's highly passionate about machine learning and you want to begin working on ML assignments, this book is for you. Prior knowledge of Python coding is assumed and basic familiarity with statistical concepts will be beneficial, although this is not necessary.

What this book covers

Chapter 1, Getting Started with Machine Learning and Python, will kick off your Python machine learning journey. It starts with what machine learning is, why we need it, and its evolution over the last few decades. It then discusses typical machine learning tasks and explores several essential techniques of working with data and working with models, in a practical and fun way. You will also set up the software and tools needed for examples and projects in the upcoming chapters.

Chapter 2, Building a Movie Recommendation Engine with Naïve Bayes, focuses on classification, specifically binary classification and Naïve Bayes. The goal of the chapter is to build a movie recommendation system. You will learn the fundamental concepts of classification, and about Naïve Bayes, a simple yet powerful algorithm. It also demonstrates how to fine-tune a model, which is an important skill for every data science or machine learning practitioner to learn.

Chapter 3, *Predicting Online Ad Click-Through with Tree-Based Algorithms*, introduces and explains in depth tree-based algorithms (including decision trees, random forests, and boosted trees) throughout the course of solving the advertising click-through rate problem. You will explore decision trees from the root to the leaves, and work on implementations of tree models from scratch, using scikit-learn and XGBoost. Feature importance, feature selection, and ensemble will be covered alongside.

Chapter 4, *Predicting Online Ad Click-Through with Logistic Regression*, is a continuation of the ad click-through prediction project, with a focus on a very scalable classification model—logistic regression. You will explore how logistic regression works, and how to work with large datasets. The chapter also covers categorical variable encoding, L1 and L2 regularization, feature selection, online learning, and stochastic gradient descent.

Chapter 5, *Predicting Stock Prices with Regression Algorithms*, focuses on several popular regression algorithms, including linear regression, regression tree and regression forest. It will encourage you to utilize them to tackle a billion (or trillion) dollar problem—stock price prediction. You will practice solving regression problems using scikit-learn and TensorFlow.

Chapter 6, *Predicting Stock Prices with Artificial Neural Networks*, introduces and explains in depth neural network models. It covers the building blocks of neural networks, and important concepts such as activation functions, feedforward, and backpropagation. You will start by building the simplest neural network and go deeper by adding more layers to it. We will implement neural networks from scratch, use TensorFlow and PyTorch, and train a neural network to predict stock prices.

Chapter 7, *Mining the 20 Newsgroups Dataset with Text Analysis Techniques*, will start the second step of your learning journey—unsupervised learning. It explores a natural language processing problem—exploring newsgroups data. You will gain hands-on experience in working with text data, especially how to convert words and phrases into machine-readable values and how to clean up words with little meaning. You will also visualize text data using a dimension reduction technique called t-SNE. Finally, you will learn how to represent words with embedding vectors.

Chapter 8, *Discovering Underlying Topics in the Newsgroups Dataset with Clustering and Topic Modeling*, talks about identifying different groups of observations from data in an unsupervised manner. You will cluster the newsgroups data using the K-means algorithm, and detect topics using non-negative matrix factorization and latent Dirichlet allocation. You will be amused by how many interesting themes you are able to mine from the 20 newsgroups dataset!

Chapter 9, *Recognizing Faces with Support Vector Machine*, continues the journey of supervised learning and classification. Specifically, it focuses on multiclass classification and support vector machine classifiers. It discusses how the support vector machine algorithm searches for a decision boundary in order to separate data from different classes. You will implement the algorithm with scikit-learn, and apply it to solve various real-life problems including face recognition.

Chapter 10, *Machine Learning Best Practices*, aims to fully prove your learning and get you ready for real-world projects. It includes 21 best practices to follow throughout the entire machine learning workflow.

Chapter 11, Categorizing Images of Clothing with Convolutional Neural Networks, is about using **Convolutional Neural Networks (CNNs)**, a very powerful modern machine learning model, to classify images of clothing. It covers the building blocks and architecture of CNNs, and their implementation using PyTorch. After exploring the data of clothing images, you will develop CNN models to categorize the images into ten classes, and utilize data augmentation and transfer learning techniques to boost the classifier.

Chapter 12, Making Predictions with Sequences using Recurrent Neural Networks, starts by defining sequential learning, and exploring how **Recurrent Neural Networks (RNNs)** are well suited for it. You will learn about various types of RNNs and their common applications. You will implement RNNs with PyTorch, and apply them to solve three interesting sequential learning problems: sentiment analysis on IMDb movie reviews, stock price forecasting, and text auto-generation.

Chapter 13, Advancing Language Understanding and Generation with the Transformer Models, dives into the Transformer neural network, designed for sequential learning. It focuses on crucial parts of the input sequence and captures long-range relationships better than RNNs. You will explore two cutting-edge Transformer models BERT and GPT, and use them for sentiment analysis and text generation, which surpass the performance achieved in the previous chapter.

Chapter 14, Building an Image Search Engine Using CLIP: A Multimodal Approach, explores a multimodal model, CLIP, that merges visual and textual data. This powerful model can understand connections between images and text. You will dive into its architecture and how it learns, then build an image search engine. Finally, you will cap it all off with a zero-shot image classification project, pushing the boundaries of what this model can do.

Chapter 15, Making Decisions in Complex Environments with Reinforcement Learning, is about learning from experience, and interacting with the environment. After exploring the fundamentals of reinforcement learning, you will explore the FrozenLake environment with a simple dynamic programming algorithm. You will learn about Monte Carlo learning and use it for value approximation and control. You will also develop temporal difference algorithms and use Q-learning to solve the taxi problem.

To get the most out of this book

A basic foundation of Python knowledge, basic machine learning algorithms, and some basic Python libraries, such as NumPy and pandas, is assumed in order to create smart cognitive actions for your projects.

Download the example code files

The code bundle for the book is hosted on GitHub at `https://github.com/packtjaniceg/Python-Machine-Learning-by-Example-Fourth-Edition/`. We also have other code bundles from our rich catalog of books and videos available at `https://github.com/PacktPublishing/`. Check them out!

Download the color images

We also provide a PDF file that has color images of the screenshots/diagrams used in this book. You can download it here: `https://packt.link/gbp/9781835085622`.

Conventions used

There are a number of text conventions used throughout this book.

`Code in text`: Indicates code words in text, database table names, folder names, filenames, file extensions, pathnames, dummy URLs, user input, and Twitter (X) handles. Here is an example: "Besides the rating matrix `data`, we also record the `movie ID` to column index mapping."

A block of REPL code is set as follows:

```
>>> smoothing = 1
>>> likelihood = get_likelihood(X_train, label_indices, smoothing)
>>> print('Likelihood:\n', likelihood)
```

Any output from the code will appear like this:

```
Likelihood:
 {'Y': array([0.4, 0.6, 0.4]), 'N': array([0.33333333, 0.33333333,
0.66666667])}
```

Bold: Indicates a new term, an important word, or words that you see onscreen. For example, words in menus or dialog boxes appear in the text like this. Here is an example: "There are three types of classification based on the possibility of class output—**binary**, **multiclass**, and **multi-label classification**."

Warnings or important notes appear like this.

Tips and tricks appear like this.

Get in touch

Feedback from our readers is always welcome.

General feedback: Email feedback@packtpub.com and mention the book's title in the subject of your message. If you have questions about any aspect of this book, please email us at questions@packtpub.com.

Errata: Although we have taken every care to ensure the accuracy of our content, mistakes do happen. If you have found a mistake in this book, we would be grateful if you reported this to us. Please visit http://www.packtpub.com/submit-errata, click **Submit Errata**, and fill in the form.

Piracy: If you come across any illegal copies of our works in any form on the internet, we would be grateful if you would provide us with the location address or website name. Please contact us at copyright@packtpub.com with a link to the material.

If you are interested in becoming an author: If there is a topic that you have expertise in and you are interested in either writing or contributing to a book, please visit http://authors.packtpub.com.

Share your thoughts

Once you've read *Python Machine Learning By Example - Fourth Edition*, we'd love to hear your thoughts! Scan the QR code below to go straight to the Amazon review page for this book and share your feedback.

https://packt.link/r/1835085628

Your review is important to us and the tech community and will help us make sure we're delivering excellent quality content.

Making the Most Out of This Book — Get to Know Your Free Benefits

Unlock exclusive free benefits that come with your purchase, thoughtfully crafted to supercharge your learning journey and help you learn without limits.

UNLOCK NOW

Note: Have your purchase invoice ready before you begin.

https://www.packtpub.com/unlock/9781835085622

Figure 0.1: Next-Gen Reader, AI Assistant (Beta), and Free PDF access

Enhanced reading experience with our Next-gen Reader:

⟳ Multi-device progress sync: Learn from any device with seamless progress sync.

📖 Highlighting and Notetaking: Turn your reading into lasting knowledge.

🔖 Bookmarking: Revisit your most important learnings anytime.

☀ Dark mode: Focus with minimal eye strain by switching to dark or sepia modes.

Learn smarter using our AI assistant (Beta):

✦ Summarize it: Summarize key sections or an entire chapter.

✦ AI code explainers: In Packt Reader, click the "Explain" button above each code block for AI-powered code explanations.

Note: AI Assistant is part of next-gen Packt Reader and is still in beta.

Learn anytime, anywhere:

📄📘: Access your content offline with DRM-free PDF and ePub versions—compatible with your favorite e-readers.

Unlock Your Book's Exclusive Benefits

Your copy of this book comes with the following exclusive benefits:

- ☁ Next-gen Packt Reader
- ✦ AI assistant (beta)
- 🗐 DRM-free PDF/ePub downloads

Use the following guide to unlock them if you haven't already. The process takes just a few minutes and needs to be done only once.

How to unlock these benefits in three easy steps

Step 1

Have your purchase invoice for this book ready, as you'll need it in *Step 3*. If you received a physical invoice, scan it on your phone and have it ready as either a PDF, JPG, or PNG.

For more help on finding your invoice, visit `https://www.packtpub.com/unlock-benefits/help`.

 Note: Bought this book directly from Packt? You don't need an invoice. After completing Step 2, you can jump straight to your exclusive content.

Step 2

Scan the following QR code or visit
`https://www.packtpub.com/unlock/9781835085622`:

Step 3

Sign in to your Packt account or create a new one for free. Once you're logged in, upload your invoice. It can be in PDF, PNG, or JPG format and must be no larger than 10 MB. Follow the rest of the instructions on the screen to complete the process.

Need help?

If you get stuck and need help, visit `https://www.packtpub.com/unlock-benefits/help` for a detailed FAQ on how to find your invoices and more. The following QR code will take you to the help page directly:

 Note: If you are still facing issues, reach out to `customercare@packt.com`.

1

Getting Started with Machine Learning and Python

The concept of **artificial intelligence** (AI) outpacing human knowledge is often referred to as the "technological singularity." Some predictions in the AI research community and other fields suggest that the singularity could happen within the next 30 years. Regardless of its time horizon, one thing is clear: the rise of AI highlights the growing importance of analytical and machine learning skills. Mastering these disciplines equips us to not only understand and interact with increasingly complex AI systems but also actively participate in shaping their development and application, ensuring they benefit humanity.

In this chapter, we will kick off our machine learning journey with the basic, yet important, concepts of machine learning. We will start with what machine learning is all about, why we need it, and its evolution over a few decades. We will then discuss typical machine learning tasks and explore several essential techniques to work with data and models.

At the end of the chapter, we will set up the software for Python, the most popular language for machine learning and data science, and the libraries and tools that are required for this book.

We will go into detail on the following topics:

- An introduction to machine learning
- Knowing the prerequisites
- Getting started with three types of machine learning
- Digging into the core of machine learning
- Data preprocessing and feature engineering
- Combining models
- Installing software and setting up

An introduction to machine learning

In this first section, we will kick off our machine learning journey with a brief introduction to machine learning, why we need it, how it differs from automation, and how it improves our lives.

Machine learning is a term that was coined around 1960, consisting of two words—**machine**, which corresponds to a computer, robot, or other device, and **learning**, which refers to an activity intended to acquire or discover event patterns, which we humans are good at. Interesting examples include facial recognition, language translation, responding to emails, making data-driven business decisions, and creating various types of content. You will see many more examples throughout this book.

Understanding why we need machine learning

Why do we need machine learning, and why do we want a machine to learn the same way as a human? We can look at it from three main perspectives: maintenance, risk mitigation, and enhanced performance.

First and foremost, of course, computers and robots can work 24/7 and don't get tired. Machines cost a lot less in the long run. Also, for sophisticated problems that involve a variety of huge datasets or complex calculations, it's much more justifiable, not to mention intelligent, to let computers do all the work. Machines driven by algorithms that are designed by humans can learn latent rules and inherent patterns, enabling them to carry out tasks effectively.

Learning machines are better suited than humans for tasks that are routine, repetitive, or tedious. Beyond that, automation by machine learning can mitigate risks caused by fatigue or inattention. Self-driving cars, as shown in *Figure 1.1*, are a great example: a vehicle is capable of navigating by sensing its environment and making decisions without human input. Another example is the use of robotic arms in production lines, which are capable of causing a significant reduction in injuries and costs.

Figure 1.1: An example of a self-driving car

Let's assume that humans don't fatigue or we have the resources to hire enough shift workers; would machine learning still have a place? Of course it would! There are many cases, reported and unreported, where machines perform comparably, or even better, than domain experts. As algorithms are designed to learn from the ground truth and the best thought-out decisions made by human experts, machines can perform just as well as experts.

In reality, even the best expert makes mistakes. Machines can minimize the chance of making wrong decisions by utilizing collective intelligence from individual experts. A major study that identified that machines are better than doctors at diagnosing certain types of cancer is proof of this philosophy (https://www.nature.com/articles/d41586-020-00847-2). **AlphaGo** (https://deepmind.com/research/case-studies/alphago-the-story-so-far) is probably the best-known example of machines beating humans—an AI program created by DeepMind defeated Lee Sedol, a world champion Go player, in a five-game Go match.

Also, it's much more scalable to deploy learning machines than to train individuals to become experts, from the perspective of economic and social barriers. Current diagnostic devices can achieve a level of performance similar to that of qualified doctors. We can distribute thousands of diagnostic devices across the globe within a week, but it's almost impossible to recruit and assign the same number of qualified doctors within the same period.

You may argue against this: what if we have sufficient resources and the capacity to hire the best domain experts and later aggregate their opinions—would machine learning still have a place? Probably not (at least right now)—learning machines might not perform better than the joint efforts of the most intelligent humans. However, individuals equipped with learning machines can outperform the best group of experts. This is an emerging concept called **AI-based assistance** or **AI plus human intelligence**, which advocates for combining the efforts of machines and humans. It provides support, guidance, or solutions to users. And more importantly, it can adapt and learn from user interactions to improve performance over time.

We can summarize the previous statement in the following inequality:

human + machine learning → most intelligent tireless human ≥ machine learning > human

 Artificial intelligence-generated content (AIGC) is one of the recent breakthroughs. It uses AI technologies to create or assist in creating various types of content, such as articles, product descriptions, music, images, and videos.

A medical operation involving robots is one great example of human and machine learning synergy. *Figure 1.2* shows robotic arms in an operation room alongside a surgeon:

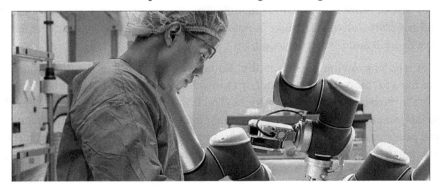

Figure 1.2: AI-assisted surgery

Differentiating between machine learning and automation

So does machine learning simply equate to automation that involves the programming and execution of human-crafted or human-curated rule sets? A popular myth says that machine learning is the same as automation because it performs instructive and repetitive tasks and thinks no further. If the answer to that question is *yes*, why can't we just hire many software programmers and continue programming new rules or extending old rules?

One reason is that defining, maintaining, and updating rules becomes increasingly expensive over time. The number of possible patterns for an activity or event could be enormous, and therefore, exhausting all enumeration isn't practically feasible. It gets even more challenging when it comes to events that are dynamic, ever-changing, or evolve in real time. It's much easier and more efficient to develop learning algorithms that command computers to learn, extract patterns, and figure things out themselves from abundant data.

The difference between machine learning and traditional programming can be seen in *Figure 1.3*:

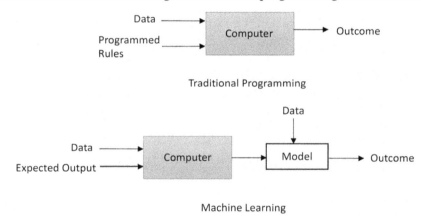

Figure 1.3: Machine learning versus traditional programming

In traditional programming, the computer follows a set of predefined rules to process the input data and produce the outcome. In machine learning, the computer tries to mimic human thinking. It interacts with the input data, expected output, and the environment, and it derives patterns that are represented by one or more mathematical models. The models are then used to interact with future input data and generate outcomes. Unlike in automation, the computer in a machine learning setting doesn't receive explicit and instructive coding.

The volume of data is growing exponentially. Nowadays, the floods of textual, audio, image, and video data are hard to fathom. The **Internet of Things (IoT)** is a recent development of a new kind of internet, which interconnects everyday devices. The IoT will bring data from household appliances and autonomous cars to the fore. This trend is likely to continue, and we will have more data that is generated and processed. Besides the quantity, the quality of data available has kept increasing in the past few years, partly due to cheaper storage. This has empowered the evolution of machine learning algorithms and data-driven solutions.

Machine learning applications

Jack Ma, co-founder of the e-commerce company Alibaba, explained in a speech in 2018 that IT was the focus of the past 20 years, but for the next 30 years, we will be in the age of **data technology (DT)** (`https://www.alizila.com/jack-ma-dont-fear-smarter-computers/`). During the age of IT, companies grew larger and stronger thanks to computer software and infrastructure. Now that businesses in most industries have already gathered enormous amounts of data, it's presently the right time to exploit DT to unlock insights, derive patterns, and boost new business growth. Broadly speaking, machine learning technologies enable businesses to better understand customer behavior, engage with customers, and optimize operations management.

As for us individuals, machine learning technologies are already making our lives better every day. One application of machine learning with which we're all familiar is spam email filtering. Another is online advertising, where adverts are served automatically based on information advertisers have collected about us. Stay tuned for the next few chapters, where you will learn how to develop algorithms to solve these two problems and more.

A search engine is an application of machine learning we can't imagine living without. It involves information retrieval, which parses what we look for, queries the related top records, and applies contextual ranking and personalized ranking, which sorts pages by topical relevance and user preference. E-commerce and media companies have been at the forefront of employing recommendation systems, which help customers find products, services, and articles faster.

The application of machine learning is boundless, and we just keep hearing new examples everyday: credit card fraud detection, presidential election prediction, instant speech translation, robo advisors, AI-generated art, chatbots for customer support, and medical or legal advice provided by generative AI technologies—you name it!

In the 1983 *War Games* movie, a computer made life-and-death decisions that could have resulted in World War III. As far as we know, technology wasn't able to pull off such feats at the time. However, in 1997, the Deep Blue supercomputer did manage to beat a world chess champion (`https://en.wikipedia.org/wiki/Deep_Blue_(chess_computer)`). In 2005, a Stanford self-driving car drove by itself for more than 130 miles in a desert (`https://en.wikipedia.org/wiki/DARPA_Grand_Challenge_(2005)`). In 2007, the car of another team drove through regular urban traffic for more than 60 miles (`https://en.wikipedia.org/wiki/DARPA_Grand_Challenge_(2007)`). In 2011, the Watson computer won a quiz against human opponents (`https://en.wikipedia.org/wiki/Watson_(computer)`). As mentioned earlier, the AlphaGo program beat one of the best Go players in the world in 2016. As of 2023, ChatGPT has been widely used across various industries, such as customer support, content generation, market research, and training and education (`https://www.forbes.com/sites/bernardmarr/2023/05/30/10-amazing-real-world-examples-of-how-companies-are-using-chatgpt-in-2023`).

If we assume that computer hardware is the limiting factor, then we can try to extrapolate into the future. A famous American inventor and futurist, Ray Kurzweil, did just that, predicting in 2017 that we can expect AI to gain human-level intelligence around 2029 (`https://aibusiness.com/responsible-ai/ray-kurzweil-predicts-that-the-singularity-will-take-place-by-2045`). What's next?

Can't wait to launch your own machine learning journey? Let's start with the prerequisites and the basic types of machine learning.

Knowing the prerequisites

Machine learning mimicking human intelligence is a subfield of AI—a field of computer science concerned with creating systems. Software engineering is another field in computer science. Generally, we can label Python programming as a type of software engineering. Machine learning is also closely related to linear algebra, probability theory, statistics, and mathematical optimization. We usually build machine learning models based on statistics, probability theory, and linear algebra, and then optimize the models using mathematical optimization.

Most of you reading this book should have a good, or at least sufficient, command of Python programming. Those who aren't feeling confident about mathematical knowledge might be wondering how much time should be spent learning or brushing up on the aforementioned subjects. Don't panic; we will get machine learning to work for us without going into any deep mathematical details in this book. It just requires some basic 101 knowledge of probability theory and linear algebra, which helps us to understand the mechanics of machine learning techniques and algorithms. And it gets easier, as we will build models both from scratch and with popular packages in Python, a language we like and are familiar with.

For those who want to learn or brush up on probability theory and linear algebra, feel free to search for basic probability theory and basic linear algebra. There are a lot of resources available online, for example, `https://people.ucsc.edu/~abrsvn/intro_prob_1.pdf`, the online course *Introduction to Probability* by Harvard University (`https://pll.harvard.edu/course/introduction-probability-edx`) regarding *probability 101*, and the following paper regarding basic linear algebra: `http://www.maths.gla.ac.uk/~ajb/dvi-ps/2w-notes.pdf`.

Those who want to study machine learning systematically can enroll in computer science, AI, and, more recently, data science and AI master's programs. There are also various data science boot camps. However, the selection for boot camps is usually stricter, as they're more job-oriented and the program duration is often short, ranging from 4 to 10 weeks. Another option is free **Massive Open Online Courses** (**MOOCs**), such as Andrew Ng's popular course on machine learning. Last but not least, industry blogs and websites are great resources for us to keep up with the latest developments.

Machine learning is not only a skill but also a bit of a sport. We can compete in several machine learning competitions, such as Kaggle (`www.kaggle.com`)—sometimes for decent cash prizes, sometimes for joy, but most of the time to play to our strengths. However, to win these competitions, we may need to utilize certain techniques, which are only useful in the context of competitions and not in the context of trying to solve a business problem. That's right—the **no free lunch** theorem (`https://en.wikipedia.org/wiki/No_free_lunch_theorem`) applies here. In the context of machine learning, this theorem suggests that no single algorithm is universally superior across all possible datasets and problem domains.

Next, we'll take a look at the three types of machine learning.

Getting started with three types of machine learning

A machine learning system is fed with input data—this can be numerical, textual, visual, or audiovisual. The system usually has an output—this can be a floating-point number, for instance, the acceleration of a self-driving car, or an integer representing a category (also called a **class**), for example, a cat or tiger from image recognition.

The main task of machine learning is to explore and construct algorithms that can learn from historical data and make predictions on new input data. For a data-driven solution, we need to define (or have it defined by an algorithm) an evaluation function called a **loss** or **cost function**, which measures how well the models learn. In this setup, we create an optimization problem with the goal of learning most efficiently and effectively.

Depending on the nature of the learning data, machine learning tasks can be broadly classified into the following three categories:

- **Unsupervised learning**: When the learning data only contains indicative signals without any description attached (we call this **unlabeled data**), it's up to us to find the structure of the data underneath, discover hidden information, or determine how to describe the data. Unsupervised learning can be used to detect anomalies, such as fraud or defective equipment, or group customers with similar online behaviors for a marketing campaign. Data visualization that makes data more digestible, as well as dimensionality reduction that distills relevant information from noisy data, are also in the family of unsupervised learning.

- **Supervised learning**: When learning data comes with a description, targets, or desired output besides indicative signals (we call this **labeled data**), the learning goal is to find a general rule that maps input to output. The learned rule is then used to label new data with unknown output. The labels are usually provided by event-logging systems or evaluated by human experts. Also, if feasible, they may be produced by human raters, through crowd-sourcing, for instance.

  Supervised learning is commonly used in daily applications, such as face and speech recognition, product or movie recommendations, sales forecasting, and spam email detection.

- **Reinforcement learning**: Learning data provides feedback so that a system adapts to dynamic conditions in order to ultimately achieve a certain goal. The system evaluates its performance based on the feedback responses and reacts accordingly. The best-known instances include robotics for industrial automation, self-driving cars, and the chess master AlphaGo. The key difference between reinforcement learning and supervised learning is the interaction with the environment.

The following diagram depicts the types of machine learning tasks:

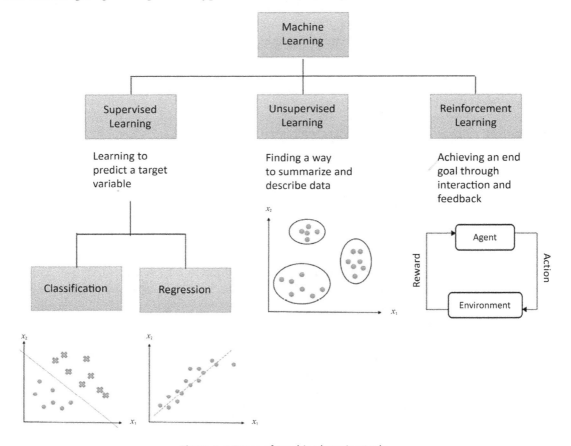

Figure 1.4: Types of machine learning tasks

As shown in the diagram, we can further subdivide supervised learning into regression and classification. **Regression** trains on and predicts continuous-valued responses, for example, predicting house prices, while **classification** attempts to find the appropriate class label, such as analyzing a positive/ negative sentiment and predicting a loan default.

If not all learning samples are labeled, but some are, we have **semi-supervised learning**. This makes use of unlabeled data (typically a large amount) for training, besides a small amount of labeled data. Semi-supervised learning is applied in cases where it is expensive to acquire a fully labeled dataset and more practical to label a small subset. For example, it often requires skilled experts to label hyperspectral remote sensing images, while acquiring unlabeled data is relatively easy.

Feeling a little bit confused by the abstract concepts? Don't worry. We will encounter many concrete examples of these types of machine learning tasks later in this book. For example, in *Chapter 2, Building a Movie Recommendation Engine with Naïve Bayes*, we will dive into supervised learning classification and its popular algorithms and applications. Similarly, in *Chapter 5, Predicting Stock Prices with Regression Algorithms*, we will explore supervised learning regression.

We will focus on unsupervised techniques and algorithms in *Chapter 8, Discovering Underlying Topics in the Newsgroups Dataset with Clustering and Topic Modeling*. Last but not least, the third machine learning task, reinforcement learning, will be covered in *Chapter 15, Making Decisions in Complex Environments with Reinforcement Learning*.

Besides categorizing machine learning based on the learning task, we can categorize it chronologically.

A brief history of the development of machine learning algorithms

In fact, we have a whole zoo of machine learning algorithms that have experienced varying popularity over time. We can roughly categorize them into five main approaches: logic-based learning, statistical learning, artificial neural networks, genetic algorithms, and deep learning.

The **logic-based** systems were the first to be dominant. They used basic rules specified by human experts, and with these rules, systems tried to reason using formal logic, background knowledge, and hypotheses.

Statistical learning theory attempts to find a function to formalize the relationships between variables. In the mid-1980s, **artificial neural networks** (ANNs) came to the fore. ANNs imitate animal brains and consist of interconnected neurons that are also an imitation of biological neurons. They try to model complex relationships between input and output values and capture patterns in data. ANNs were superseded by statistical learning systems in the 1990s.

Genetic algorithms (GA) were popular in the 1990s. They mimic the biological process of evolution and try to find optimal solutions, using methods such as mutation and crossover.

In the 2000s, ensemble learning methods gained attention, which combined multiple models to improve performance.

We have seen **deep learning** become a dominant force since the late 2010s. The term deep learning was coined around 2006 and refers to deep neural networks with many layers. The breakthrough in deep learning was the result of the integration and utilization of **Graphical Processing Units (GPUs)**, which massively speed up computation. The availability of large datasets also fuels the deep learning revolution.

GPUs were originally developed to render video games and are very good in parallel matrix and vector algebra. It's believed that deep learning resembles the way humans learn. Therefore, it may be able to deliver on the promise of sentient machines. Of course, in this book, we will dig deep into deep learning in *Chapter 11, Categorizing Images of Clothing with Convolutional Neural Networks*, and *Chapter 12, Making Predictions with Sequences Using Recurrent Neural Networks*, after touching on it in *Chapter 6, Predicting Stock Prices with Artificial Neural Networks*.

Machine learning algorithms continue to evolve rapidly, with ongoing research in areas including **transfer learning, generative models**, and reinforcement learning, which are the backbone of AIGC. We will explore the latest developments in *Chapter 13, Advancing Language Understanding and Generation with the Transformer Models*, and *Chapter 14, Building an Image Search Engine Using CLIP: a Multimodal Approach*.

Some of us may have heard of **Moore's law**—an empirical observation claiming that computer hardware improves exponentially with time. The law was first formulated by Gordon Moore, the co-founder of Intel, in 1965. According to the law, the number of transistors on a chip should double every two years. In the following diagram, you can see that the law holds up nicely (the size of the bubbles corresponds to the average transistor count in GPUs):

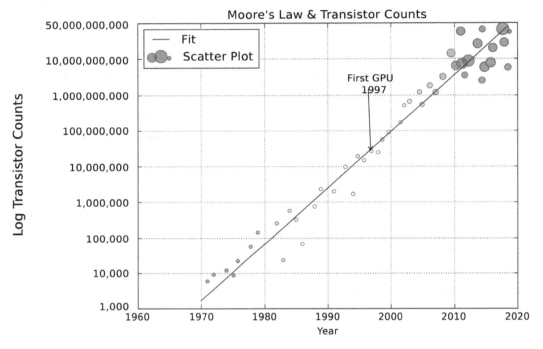

Figure 1.5: Transistor counts over the past decades

The consensus seems to be that Moore's law should continue to be valid for a couple of decades. This gives some credibility to Ray Kurzweil's predictions of achieving true machine intelligence by 2029.

Digging into the core of machine learning

After discussing the categorization of machine learning algorithms, we are now going to dig into the core of machine learning—generalizing with data, the different levels of generalization, as well as the approaches to attain the right level of generalization.

Generalizing with data

The good thing about data is that there's a lot of it in the world. The bad thing is that it's hard to process this data. The challenge stems from the diversity and noisiness of it. We humans usually process data coming into our ears and eyes. These inputs are transformed into electrical or chemical signals. On a very basic level, computers and robots also work with electrical signals.

These electrical signals are then translated into ones and zeros. However, we program in Python in this book, and on that level, normally we represent the data either as numbers or texts. However, text isn't very convenient, so we need to transform this into numerical values.

Especially in the context of supervised learning, we have a scenario similar to studying for an exam. We have a set of practice questions and the actual exams. We should be able to answer exam questions without being exposed to identical questions beforehand. This is called **generalization**—we learn something from our practice questions and, hopefully, can apply this knowledge to other similar questions. In machine learning, these practice questions are called **training sets** or **training samples**. This is where the machine learning models derive patterns from. And the actual exams are **testing sets** or **testing samples**. They are where the models are eventually applied. Learning effectiveness is measured by the compatibility of the learning models and the testing.

Sometimes, between practice questions and actual exams, we have mock exams to assess how well we will do in actual exams and to aid revision. These mock exams are known as **validation sets** or **validation samples** in machine learning. They help us to verify how well the models will perform in a simulated setting, and then we fine-tune the models accordingly in order to achieve greater accuracy.

An old-fashioned programmer would talk to a business analyst or other expert, and then implement a tax rule that adds a certain value multiplied by another corresponding value, for instance. In a machine learning setting, we can give the computer a bunch of input and output examples; alternatively, if we want to be more ambitious, we can feed the program the actual tax texts. We can let the machine consume the data and figure out the tax rule, just as an autonomous car doesn't need a lot of explicit human input.

In physics, we have almost the same situation. We want to know how the universe works and formulate laws in a mathematical language. Since we don't know how it works, all we can do is measure the error produced in our attempt at law formulation and try to minimize it. In supervised learning tasks, we compare our results against the expected values. In unsupervised learning, we measure our success with related metrics. For instance, we want data points to be grouped based on similarities, forming clusters; the metrics could be how similar the data points within one cluster are, or how different the data points from two clusters are. In reinforcement learning, a program evaluates its moves, for example, by using a predefined function in a chess game.

Aside from correct generalization with data, there are two levels of generalization, overfitting and underfitting, which we will explore in the next section.

Overfitting, underfitting, and the bias-variance trade-off

In this section, let's take a look at both levels of generalization in detail and explore the bias-variance trade-off.

Overfitting

Reaching the right fit model is the goal of a machine learning task. What if the model overfits? **Overfitting** means a model fits the existing observations **too well** but fails to predict future new observations. Let's look at the following analogy.

If we go through many practice questions for an exam, we may start to find ways to answer questions that have nothing to do with the subject material. For instance, given only five practice questions, we might find that if there are two occurrences of *potatoes*, one of *tomato*, and three of *banana* in a multiple-choice question, the answer is always *A*, and if there is one occurrence of *potato*, three of *tomato*, and two of *banana* in a question, the answer is always *B*. We could then conclude that this is always true and apply such a theory later, even though the subject or answer may not be relevant to potatoes, tomatoes, or bananas. Or, even worse, we might memorize the answers to each question verbatim. We would then score highly on the practice questions, leading us to hope that the questions in the actual exams would be the same as the practice questions. However, in reality, we would score very low on the exam questions, as it's rare that the exact same questions occur in exams.

The phenomenon of memorization can cause overfitting. This can occur when we're over-extracting too much information from the training sets and making our model just work well with them. At the same time, however, overfitting won't help us to generalize it to new data and derive true patterns from it. The model, as a result, will perform poorly on datasets that weren't seen before. We call this situation **high variance** in machine learning. Let's quickly recap variance: *variance* measures the spread of the prediction, which is the variability of the prediction. It can be calculated as follows:

$$Variance = E[\hat{y}^2] - E[\hat{y}]^2$$

Here, $\hat{y}$ is the prediction, and E[] is the expectation or expected value that represents the average value of a random variable, based on its probability distribution in statistics.

The following example demonstrates what a typical instance of overfitting looks like, where the regression curve tries to flawlessly accommodate all observed samples:

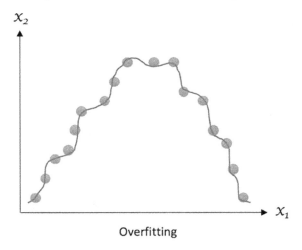

Overfitting

Figure 1.6: Example of overfitting

Overfitting occurs when we try to describe the learning rules based on too many parameters relative to the small number of observations, instead of the underlying relationship, such as the preceding potato, tomato, and banana example, where we deduced three parameters from only five learning samples. Overfitting also takes place when we make the model so excessively complex that it fits every training sample, such as memorizing the answers for all questions, as mentioned previously.

Underfitting

The opposite scenario is **underfitting**. When a model is underfit, it doesn't perform well on the training sets and won't do so on the testing sets, which means it fails to capture the underlying trend of the data. Underfitting may occur if we don't use enough data to train the model, just like we will fail the exam if we don't review enough material; this may also happen if we try to fit a wrong model to the data, just as we will score low in any exercises or exams if we take the wrong approach and learn in the wrong way. We describe any of these situations as **high bias** in machine learning, although its variance is low, as the performance in training and test sets is consistent, in a bad way. If you need a quick recap of bias, here it is: **bias** is the difference between the average prediction and the true value. It is computed as follows:

$$Bias[\hat{y}] = E[\hat{y} - y]$$

Here, $\hat{y}$ is the prediction and y is the ground truth.

The following example shows what typical underfitting looks like, where the regression curve doesn't fit the data well enough or capture enough of the underlying pattern of the data:

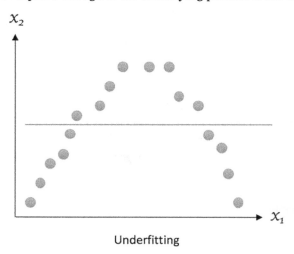

Underfitting

Figure 1.7: Example of underfitting

Now, let's look at what a well-fitting example should look like:

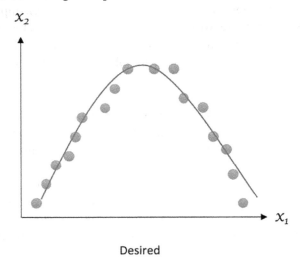

Desired

Figure 1.8: Example of desired fitting

The bias-variance trade-off

Obviously, we want to avoid both overfitting and underfitting. Recall that **bias** is the error stemming from incorrect assumptions in the learning algorithm; high bias results in underfitting. **Variance** measures how sensitive the model prediction is to variations in the datasets. Hence, we need to avoid cases where either bias or variance gets high. So, does it mean we should always make both bias and variance as low as possible? The answer is yes, if we can. But, in practice, there is an explicit trade-off between them, where decreasing one increases the other. This is the so-called **bias-variance trade-off**. Sounds abstract? Let's look at the next example.

Let's say we're asked to build a model to predict the probability of a candidate being the next president of America based on phone poll data. The poll is conducted using zip codes. We randomly choose samples from one zip code, and we estimate there's a 61% chance the candidate will win. However, it turns out they lost the election. Where did our model go wrong? The first thing we might think of is the small size of samples from only one zip code. It's a source of high bias also, as people in a geographic area tend to share similar demographics, although it results in a low variance of estimates. So can we fix it simply by using samples from a large number of zip codes? Yes, but don't get happy too soon. This might cause an increased variance of estimates at the same time. We need to find the optimal sample size—the best number of zip codes to achieve the lowest overall bias and variance.

Minimizing the total error of a model requires a careful balancing of bias and variance. Given a set of training samples, $x_1, x_2, ..., x_n$, and their targets, $y_1, y_2, ..., y_n$, we want to find a regression function $\hat{y}(x)$ that estimates the true relation $y(x)$ as correctly as possible. We measure the error of estimation, i.e., how good (or bad) the regression model is, in **mean squared error (MSE)**:

$$MSE = E[(y(x) - \hat{y}(x))^2]$$

The E denotes the expectation. This error can be decomposed into bias and variance components following the analytical derivation, as shown in the following formula (although it requires a bit of basic probability theory to understand):

$$MSE = E[(y - \hat{y}^2)]$$

$$= E[(y - E[\hat{y}] + E[\hat{y}] - \hat{y}^2)]$$

$$= E[(y - E[\hat{y}]^2)] + E[(E[\hat{y}] - \hat{y})^2] + E[2(y - E[\hat{y}])(E[\hat{y}] - \hat{y})]$$

$$= E[(y - E[\hat{y}]^2)] + E[(E[\hat{y}] - \hat{y})^2] + 2(y - E[\hat{y}])(E[\hat{y}] - \hat{y})]$$

$$= (E[\hat{y} - y])^2 + E[\hat{y}^2] - E[\hat{y}]^2$$

$$= Bias[\hat{y}]^2 + Variance[\hat{y}]$$

The term *Bias* measures the error of estimations, and the term *Variance* describes how much the estimation, $\hat{y}$, moves around its mean, $E[\hat{y}]$. The more complex the learning model $\hat{y}(x)$ is, and the larger the size of the training samples is, the lower the bias will become. However, this will also create more adjustments to the model to better fit the increased data points. As a result, the variance will be lifted.

We usually employ the cross-validation technique, as well as regularization and feature reduction, to find the optimal model balancing bias and variance and diminish overfitting. We will discuss these next.

You may ask why we only want to deal with overfitting: how about underfitting? This is because underfitting can be easily recognized: it occurs if a model doesn't work well on a training set. When this occurs, we need to find a better model or tweak some parameters to better fit the data, which is a must under all circumstances. On the other hand, overfitting is hard to spot. Oftentimes, when we achieve a model that performs well on a training set, we are overly happy and think it is ready for production right away. This can be very dangerous. We should instead take extra steps to ensure that the great performance isn't due to overfitting and that the great performance applies to data that excludes the training data.

Avoiding overfitting with cross-validation

You will see cross-validation in action multiple times later in this book. So don't panic if you find this section difficult to understand, as you will become an expert on cross-validation very soon.

Recall that between practice questions and actual exams, there are mock exams where we can assess how well we will perform in actual exams and use that information to conduct the necessary revision. In machine learning, the validation procedure helps to evaluate how models will generalize to independent or unseen datasets in a simulated setting. In a conventional validation setting, the original data is partitioned into three subsets, usually 60% for the training set, 20% for the validation set, and the rest (20%) for the testing set. This setting suffices if we have enough training samples after partitioning and we only need a rough estimate of simulated performance. Otherwise, cross-validation is preferable. Cross-validation helps to reduce variability and, therefore, limit overfitting.

In one round of cross-validation, the original data is divided into two subsets, for **training** and **testing** (or **validation**), respectively. The testing performance is recorded. Similarly, multiple rounds of cross-validation are performed under different partitions. Testing results from all rounds are finally averaged to generate a more reliable estimate of model prediction performance.

When the training size is very large, it's often sufficient to split it into training, validation, and testing (three subsets) and conduct a performance check on the latter two. Cross-validation is less preferable in this case, since it's computationally costly to train a model for each single round. But if you can afford it, there's no reason not to use cross-validation. When the size isn't so large, cross-validation is definitely a good choice.

There are mainly two cross-validation schemes in use: exhaustive and non-exhaustive. In the **exhaustive scheme**, we leave out a fixed number of observations in each round as testing (or validation) samples and use the remaining observations as training samples. This process is repeated until all possible different subsets of samples are used for testing once. For instance, we can apply **Leave-One-Out-Cross-Validation (LOOCV)**, which lets each sample be in the testing set once. For a dataset of the size n, LOOCV requires n rounds of cross-validation. This can be slow when n gets large. The following diagram presents the workflow of LOOCV:

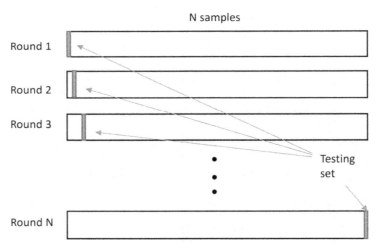

Figure 1.9: Workflow of leave-one-out-cross-validation

A **non-exhaustive scheme,** on the other hand, as the name implies, doesn't try out all possible partitions. The most widely used type of this scheme is **k-fold cross-validation**. First, we randomly split the original data into **k equal-sized** folds. In each trial, one of these folds becomes the testing set, and the rest of the data becomes the training set.

We repeat this process k times, with each fold being the designated testing set once. Finally, we average the k sets of test results for the purpose of evaluation. Common values for k are 3, 5, and 10. The following table illustrates the setup for five-fold cross-validation:

Round	Fold 1	Fold 2	Fold 3	Fold 4	Fold 5
1	**Testing**	Training	Training	Training	Training
2	Training	**Testing**	Training	Training	Training
3	Training	Training	**Testing**	Training	Training
4	Training	Training	Training	**Testing**	Training
5	Training	Training	Training	Training	**Testing**

Table 1.1: Setup for five-fold cross-validation

K-fold cross-validation often has a lower variance compared to LOOCV, since we're using a chunk of samples instead of a single one for validation.

We can also randomly split the data into training and testing sets numerous times. This is formally called the **holdout** method. The problem with this algorithm is that some samples may never end up in the testing set, while some may be selected multiple times in the testing set.

Last but not least, **nested cross-validation** is a combination of cross-validations. It consists of the following two phases:

- **Inner cross-validation:** This phase is conducted to find the best fit and can be implemented as a k-fold cross-validation
- **Outer cross-validation:** This phase is used for performance evaluation and statistical analysis

We will apply cross-validation very intensively throughout this entire book. Before that, let's look at cross-validation with an analogy next, which will help us to better understand it.

A data scientist plans to take his car to work, and his goal is to arrive before 9 a.m. every day. He needs to decide the departure time and the route to take. He tries out different combinations of these two parameters on certain Mondays, Tuesdays, and Wednesdays and records the arrival time for each trial. He then figures out the best schedule and applies it every day. However, it doesn't work quite as well as expected.

It turns out the scheduling **model** is overfitted to the data points gathered in the first three days and may not work well on Thursdays and Fridays. A better solution would be to test the best combination of parameters derived from Mondays to Wednesdays on Thursdays and Fridays and similarly repeat this process, based on different sets of learning days and testing days of the week. This analogized cross-validation ensures that the selected schedule works for the whole week.

In summary, cross-validation derives a more accurate assessment of model performance by combining measures of prediction performance on different subsets of data. This technique not only reduces variance and avoids overfitting but also gives an insight into how a model will generally perform in practice.

Avoiding overfitting with regularization

Another way of preventing overfitting is **regularization**. Recall that the unnecessary complexity of a model is a source of overfitting. Regularization adds extra parameters to the error function we're trying to minimize, in order to penalize complex models.

According to the principle of Occam's razor, simpler methods are to be favored. William Occam was a monk and philosopher who, around the year 1320, came up with the idea that the simplest hypothesis that fits data should be preferred. One justification for this is that we can invent fewer simple models than complex models. For instance, intuitively, we know that there are more high-polynomial models than linear ones. The reason is that a line ($y = ax + b$) is governed by only two parameters—the intercept, b, and slope, a. The possible coefficients for a line span two-dimensional space. A quadratic polynomial adds an extra coefficient for the quadratic term, and we can span a three-dimensional space with the coefficients. Therefore, it is much easier to find a model that perfectly captures all training data points with a **high-order polynomial function**, as its search space is much larger than that of a linear function. However, these easily obtained models generalize worse than linear models, which are more prone to overfitting. Also, of course, simpler models require less computation time. The following diagram displays how we try to fit a linear function and a high order polynomial function, respectively, to the data:

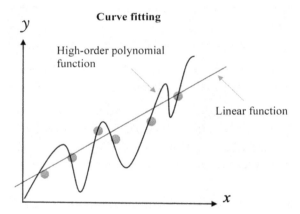

Figure 1.10: Fitting data with a linear function and a polynomial function

The linear model is preferable, as it may generalize better to more data points drawn from the underlying distribution. We can use regularization to reduce the influence of the high orders of a polynomial by imposing penalties on them. This will discourage complexity, even though a less accurate and less strict rule is learned from the training data.

We will employ regularization quite often in this book, starting from *Chapter 4, Predicting Online Ad Click-Through with Logistic Regression*. For now, let's look at an analogy that can help you better understand regularization.

A data scientist wants to equip his robotic guard dog with the ability to identify strangers and his friends. He feeds it with the following learning samples:

Male	Young	Tall	With glasses	In grey	**Friend**
Female	Middle	Average	Without glasses	In black	**Stranger**
Male	Young	Short	With glasses	In white	**Friend**
Male	Senior	Short	Without glasses	In black	**Stranger**
Female	Young	Average	With glasses	In white	**Friend**
Male	Young	Short	Without glasses	In red	**Friend**

Table 1.2: Training samples for the robotic guard dog

The robot may quickly learn the following rules:

- Any middle-aged female of average height without glasses and dressed in black is a stranger
- Any senior short male without glasses and dressed in black is a stranger
- Anyone else is his friend

Although these perfectly fit the training data, they seem too complicated and unlikely to generalize well to new visitors. In contrast, the data scientist limits the learning aspects. A loose rule that can work well for hundreds of other visitors could be as follows: anyone without glasses dressed in black is a stranger.

Besides penalizing complexity, we can also stop a training procedure early as a technique to prevent overfitting. If we limit the time a model spends learning or set some internal stopping criteria, it's more likely to produce a simpler model. The model complexity will be controlled in this way; hence, overfitting becomes less probable. This approach is called **early stopping** in machine learning.

Last but not least, it's worth noting that regularization should be kept at a moderate level or, to be more precise, fine-tuned to an optimal level. Too small a regularization doesn't make any impact; too large a regularization will result in underfitting, as it moves the model away from the ground truth. We will explore how to achieve optimal regularization in *Chapter 4, Predicting Online Ad Click-Through with Logistic Regression, Chapter 5, Predicting Stock Prices with Regression Algorithms*, and *Chapter 6, Predicting Stock Prices with Artificial Neural Networks*.

Avoiding overfitting with feature selection and dimensionality reduction

We typically represent data as a grid of numbers (a **matrix**). Each column represents a variable, which we call a **feature** in machine learning. In supervised learning, one of the variables is actually not a feature but the label that we're trying to predict. And in supervised learning, each row is an example that we can use for training or testing.

The number of features corresponds to the dimensionality of the data. Our machine learning approach depends on the number of dimensions versus the number of examples. For instance, text and image data are very high dimensional, while sensor data (such as temperature, pressure, or GPS) has relatively fewer dimensions.

Fitting high-dimensional data is computationally expensive and prone to overfitting, due to the high complexity. Higher dimensions are also impossible to visualize, and therefore, we can't use simple diagnostic methods.

Not all of the features are useful, and they may only add randomness to our results. Therefore, it's often important to do good feature selection. **Feature selection** is the process of picking a subset of significant features for use in better model construction. In practice, not every feature in a dataset carries information useful for discriminating samples; some features are either redundant or irrelevant and, hence, can be discarded with little loss.

In principle, feature selection boils down to multiple binary decisions about whether to include a feature. For n features, we get 2^n feature sets, which can be a very large number for a large number of features. For example, for 10 features, we have 1,024 possible feature sets (for instance, if we're deciding what clothes to wear, the features can be temperature, rain, the weather forecast, and where we're going). Basically, we have two options: we either start with all of the features and remove features iteratively, or we start with a minimum set of features and add features iteratively. We then take the best feature sets for each iteration and compare them. At a certain point, brute-force evaluation becomes infeasible. Hence, more advanced feature selection algorithms were invented to distill the most useful features/signals. We will discuss in detail how to perform feature selection in *Chapter 4, Predicting Online Ad Click-Through with Logistic Regression*.

Another common approach to reducing dimensionality is to transform high-dimensional data into lower-dimensional space. This is known as **dimensionality reduction** or **feature projection**. We will get into this in detail in *Chapter 7, Mining the 20 Newsgroups Dataset with Text Analysis Techniques*, where we will encode text data into two dimensions, and *Chapter 9, Recognizing Faces with Support Vector Machine*, where we will talk about projecting high-dimensional image data into low-dimensional space.

In this section, we talked about how the goal of machine learning is to find the optimal generalization to the data, and how to avoid ill-generalization. In the next two sections, we will explore tricks to get closer to the goal throughout individual phases of machine learning, including data preprocessing and feature engineering in the next section, and modeling in the section after that.

Data preprocessing and feature engineering

Data preprocessing and feature engineering play a crucial and foundational role in machine learning. It's like laying the groundwork for a building – the stronger and better prepared the foundation, the better the final structure (machine learning model) will be. Here is a breakdown of their relationship:

- **Preprocessing prepares data for efficient learning:** Raw data from various sources often contains inconsistencies, errors, and irrelevant information. Preprocessing cleans, organizes, and transforms the data into a format suitable for the chosen machine learning algorithm. This allows the algorithm to understand the data more easily and efficiently, leading to better model performance.

- **Preprocessing helps improve model accuracy and generalizability:** By handling missing values, outliers, and inconsistencies, preprocessing reduces noise in data. This enables a model to focus on the true patterns and relationships within the data, leading to more accurate predictions and better generalization on unseen data.

- **Feature engineering provides meaningful input variables:** Raw data is transformed and manipulated to create new features or select relevant ones. New features potentially improve model performance and insight generation.

Overall, data preprocessing and feature engineering is an essential step in the machine learning workflow. By dedicating time and effort to proper preprocessing and feature engineering, you lay the foundation to build reliable, accurate, and generalizable machine learning models. We will cover the preprocessing phase first in this section.

Preprocessing and exploration

When we learn, we require high-quality learning material. We can't learn from gibberish, so we automatically ignore anything that doesn't make sense. A machine learning system isn't able to recognize gibberish, so we need to help it by cleaning the input data. It's often claimed that cleaning the data forms a large part of machine learning. Sometimes, the cleaning is already done for us, but you shouldn't count on it.

To decide how to clean data, we need to be familiar with it. There are some projects that try to automatically explore the data and do something intelligent, such as producing a report. For now, unfortunately, we don't have a solid solution in general, so you need to do some work.

We can do two things, which aren't mutually exclusive: first, scan the data, and second, visualize the data. This also depends on the type of data we're dealing with—whether we have a grid of numbers, images, audio, text, or something else.

Ultimately, a grid of numbers is the most convenient form, and we will always work toward having numerical features. Let's pretend that we have a table of numbers in the rest of this section.

We want to know whether features have missing values, how the values are distributed, and what type of features we have. Values can approximately follow a normal distribution, a binomial distribution, a Poisson distribution, or another distribution altogether. Features can be binary: either yes or no, positive or negative, and so on. They can also be categorical: pertaining to a category, such as continents (Africa, Asia, Europe, South America, North America, and so on). Categorical variables can also be ordered, for instance, high, medium, and low. Features can also be quantitative, for example, the temperature in degrees or the price in dollars. Now, let's dive into how we can cope with each of these situations.

Dealing with missing values

Quite often, we miss values for certain features. This could happen for various reasons. It can be inconvenient, expensive, or even impossible to always have a value. Maybe we weren't able to measure a certain quantity in the past because we didn't have the right equipment or just didn't know that the feature was relevant. However, we're stuck with missing values from the past.

Sometimes, it's easy to figure out that we're missing values, and we can discover this just by scanning the data or counting the number of values we have for a feature and comparing this figure with the number of values we expect, based on the number of rows. Certain systems encode missing values with, for example, values such as 999,999 or -1. This makes sense if the valid values are much smaller than 999,999. If you're lucky, you'll have information about the features provided by whoever created the data in the form of a data dictionary or metadata.

Once we know that we're missing values, the question arises of how to deal with them. The simplest answer is to just ignore them. However, some algorithms can't deal with missing values, and the program will just refuse to continue. In other circumstances, ignoring missing values will lead to inaccurate results. The second solution is to substitute missing values with a fixed value—this is called **imputing**. We can impute the arithmetic **mean, median,** or **mode** of the valid values of a certain feature. Ideally, we will have some prior knowledge of a variable that is somewhat reliable. For instance, we may know the seasonal averages of temperature for a certain location and be able to impute guesses for missing temperature values, given a date. We will talk about dealing with missing data in detail in *Chapter 10, Machine Learning Best Practices*. Similarly, techniques in the following sections will be discussed and employed in later chapters, just in case you feel uncertain about how they can be used.

Label encoding

Humans are able to deal with various types of values. Machine learning algorithms (with some exceptions) require numerical values. If we offer a string such as Ivan, unless we're using specialized software, the program won't know what to do. In this example, we're dealing with a categorical feature—names, probably. We can consider each unique value to be a label. (In this particular example, we also need to decide what to do with the case—is Ivan the same as ivan?). We can then replace each label with an integer—**label encoding**.

The following example shows how label encoding works:

Label	Encoded Label
Africa	1
Asia	2
Europe	3
South America	4
North America	5
Other	6

Table 1.3: Example of label encoding

This approach can be problematic in some cases because the learner may conclude that there is an order (unless it is expected, for example, *bad=0, ok=1, good=2,* and *excellent=3*). In the preceding mapping table, Asia and North America in the preceding case differ by 4 after encoding, which is a bit counterintuitive, as it's hard to quantify them. One-hot encoding in the next section takes an alternative approach.

One-hot encoding

The **one-of-K**, or **one-hot encoding**, scheme uses dummy variables to encode categorical features. Originally, it was applied to digital circuits. The dummy variables have binary values such as bits, so they take the values zero or one (equivalent to true or false). For instance, if we want to encode continents, we will have dummy variables, such as `is_asia`, which will be true if the continent is `Asia` and false otherwise. In general, we need as many dummy variables as there are unique values minus one (or sometimes the exact number of unique values). We can determine one of the labels automatically from the dummy variables because they are exclusive.

If the dummy variables all have a false value, then the correct label is the label for which we don't have a dummy variable. The following table illustrates the encoding for continents:

Continent	Is_africa	Is_asia	Is_europe	Is_sam	Is_nam
Africa	1	0	0	0	0
Asia	0	1	0	0	0
Europe	0	0	1	0	0
South America	0	0	0	1	0
North America	0	0	0	0	1
Other	0	0	0	0	0

Table 1.4: Example of one-hot encoding

The encoding produces a matrix (grid of numbers) with lots of zeros (false values) and occasional ones (true values). This type of matrix is called a **sparse matrix**. The sparse matrix representation is handled well by the `scipy` package, which we will discuss later in this chapter.

Dense embedding

While one-hot encoding is a simple and sparse representation of categorical features, **dense embedding** provides a compact, continuous representation that captures semantic relationships based on the co-occurrence patterns in data. For example, using dense embedding, the continent categories might be represented by 3-dimensional continuous vectors like:

- Africa: [0.9, -0.2, 0.5]
- Asia: [-0.1, 0.8, 0.6]
- Europe: [0.6, 0.3, -0.7]
- South America: [0.5, 0.2, 0.1]
- North America: [0.4, 0.3, 0.2]
- Other: [-0.8, -0.5, 0.4]

In this example, you may notice the vectors of South America and North America are closer together than those of Africa and Asia. Dense embedding can capture the similarities between categories. In another example, you may see more closeness of the vectors of Europe and North America, based on cultural similarity.

We will explore dense embedding further in *Chapter 7, Mining the 20 Newsgroups Dataset with Text Analysis Techniques*.

Scaling

Values of different features can differ by orders of magnitude. Sometimes, this can mean that the larger values dominate the smaller values. This depends on the algorithm we use. For certain algorithms to work properly, we're required to scale data.

There are the following several common strategies that we can apply:

- Standardization removes the mean of a feature and divides it by the standard deviation. If the feature values are normally distributed, we will get a **Gaussian**, which is centered around zero with a variance of one.

- If the feature values aren't normally distributed, we can remove the median and divide by the interquartile range. The **interquartile range** is the range between the first and third quartile (or 25th and 75th percentile).

- A range between zero and one is a common choice of range for feature scaling.

We will use scaling in many projects throughout the book.

An advanced version of data preprocessing is usually called feature engineering. We will cover that next.

Feature engineering

Feature engineering is the process of creating or improving features. Features are often created based on common sense, domain knowledge, or prior experience. There are certain common techniques for feature creation; however, there is no guarantee that creating new features will improve your results. We are sometimes able to use the clusters found by unsupervised learning as extra features. **Deep neural networks** are often able to derive features automatically.

We will briefly look at some feature engineering techniques: polynomial transformation and binning.

Polynomial transformation

If we have two features, a and b, we can suspect that there is a polynomial relationship, such as $a^2 + ab + b^2$. We can consider a new feature an **interaction** between a and b, such as the product ab. An interaction doesn't have to be a product—although this is the most common choice—it can also be a sum, a difference, or a ratio. If we use a ratio to avoid dividing by zero, we should add a small constant to the divisor and dividend.

The number of features and the order of the polynomial for a polynomial relationship aren't limited. However, if we follow the Occam's razor principle, we should avoid higher-order polynomials and interactions of many features. In practice, complex polynomial relations tend to be more difficult to compute and tend to overfit, but if you really need better results, they may be worth considering. We will see polynomial transformation in action in *Best practice 12 – performing feature engineering without domain expertise* section in *Chapter 10, Machine Learning Best Practices*.

Binning

Sometimes, it's useful to separate feature values into several bins. For example, we may only be interested in whether it rained on a particular day. Given the precipitation values, we can binarize the values so that we get a true value if the precipitation value isn't zero, and a false value otherwise. We can also use statistics to divide values into high, low, and medium bins. In marketing, we often care more about the age group, such as 18 to 24, than a specific age, such as 23.

The binning process inevitably leads to a loss of information. However, depending on your goals, this may not be an issue, actually reducing the chance of overfitting. Certainly, there will be improvements in speed and a reduction of memory or storage requirements and redundancy.

Any real-world machine learning system should have two modules: a data preprocessing module, which we just covered in this section, and a modeling module, which will be covered next.

Combining models

A model takes in data (usually preprocessed) and produces predictive results. What if we employ multiple models? Will we make better decisions by combining predictions from individual models? We will talk about this in this section.

Let's start with an analogy. In high school, we sit together with other students and learn together, but we aren't supposed to work together during the exam. The reason is, of course, that teachers want to know what we've learned, and if we just copy exam answers from friends, we may not have learned anything. Later in life, we discover that teamwork is important. For example, this book is the product of a whole team, or possibly a group of teams.

Clearly, a team can produce better results than a single person. However, this goes against Occam's razor, since a single person can come up with simpler theories compared to what a team will produce. In machine learning, we nevertheless prefer to have our models cooperate with the following model combination schemes:

- Voting and averaging
- Bagging
- Boosting
- Stacking

Let's dive into each of them now.

Voting and averaging

This is probably the most understandable type of model aggregation. It just means the final output will be the **majority** or **average** of prediction output values from multiple models. It is also possible to assign different weights to individual models in the ensemble; for example, some models that are more reliable might be given two votes.

Nonetheless, combining the results of models that are highly correlated to each other doesn't guarantee a spectacular improvement. It is better to somehow diversify the models by using different features or different algorithms. If you find two models are strongly correlated, you may, for example, decide to remove one of them from the ensemble and increase proportionally the weight of the other model.

Bagging

Bootstrap aggregating, or **bagging,** is an algorithm introduced by Leo Breiman, a distinguished statistician at the University of California, Berkeley, in 1994, which applies **bootstrapping** to machine learning problems. Bootstrapping is a statistical procedure that creates multiple datasets from an existing one by sampling data with replacement. Bootstrapping can be used to measure the properties of a model, such as bias and variance.

In general, a bagging algorithm follows these steps:

1. We generate new training sets from input training data by sampling with replacement.
2. For each generated training set, we fit a new model.
3. We combine the results of the models by averaging or majority voting.

The following diagram illustrates the steps for bagging, using classification as an example (the circles and crosses represent samples from two classes):

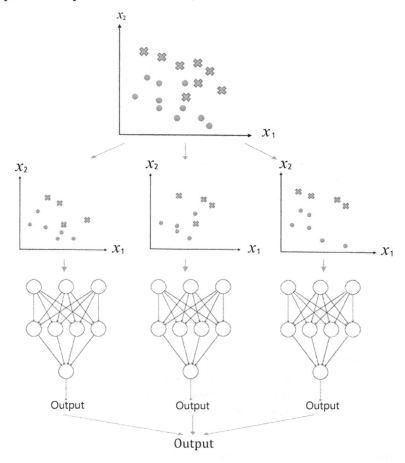

Figure 1.11: Workflow of bagging for classification

As you can imagine, bagging can reduce the chance of overfitting.

We will study bagging in depth in *Chapter 3, Predicting Online Ad Click-Through with Tree-Based Algorithms*.

Boosting

In the context of supervised learning, we define **weak learners** as learners who are just a little better than a baseline, such as randomly assigning classes or average values. Much like ants, weak learners are weak individually, but together, they have the power to do amazing things.

It makes sense to take into account the strength of each individual learner using weights. This general idea is called **boosting**. In boosting, all models are trained in sequence, instead of in parallel as in bagging. Each model is trained on the same dataset, but each data sample has a different weight, factoring in the previous model's success. The weights are reassigned after a model is trained, which will be used for the next training round. In general, weights for mispredicted samples are increased to stress their prediction difficulty.

The following diagram illustrates the steps for boosting, again using classification as an example (the circles and crosses represent samples from two classes, and the size of a circle or cross indicates the weight assigned to it):

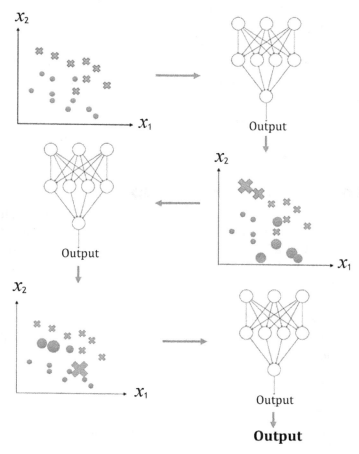

Figure 1.12: Workflow of boosting for classification

There are many boosting algorithms; boosting algorithms differ mostly in their weighting scheme. If you've studied for an exam, you may have applied a similar technique by identifying the type of practice questions you had trouble with and focusing on the hard problems.

Viola-Jones, a popular face detection framework, leverages the boosting algorithm to efficiently identify faces in images. Detecting faces in images or videos is supervised learning. We give the learner examples of regions containing faces. There's an imbalance, since we usually have far more regions that don't have faces than those that do (about 10,000 times more).

A cascade of classifiers progressively filters out these negative image areas stage by stage. In each progressive stage, the classifiers use progressively more features on fewer image windows. The idea is to spend the majority of time on image patches that contain faces. In this context, boosting is used to select features and combine results.

Stacking

Stacking takes the output values of machine learning models and then uses them as input values for another algorithm. You can, of course, feed the output of the higher-level algorithm to another predictor. It's possible to use any arbitrary topology, but for practical reasons, you should try a simple setup first, as also dictated by Occam's razor.

A fun fact is that stacking is commonly used in the winning models in the Kaggle competition. For instance, the first place for the Otto Group Product Classification Challenge (`www.kaggle.com/c/otto-group-product-classification-challenge`) went to a stacking model composed of more than 30 different models.

So far, we have covered the tricks required to more easily reach the right generalization for a machine learning model throughout the data preprocessing and modeling phase. I know you can't wait to start working on a machine learning project. Let's get ready by setting up the working environment.

Installing software and setting up

As the book title says, Python is the language we will use to implement all machine learning algorithms and techniques throughout the entire book. We will also exploit many popular Python packages and tools, such as NumPy, SciPy, scikit-learn, TensorFlow, and PyTorch. By the end of this initial chapter, make sure you have set up the tools and working environment properly, even if you are already an expert in Python or familiar with some of the aforementioned tools.

Setting up Python and environments

We will use Python 3 in this book. The Anaconda Python 3 distribution is one of the best options for data science and machine learning practitioners.

Anaconda is a free Python distribution for data analysis and scientific computing. It has its own package manager, `conda`. The distribution (`https://docs.anaconda.com/free/anaconda/`, depending on your OS, or Python version 3.7 to 3.11) includes around 700 Python packages (as of 2023), which makes it very convenient. For casual users, the **Miniconda** (`https://conda.io/miniconda.html`) distribution may be the better choice. Miniconda contains the `conda` package manager and Python. Obviously, Miniconda takes up much less disk space than Anaconda.

The procedures to install Anaconda and Miniconda are similar. You can follow the instructions from `https://docs.conda.io/projects/conda/en/latest/user-guide/install/`. First, you must download the appropriate installer for your OS and Python version, as follows:

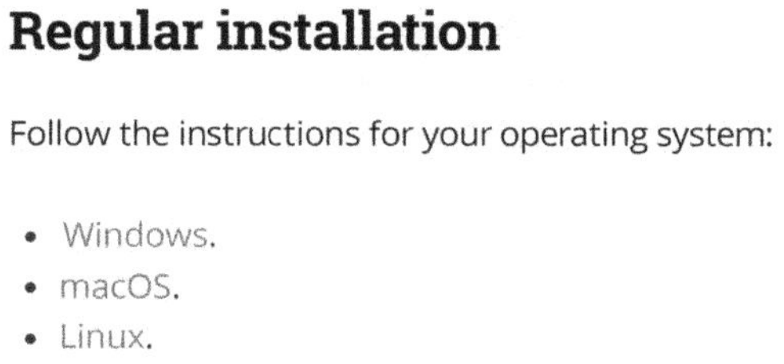

Regular installation

Follow the instructions for your operating system:

- Windows.
- macOS.
- Linux.

Figure 1.13: Installation entry based on your OS

Follow the steps listed in your OS. You can choose between a GUI and a CLI. I personally find the latter easier.

Anaconda comes with its own Python installation. On my machine, the Anaconda installer created an `anaconda` directory in my home directory and required about 900 MB. Similarly, the `Miniconda` installer installs a `miniconda` directory in your home directory.

Feel free to play around with it after you set it up. One way to verify that you have set up **Anaconda** properly is by entering the following command line in your terminal on Linux/Mac or Command Prompt on Windows (from now on, we will just mention Terminal):

```
python
```

The preceding command line will display your Python running environment, as shown in the following screenshot:

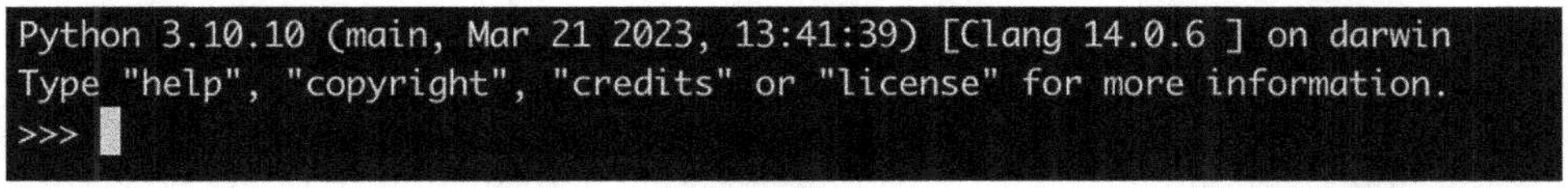

```
Python 3.10.10 (main, Mar 21 2023, 13:41:39) [Clang 14.0.6 ] on darwin
Type "help", "copyright", "credits" or "license" for more information.
>>>
```

Figure 1.14: Screenshot after running "python" in the terminal

If you don't see this, please check the system path or the path Python is running from.

To wrap up this section, I want to emphasize the reasons why Python is the most popular language for machine learning and data science. First of all, Python is famous for its high readability and simplicity, which makes it easy to build machine learning models. We spend less time worrying about getting the right syntax and compilation and, as a result, have more time to find the right machine learning solution. Second, we have an extensive selection of Python libraries and frameworks for machine learning:

Tasks	Python libraries
Data analysis	NumPy, SciPy, and pandas
Data visualization	Matplotlib, and Seaborn
Modeling	scikit-learn, TensorFlow, Keras, and PyTorch

Table 1.5: Popular Python libraries for machine learning

The next step involves setting up some of the packages that we will use throughout this book.

Installing the main Python packages

For most projects in this book, we will use NumPy (http://www.numpy.org/), SciPy (https://scipy.org/), the pandas library (https://pandas.pydata.org/), scikit-learn (http://scikit-learn.org/stable/), TensorFlow (https://www.tensorflow.org/), and PyTorch (https://pytorch.org/).

In the sections that follow, we will cover the installation of several Python packages that we will mainly use in this book.

Conda environments provide a way to isolate dependencies and packages for different projects. So it is recommended to create and use an environment for a new project. Let's create one using the following command to create an environment called "pyml":

```
conda create --name pyml python=3.10
```

Here, we also specify the Python version, 3.10, which is optional but highly recommended. This is to avoid using the latest Python version by default, which may not be compatible with many Python packages. For example, at the time of writing (late 2023), PyTorch does not support Python 3.11.

To activate the newly created environment, we use the following command:

```
conda activate pyml
```

The activated environment is displayed in front of your prompt like this:

```
(pyml) hayden@haydens-Air ~ %
```

NumPy

NumPy is the fundamental package for machine learning with Python. It offers powerful tools including the following:

- The *N*-dimensional array (ndarray) class and several subclasses representing matrices and arrays
- Various sophisticated array functions
- Useful linear algebra capabilities

Installation instructions for NumPy can be found at https://numpy.org/install/. Alternatively, an easier method involves installing it with conda or pip in the command line, as follows:

```
conda install numpy
```

or

```
pip install numpy
```

A quick way to verify your installation is to import it in Python, as follows:

```
>>> import numpy
```

It is installed correctly if no error message is visible.

SciPy

In machine learning, we mainly use NumPy arrays to store data vectors or matrices composed of feature vectors. SciPy (https://scipy.org/) uses NumPy arrays and offers a variety of scientific and mathematical functions. Installing SciPy in the terminal is similar, again as follows:

```
conda install scipy
```

or

```
pip install scipy
```

pandas

We also use the pandas library (https://pandas.pydata.org/) for data wrangling later in this book. The best way to get pandas is via pip or conda, for example:

```
conda install pandas
```

scikit-learn

The scikit-learn library is a Python machine learning package optimized for performance, as a lot of its code runs almost as fast as equivalent C code. The same statement is true for NumPy and SciPy. scikit-learn requires both NumPy and SciPy to be installed. As the installation guide in http://scikit-learn.org/stable/install.html states, the easiest way to install scikit-learn is to use pip or conda, as follows:

```
pip install -U scikit-learn
```

or

```
conda install -c conda-forge scikit-learn
```

Here, we use the "-c conda-forge" option to tell conda to search for packages in the conda-forge channel, which is a community-driven channel with a wide range of open-source packages.

TensorFlow

TensorFlow is a Python-friendly open-source library invented by the Google Brain team for high-performance numerical computation. It makes machine learning faster and deep learning easier, with the Python-based convenient frontend API and high-performance C++-based backend execution. TensorFlow 2 was largely a redesign of its first mature version, 1.0, and was released at the end of 2019.

TensorFlow has been widely known for its deep learning modules. However, its most powerful point is **computation graphs**, which algorithms are built on. Basically, a computation graph is used to convey relationships between the input and the output via tensors.

For instance, if we want to evaluate a linear relationship, *y = 3 * a + 2 * b*, we can represent it in the following computation graph:

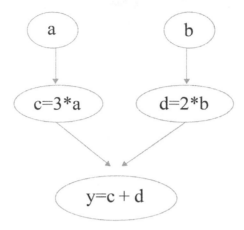

Figure 1.15: Computation graph for a y = 3 * a + 2 * b machine

Here, *a* and *b* are the input tensors, *c* and *d* are the intermediate tensors, and y is the output.

You can think of a computation graph as a network of nodes connected by edges. Each node is a tensor, and each edge is an operation or function that takes its input node and returns a value to its output node. To train a machine learning model, TensorFlow builds the computation graph and computes the **gradients** accordingly (gradients are vectors that provide the steepest direction where an optimal solution is reached). In the upcoming chapters, you will see some examples of training machine learning models using TensorFlow.

 We highly recommend you go through https://www.tensorflow.org/guide/data if you are interested in exploring more about TensorFlow and computation graphs.

TensorFlow allows easy deployment of computation across CPUs and GPUs, which empowers expensive and large-scale machine learning. In this book, we will focus on the CPU as our computation platform. Hence, according to https://www.tensorflow.org/install/, installing TensorFlow 2 is done via the following command line:

```
conda install -c conda-forge tensorflow
```

or

```
pip install tensorflow
```

You can always verify the installation by importing it in Python.

PyTorch

PyTorch is an open-source machine learning library primarily used to develop deep learning models. It provides a flexible and efficient framework to build neural networks and perform computations on GPUs. PyTorch was developed by Facebook's AI Research lab and is widely used in both research and industry.

Similar to TensorFlow, PyTorch performs its computations based on a **directed acyclic graph** (**DAG**). The difference is that PyTorch utilizes a **dynamic computational graph**, which allows for on-the-fly graph construction during runtime, while TensorFlow uses a **static** computational graph, where the graph structure is defined upfront and then executed. This dynamic nature enables greater flexibility in model design and easier debugging, and also facilitates dynamic control flow, making it suitable for a wide range of applications.

PyTorch has become a popular choice among researchers and practitioners in the field of deep learning, due to its flexibility, ease of use, and efficient computational capabilities. Its intuitive interface and strong community support make it a powerful tool for various applications, including computer vision, natural language processing, reinforcement learning, and more.

To install PyTorch, it is recommended to look up the command in the latest instructions on `https://pytorch.org/get-started/locally/`, based on the system and method.

As an example, we install the latest stable version (`2.2.0` as of late 2023) via `conda` on a Mac using the following command:

```
conda install pytorch::pytorch torchvision  -c pytorch
```

Best practice

If you encounter issues in installation, please read more about the platform and package-specific recommendations provided on the instructions page. All PyTorch code in this book can be run on your CPU, unless specifically indicated for a GPU only. However, using a GPU is recommended if you want to expedite training neural network models and fully enjoy the benefits of PyTorch. If you have a graphics card, refer to the instructions and set up PyTorch with the appropriate compute platform. For example, I install it on Windows with a GPU using the following command:

```
conda install pytorch torchvision pytorch-cuda=11.8 -c pytorch -c
nvidia
```

To check if PyTorch with GPU support is installed correctly, run the following Python code:

```
>>> import torch
>>> torch.cuda.is_available()
True
```

Alternatively, you can use Google Colab (`https://colab.research.google.com/`) to train some neural network models using GPUs for free.

There are many other packages we will use intensively, for example, **Matplotlib** for plotting and visualization, **Seaborn** for visualization, **NLTK** for natural language processing tasks, **transformers** for state-of-the-art models pretrained on large datasets, and **OpenAI Gym** for reinforcement learning. We will provide installation details for any package when we first encounter it in this book.

Summary

We just finished our first mile on the Python and machine learning journey! Throughout this chapter, we became familiar with the basics of machine learning. We started with what machine learning is all about, the importance of machine learning and its brief history, and looked at recent developments as well. We also learned typical machine learning tasks and explored several essential techniques to work with data and models. Now that we're equipped with basic machine learning knowledge and have set up the software and tools, let's get ready for the real-world machine learning examples ahead.

In the next chapter, we will build a movie recommendation engine as our first machine learning project!

Exercises

1. Can you tell the difference between machine learning and traditional programming (rule-based automation)?
2. What's overfitting, and how do we avoid it?
3. Name two feature engineering approaches.
4. Name two ways to combine multiple models.
5. Install Matplotlib (`https://matplotlib.org/`) if this is of interest to you. We will use it for data visualization throughout the book.

Join our book's Discord space

Join our community's Discord space for discussions with the authors and other readers:

`https://packt.link/yuxi`

2

Building a Movie Recommendation Engine with Naïve Bayes

As promised, in this chapter, we will kick off our supervised learning journey with machine learning classification, and specifically, binary classification. The goal of the chapter is to build a movie recommendation system, which is a good starting point for learning classification from a real-life example—movie streaming service providers are already doing this, and we can do the same.

In this chapter, you will learn the fundamental concepts of classification, including what it does and its various types and applications, with a focus on solving a binary classification problem using a simple, yet powerful, algorithm, Naïve Bayes. Finally, the chapter will demonstrate how to fine-tune a model, which is an important skill that every data science or machine learning practitioner should learn.

We will go into detail on the following topics:

- Getting started with classification
- Exploring Naïve Bayes
- Implementing Naïve Bayes
- Building a movie recommender with Naïve Bayes
- Evaluating classification performance
- Tuning models with cross-validation

Getting started with classification

Movie recommendation can be framed as a machine learning classification problem. If it is predicted that you'll like a movie because you've liked or watched similar movies, for example, then it will be on your recommended list; otherwise, it won't. Let's get started by learning the important concepts of machine learning classification.

Classification is one of the main instances of supervised learning. Given a training set of data containing observations and their associated categorical outputs, the goal of classification is to learn a general rule that correctly maps the **observations** (also called **features** or **predictive variables**) to the target **categories** (also called **labels** or **classes**). Putting it another way, a trained classification model will be generated after the model learns from the features and targets of training samples, as shown in the first half of *Figure 2.1*. When new or unseen data comes in, the trained model will be able to determine their desired class memberships. Class information will be predicted based on the known input features using the trained classification model, as displayed in the second half of *Figure 2.1*:

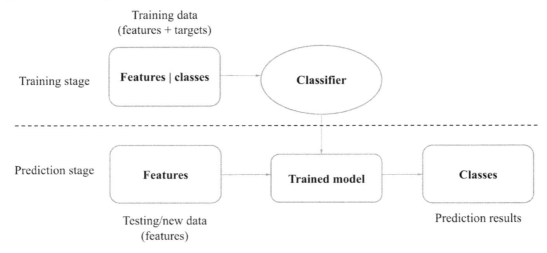

Figure 2.1: The training and prediction stages in classification

In general, there are three types of classification based on the possibility of class output—**binary**, **multiclass**, and **multi-label classification**. We will cover them one by one in this section.

Binary classification

Binary classification classifies observations into one of two possible classes. Spam email filtering we encounter every day is a typical use case of binary classification, which identifies email messages (input observations) as spam or not spam (output classes). Customer churn prediction is another frequently mentioned example, where a prediction system takes in customer segment data and activity data from **customer relationship management (CRM)** systems and identifies which customers are likely to churn.

Another application in the marketing and advertising industry is click-through prediction for online ads—that is, whether or not an ad will be clicked, given users' interest information and browsing history. Last but not least, binary classification is also being employed in biomedical science, for example, in early cancer diagnosis, classifying patients into high- or low-risk groups based on MRI images.

As demonstrated in *Figure 2.2*, binary classification tries to find a way to separate data into two classes (denoted by dots and crosses):

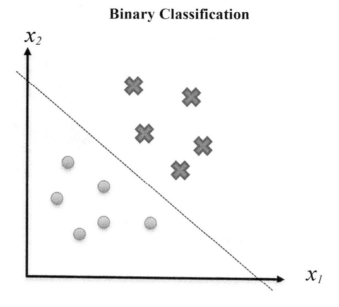

Figure 2.2: Binary classification example

Don't forget that predicting whether a person likes a movie is also a binary classification problem.

Multiclass classification

This type of classification is also referred to as **multinomial classification.** It allows more than two possible classes, as opposed to only two in binary cases. Handwritten digit recognition is a common instance of classification and has a long history of research and development since the early 1900s. A classification system, for example, can learn to read and understand handwritten ZIP codes (digits from 0 to 9 in most countries) by which envelopes are automatically sorted.

Handwritten digit recognition has become a *"Hello, World!"* in the journey of studying machine learning, and the scanned document dataset constructed by the **National Institute of Standards and Technology (NIST)**, called **Modified National Institute of Standards and Technology (MNIST)**, is a benchmark dataset frequently used to test and evaluate multiclass classification models. *Figure 2.3* shows four samples taken from the MNIST dataset, representing the digits "9," "2," "1," and "3," respectively:

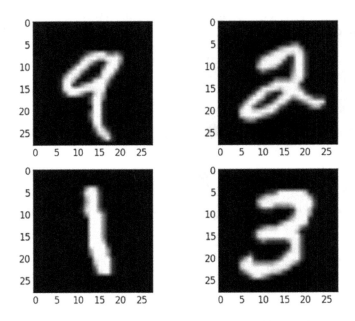

Figure 2.3: Samples from the MNIST dataset

As another example, in *Figure 2.4*, the multiclass classification model tries to find segregation boundaries to separate data into the following three different classes (denoted by dots, crosses, and triangles):

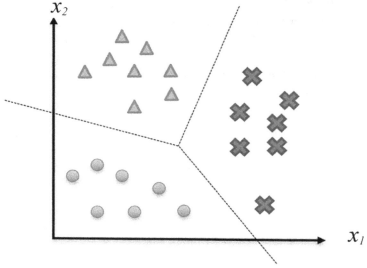

Figure 2.4: Multiclass classification example

Multi-label classification

In the first two types of classification, target classes are mutually exclusive and a sample is assigned *one, and only one*, label. It is the opposite in multi-label classification. Increasing research attention has been drawn to multi-label classification by the nature of the combination of categories in modern applications. For example, a picture that captures a sea and a sunset can simultaneously belong to both conceptual scenes, whereas it can only be an image of either a cat or dog in a binary case, or one type of fruit among oranges, apples, and bananas in a multiclass case. Similarly, adventure films are often combined with other genres, such as fantasy, science fiction, horror, and drama.

Another typical application is protein function classification, as a protein may have more than one function—storage, antibody, support, transport, and so on.

A typical approach to solving an n-label classification problem is to transform it into a set of n binary classification problems, where each binary classification problem is handled by an individual binary classifier.

Refer to *Figure 2.5* to see the restructuring of a multi-label classification problem into a multiple-binary classification problem:

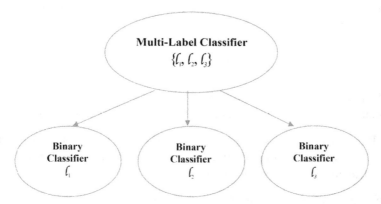

Figure 2.5: Transforming three-label classification into three independent binary classifications

Using the protein function classification example once more, we can transform it into several binary classifications, such as: Is it for storage? Is it for antibodies? Is it for support?

To solve problems like these, researchers have developed many powerful classification algorithms, among which Naïve Bayes, **Support Vector Machines** (**SVMs**), decision trees, logistic regression, and neural networks are often used.

In the following sections, we will cover the mechanics of Naïve Bayes and its in-depth implementation, along with other important concepts, including classifier tuning and classification performance evaluation. Stay tuned for upcoming chapters that cover the other classification algorithms.

Exploring Naïve Bayes

The **Naïve Bayes** classifier belongs to the family of probabilistic classifiers. It computes the probabilities of each predictive **feature** (also referred to as an **attribute** or **signal**) of the data belonging to each class in order to make a prediction of the probability distribution over all classes. Of course, from the resulting probability distribution, we can conclude the most likely class that the data sample is associated with. What Naïve Bayes does specifically, as its name indicates, is as follows:

- **Bayes**: As in, it maps the probability of observed input features given a possible class to the probability of the class given observed pieces of evidence based on Bayes' theorem.
- **Naïve**: As in, it simplifies probability computation by assuming that predictive features are mutually independent.

I will explain Bayes' theorem with examples in the next section.

Bayes' theorem by example

It is important to understand Bayes' theorem before diving into the classifier. Let A and B denote any two events. Events could be that *it will rain tomorrow, two kings are drawn from a deck of cards, or a person has cancer*. In Bayes' theorem, $P(A \mid B)$ is the probability that A occurs given that B is true. It can be computed as follows:

$$P(A \mid B) = \frac{P(B \mid A)\, P(A)}{P(B)}$$

Here, $P(B \mid A)$ is the probability of observing B given that A occurs, while $P(A)$ and $P(B)$ are the probability that A and B occur, respectively. Is that too abstract? Let's consider the following concrete examples:

- **Example 1**: Given two coins, one is unfair, with 90% of flips getting a head and 10% getting a tail, while the other one is fair. Randomly pick one coin and flip it. What is the probability that this coin is the unfair one, if we get a head?

  We can solve this by first denoting U for the event of picking the unfair coin, F for the fair coin, and H for the event of getting a head. So, the probability that the unfair coin has been picked when we get a head, $P(U \mid H)$, can be calculated with the following:

$$P(U|H) = \frac{P(H|U)P(U)}{P(H)}$$

  As we know, $P(H \mid U)$ is 0.9. $P(U)$ is 0.5 because we randomly pick a coin out of two. However, deriving the probability of getting a head, $P(H)$, is not that straightforward, as two events can lead to the following, where U is when the unfair coin is picked, and F is when the fair coin is picked:

$$P(H) = P(H|U)P(U) + P(H|F)P(F)$$

Now, $P(U|H)$ becomes the following:

$$P(U|H) = \frac{P(H|U)P(U)}{P(H)} = \frac{P(H|U)P(U)}{P(H|U)P(U) + P(H|F)P(F)} = \frac{0.9 * 0.5}{0.9 * 0.5 + 0.5 * 0.5} = 0.64$$

So, under Bayes' theorem, the probability that the unfair coin has been picked when we get a head is 0.64.

- **Example 2**: Suppose a physician reported the following cancer screening test scenario among 10,000 people:

	Cancer	No Cancer	Total
Test Positive	80	900	980
Test Negative	20	9000	9020
Total	100	9900	10000

Table 2.1: Example of a cancer screening result

This indicates that 80 out of 100 cancer patients are correctly diagnosed, while the other 20 are not; cancer is falsely detected in 900 out of 9,900 healthy people.

If the result of this screening test on a person is positive, what is the probability that they actually have cancer? Let's assign the event of having cancer and positive testing results as C and *Pos*, respectively. So we have $P(Pos\,|C)$ = 80/100 = 0.8, $P(C)$ = 100/10000 = 0.01, and $P(Pos)$ = 980/10000 = 0.098.

We can apply Bayes' theorem to calculate $P(C|Pos)$:

$$P(C|Pos) = \frac{P(Pos|C)P(C)}{P(Pos)} = 8.16\%$$

Given a positive screening result, the chance that the subject has cancer is 8.16%, which is significantly higher than the one under the general assumption (100/10000=1%) without the subject undergoing the screening.

- **Example 3**: Three machines, A, B, and C, in a factory account for 35%, 20%, and 45% of bulb production. The fraction of defective bulbs produced by each machine is 1.5%, 1%, and 2%, respectively. A bulb produced by this factory was identified as defective, which is denoted as event D. What are the probabilities that this bulb was manufactured by machine A, B, or C, respectively?

Again, we can simply follow Bayes' theorem:

$$P(A|D) = \frac{P(D|A)P(A)}{P(D)} = \frac{P(D|A)P(A)}{P(D|A)P(A) + P(D|B)P(B) + P(D|C)P(C)}$$

$$\frac{0.015 * 0.35}{0.015 * 0.35 + 0.01 * 0.2 + 0.02 * 0.45} = 0.323$$

$$P(B|D) = \frac{P(D|B)P(B)}{P(D)} = \frac{P(D|B)P(B)}{P(D|A)P(A) + P(D|B)P(B) + P(D|C)P(C)}$$

$$\frac{0.01 * 0.2}{0.015 * 0.35 + 0.01 * 0.2 + 0.02 * 0.45} = 0.123$$

$$P(C|D) = \frac{P(D|C)P(C)}{P(D)} = \frac{P(D|C)P(C)}{P(D|A)P(A) + P(D|B)P(B) + P(D|C)P(C)}$$

$$\frac{0.02 * 0.45}{0.015 * 0.35 + 0.01 * 0.2 + 0.02 * 0.45} = 0.554$$

So, under Bayes' theorem, the probabilities that this bulb was manufactured by machine *A*, *B*, or *C*, are `0.323`, `0.123`, and `0.554` respectively.

Also, either way, we do not even need to calculate *P(D)* since we know that the following is the case:

$$P(A|D): P(B|D): P(C|D) = P(D|A)P(A): P(D|B)P(B): P(D|C)P(C) = 21:8:36$$

We also know the following concept:

$$P(A|D) + P(B|D) + P(C|D) = 1$$

So, we have the following formula:

$$P(A|D) = \frac{21}{21 + 8 + 36} = 0.323$$

$$P(B|D) = \frac{8}{21 + 8 + 36} = 0.133$$

This shortcut approach gave us the same results as the original method, but faster. Now that you understand Bayes' theorem as the backbone of Naïve Bayes, we can easily move forward with the classifier itself.

The mechanics of Naïve Bayes

Let's start by discussing the magic behind the algorithm—how Naïve Bayes works. Given a data sample, *x*, with *n* features, $x_1, x_2,..., x_n$ (*x* represents a feature vector and $x = (x_1, x_2,..., x_n)$), the goal of Naïve Bayes is to determine the probabilities that this sample belongs to each of *K* possible classes $y_1, y_2,..., y_K$, which is $P(y_K |x)$ or $P(x_1, x_2,..., x_n)$, where *k* = 1, 2, ..., *K*.

This looks no different from what we have just dealt with: *x* or $x_1, x_2,..., x_n$. This is a joint event where a sample that has observed feature values $x_1, x_2,..., x_n$. y_K is the event that the sample belongs to class *k*. We can apply Bayes' theorem right away:

$$P(y_k|x) = \frac{P(x|y_k)P(y_k)}{P(x)}$$

Let's look at each component in detail:

- $P(y_k)$ portrays how classes are distributed, with no further knowledge of observation features. Thus, it is also called **prior** in Bayesian probability terminology. Prior can be either predetermined (usually in a uniform manner where each class has an equal chance of occurrence) or learned from a set of training samples.

- $P(y_k | x)$, in contrast to prior $P(y_k)$, is the **posterior**, with extra knowledge of observation.

- $P(x | y_k)$, or $P(x_1, x_2,..., x_n | y_k)$, is the joint distribution of n features, given that the sample belongs to class y_k. This is how likely the features with such values co-occur. This is named **likelihood** in Bayesian terminology. Obviously, the likelihood will be difficult to compute as the number of features increases. In Naïve Bayes, this is solved thanks to the feature independence assumption. The joint conditional distribution of n features can be expressed as the joint product of individual feature conditional distributions:

$$P(x|y_k) = P(x_1|y_k) * P(x_2|y_k) * ... * P(x_n|y_k)$$

Each conditional distribution can be efficiently learned from a set of training samples.

- $P(x)$, also called **evidence**, solely depends on the overall distribution of features, which is not specific to certain classes and is therefore a normalization constant. As a result, posterior is proportional to prior and likelihood:

$$P(y_k|x) \propto P(x|y_k)P(y_k) = P(x_1|y_k) * P(x_2|y_k) * ... * P(x_n|y_k) * P(y_k)$$

Figure 2.6 summarizes how a Naïve Bayes classification model is trained and applied to new data:

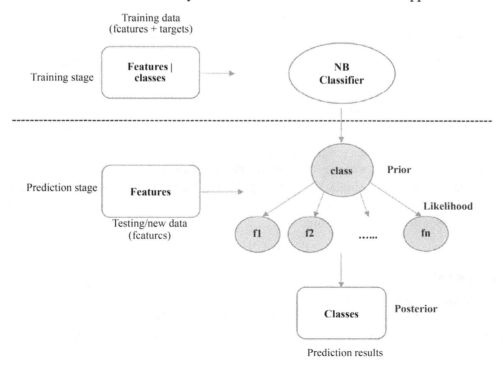

Figure 2.6: Training and prediction stages in Naïve Bayes classification

A Naïve Bayes classification model is trained using labeled data, where each instance is associated with a class label. During training, the model learns the probability distribution of the features given each class. This involves calculating the likelihood of observing each feature value given each class. Once trained, the model can be applied to new, unlabeled data. To classify a new instance, the model calculates the probability of each class given the observed features using Bayes' theorem.

Let's see a Naïve Bayes classifier in action through a simplified example of movie recommendation before we jump to the implementations of Naïve Bayes. Given four (pseudo) users, whether they like each of three movies, m_1, m_2, and m_3 (indicated as 1 or 0), and whether they like a target movie (denoted as event Y) or not (denoted as event N), as shown in the following table, we are asked to predict how likely it is that another user will like that movie:

	ID	m1	m2	m3	Whether the user likes the target movie
Training data	1	0	1	1	Y
	2	0	0	1	N
	3	0	0	0	Y
	4	1	1	0	Y
Testing case	5	1	1	0	?

Table 2.2: Toy data example for a movie recommendation

Whether users like three movies, m_1, m_2, and m_3, are features (signals) that we can utilize to predict the target class. The training data we have are the four samples with both ratings and target information.

Now, let's first compute the prior, $P(Y)$ and $P(N)$. From the training set, we can easily get the following:

$$P(Y) = \frac{3}{4} \quad P(N) = \frac{1}{4}$$

Alternatively, we can also impose an assumption of a uniform prior that $P(Y) = 50\%$, for example.

For simplicity, we will denote the event that a user likes three movies or not as f_1, f_2, and f_3, respectively. To calculate posterior $P(Y|x)$, where $x = (1, 1, 0)$, the first step is to compute the likelihoods, $P(f_1 = 1|Y)$, $P(f_2 = 1|Y)$, and $P(f_3 = 0|Y)$, and similarly, $P(f_1 = 1|N)$, $P(f_2 = 1|N)$, and $P(f_3 = 0|N)$, based on the training set. However, you may notice that since $f_1 = 1$ was not seen in the N class, we will get $P(f_1 = 1|N) = 0$. Consequently, we will have the following:

$$P(N|x) \propto P(f_1 = 1|N) * P(f_2 = 1|N) = 0$$

This means we will recklessly predict class = Y by any means.

To eliminate the zero-multiplication factor, the unknown likelihood, we usually assign an initial value of 1 to each feature, that is, we start counting each possible value of a feature from one. This technique is also known as **Laplace smoothing**. With this amendment, we now have the following:

$$P(f_1 = 1|N) = \frac{0+1}{1+2} = \frac{1}{3}$$

$$P(f_1 = 1|Y) = \frac{1+1}{3+2} = \frac{2}{5}$$

Here, given class N, $0 + 1$ means there are zero likes of m_1 plus $+ 1$ smoothing; $1 + 2$ means there is one data point (ID = 2) plus 2 (2 possible values) + 1 smoothing. Given class Y, $1 + 1$ means there is one like of m_1 (ID = 4) plus $+ 1$ smoothing; $3 + 2$ means there are 3 data points (ID = 1, 3, 4) plus 2 (2 possible values) + 1 smoothing.

Similarly, we can compute the following:

$$P(f_2 = 1|N) = \frac{0+1}{1+2} = \frac{1}{3}$$

$$P(f_2 = 1|Y) = \frac{2+1}{3+2} = \frac{3}{5}$$

$$P(f_3 = 0|N) = \frac{0+1}{1+2} = \frac{1}{3}$$

$$P(f_3 = 0|Y) = \frac{2+1}{3+2} = \frac{3}{5}$$

Now, we can compute the ratio between two posteriors as follows:

$$\frac{P(N|x)}{P(Y|x)} \propto \frac{P(N) * P(f_1 = 1|N) * P(f_2 = 1|N) * P(f_3 = 0|N)}{P(Y) * P(f_1 = 1|Y) * P(f_2 = 1|Y) * P(f_3 = 0|Y)} = \frac{125}{1458}$$

Also, remember this:

$$P(N|x) + P(Y|x) = 1$$

So, finally, we have the following:

$$P(Y|x) = 92.1\%$$

There is a 92.1% chance that the new user will like the target movie.

I hope that you now have a solid understanding of Naïve Bayes after going through the theory and a toy example. Let's get ready for its implementation in the next section.

Implementing Naïve Bayes

After calculating the movie preference example by hand, as promised, we are going to implement Naïve Bayes from scratch. After that, we will implement it using the `scikit-learn` package.

Implementing Naïve Bayes from scratch

Before we develop the model, let's define the toy dataset we just worked with:

```
>>> import numpy as np
```

```
>>> X_train = np.array([
...     [0, 1, 1],
...     [0, 0, 1],
...     [0, 0, 0],
...     [1, 1, 0]])
>>> Y_train = ['Y', 'N', 'Y', 'Y']
>>> X_test = np.array([[1, 1, 0]])
```

💡 **Quick tip:** Enhance your coding experience with the **AI Code Explainer** and **Quick Copy** features. Open this book in the next-gen Packt Reader. Click the **Copy** button (**1**) to quickly copy code into your coding environment, or click the **Explain** button (**2**) to get the AI assistant to explain a block of code to you.

```
                                                              Copy      Explain
function calculate(a, b) {                                     1          2
    return {sum: a + b};
};
```

🔒 **The next-gen Packt Reader** is included for free with the purchase of this book. Unlock it by scanning the QR code below or visiting
`https://www.packtpub.com/unlock/9781835085622`.

For the model, starting with the prior, we first group the data by label and record their indices by classes:

```
>>> def get_label_indices(labels):
...     """
...     Group samples based on their labels and return indices
...     @param labels: list of labels
...     @return: dict, {class1: [indices], class2: [indices]}
...     """
...     from collections import defaultdict
...     label_indices = defaultdict(list)
...     for index, label in enumerate(labels):
...         label_indices[label].append(index)
...     return label_indices
```

Take a look at what we get:

```
>>> label_indices = get_label_indices(Y_train)
>>> print('label_indices:\n', label_indices)
    label_indices:
    defaultdict(<class 'list'>, {'Y': [0, 2, 3], 'N': [1]})
```

With `label_indices`, we calculate the prior:

```
>>> def get_prior(label_indices):
...     """
...     Compute prior based on training samples
...     @param label_indices: grouped sample indices by class
...     @return: dictionary, with class label as key, corresponding
...                 prior as the value
...     """
...     prior = {label: len(indices) for label, indices in
...                                     label_indices.items()}
...     total_count = sum(prior.values())
...     for label in prior:
...         prior[label] /= total_count
...     return prior
```

Take a look at the computed prior:

```
>>> prior = get_prior(label_indices)
>>> print('Prior:', prior)
 Prior: {'Y': 0.75, 'N': 0.25}
```

With `prior` calculated, we continue with `likelihood`, which is the conditional probability, `P(feature|class)`:

```
>>> def get_likelihood(features, label_indices, smoothing=0):
...     """
...     Compute likelihood based on training samples
...     @param features: matrix of features
...     @param label_indices: grouped sample indices by class
...     @param smoothing: integer, additive smoothing parameter
...     @return: dictionary, with class as key, corresponding
...                 conditional probability P(feature|class) vector
...                 as value
...     """
...     likelihood = {}
...     for label, indices in label_indices.items():
...         likelihood[label] = features[indices, :].sum(axis=0)
```

```
...                                        + smoothing
...           total_count = len(indices)
...           likelihood[label] = likelihood[label] /
...                                   (total_count + 2 * smoothing)
...       return likelihood
```

We set the `smoothing` value to 1 here, which can also be 0 for no smoothing, or any other positive value, as long as a higher classification performance is achieved:

```
>>> smoothing = 1
>>> likelihood = get_likelihood(X_train, label_indices, smoothing)
>>> print('Likelihood:\n', likelihood)
Likelihood:
 {'Y': array([0.4, 0.6, 0.4]), 'N': array([0.33333333, 0.33333333,
0.66666667])}
```

If you ever find any of this confusing, feel free to check *Figure 2.7* to refresh your memory:

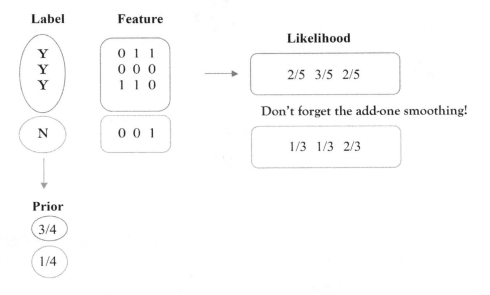

Figure 2.7: A simple example of computing prior and likelihood

With prior and likelihood ready, we can now compute posterior for the testing/new samples:

```
>>> def get_posterior(X, prior, likelihood):
...       """
...       Compute posterior of testing samples, based on prior and
...       likelihood
...       @param X: testing samples
...       @param prior: dictionary, with class label as key,
```

```
...                 corresponding prior as the value
...     @param likelihood: dictionary, with class label as key,
...                         corresponding conditional probability
...                         vector as value
...     @return: dictionary, with class label as key, corresponding
...              posterior as value
...     """
...     posteriors = []
...     for x in X:
...         # posterior is proportional to prior * likelihood
...         posterior = prior.copy()
...         for label, likelihood_label in likelihood.items():
...             for index, bool_value in enumerate(x):
...                 posterior[label] *= likelihood_label[index] if
...                 bool_value else (1 - likelihood_label[index])
...         # normalize so that all sums up to 1
...         sum_posterior = sum(posterior.values())
...         for label in posterior:
...             if posterior[label] == float('inf'):
...                 posterior[label] = 1.0
...             else:
...                 posterior[label] /= sum_posterior
...         posteriors.append(posterior.copy())
...     return posteriors
```

Now, let's predict the class of our one sample test set using this prediction function:

```
>>> posterior = get_posterior(X_test, prior, likelihood)
>>> print('Posterior:\n', posterior)
Posterior:
 [{'Y': 0.9210360075805433, 'N': 0.07896399241945673}]
```

This is exactly what we got previously. We have successfully developed Naïve Bayes from scratch and we can now move on to the implementation using scikit-learn.

Implementing Naïve Bayes with scikit-learn

Coding from scratch and implementing your own solutions is the best way to learn about machine learning models. Of course, you can take a shortcut by directly using the BernoulliNB module (https://scikit-learn.org/stable/modules/generated/sklearn.naive_bayes.BernoulliNB.html) from the scikit-learn API:

```
>>> from sklearn.naive_bayes import BernoulliNB
```

Let's initialize a model with a smoothing factor (specified as `alpha` in `scikit-learn`) of `1.0`, and `prior` learned from the training set (specified as `fit_prior=True` in `scikit-learn`):

```
>>> clf = BernoulliNB(alpha=1.0, fit_prior=True)
```

To train the Naïve Bayes classifier with the `fit` method, we use the following line of code:

```
>>> clf.fit(X_train, Y_train)
```

To obtain the predicted probability results with the `predict_proba` method, we use the following lines of code:

```
>>> pred_prob = clf.predict_proba(X_test)
>>> print('[scikit-learn] Predicted probabilities:\n', pred_prob)
[scikit-learn] Predicted probabilities:
 [[0.07896399 0.92103601]]
```

Finally, we do the following to directly acquire the predicted class with the `predict` method (0.5 is the default threshold, and if the predicted probability of class Y is greater than 0.5, class Y is assigned; otherwise, N is used):

```
>>> pred = clf.predict(X_test)
>>> print('[scikit-learn] Prediction:', pred)
[scikit-learn] Prediction: ['Y']
```

The prediction results using scikit-learn are consistent with what we got using our own solution. Now that we've implemented the algorithm both from scratch and using `scikit-learn`, why don't we use it to solve the movie recommendation problem?

Building a movie recommender with Naïve Bayes

After the toy example, it is now time to build a movie recommender (or, more specifically, movie preference classifier) using a real dataset. We herein use a movie rating dataset (https://grouplens.org/datasets/movielens/). The movie rating data was collected by the GroupLens Research group from the MovieLens website (http://movielens.org).

For demonstration purposes, we will use the stable small dataset, MovieLens 1M Dataset (which can be downloaded from https://files.grouplens.org/datasets/movielens/ml-1m.zip or https://grouplens.org/datasets/movielens/1m/) for ml-1m.zip (size: 1 MB) file. It has around 1 million ratings, ranging from 1 to 5 with half-star increments, given by 6,040 users on 3,706 movies (last updated September 2018).

Unzip the `ml-1m.zip` file and you will see the following four files:

- `movies.dat`: It contains the movie information in the format of `MovieID::Title::Genres`.
- `ratings.dat`: It contains user movie ratings in the format of `UserID::MovieID::Rating::Timestamp`. We will only be using data from this file in this chapter.
- `users.dat`: It contains user information in the format of `UserID::Gender::Age::Occupation::Zip-code`.
- `README`

Let's attempt to predict whether a user likes a particular movie based on how they rate other movies (again, ratings are from 1 to 5).

Preparing the data

First, we import all the necessary modules and read the `ratings.dat` into a pandas DataFrame object:

```
>>> import numpy as np
>>> import pandas as pd
>>> data_path = 'ml-1m/ratings.dat'
>>> df = pd.read_csv(data_path, header=None, sep='::', engine='python')
>>> df.columns = ['user_id', 'movie_id', 'rating', 'timestamp']
>>> print(df)
         user_id  movie_id  rating   timestamp
0              1      1193       5   978300760
1              1       661       3   978302109
2              1       914       3   978301968
3              1      3408       4   978300275
4              1      2355       5   978824291
...          ...       ...     ...         ...
1000204     6040      1091       1   956716541
1000205     6040      1094       5   956704887
1000206     6040       562       5   956704746
1000207     6040      1096       4   956715648
1000208     6040      1097       4   956715569

[1000209 rows x 4 columns]
```

Now, let's see how many unique users and movies are in this million-row dataset:

```
>>> n_users = df['user_id'].nunique()
>>> n_movies = df['movie_id'].nunique()
>>> print(f"Number of users: {n_users}")
Number of users: 6040
>>> print(f"Number of movies: {n_movies}")
Number of movies: 3706
```

Next, we will construct a 6,040 (the number of users) by 3,706 (the number of movies) matrix where each row contains movie ratings from a user, and each column represents a movie, using the following function:

```
>>> def load_user_rating_data(df, n_users, n_movies):
...      data = np.zeros([n_users, n_movies], dtype=np.intc)
             movie_id_mapping = {}
             for user_id, movie_id, rating in zip(df['user_id'], df['movie_
id'], df['rating']):
                 user_id = int(user_id) - 1
                 if movie_id not in movie_id_mapping:
                     movie_id_mapping[movie_id] = len(movie_id_mapping)
                 data[user_id, movie_id_mapping[movie_id]] = rating
             return data, movie_id_mapping

>>> data, movie_id_mapping = load_user_rating_data(df, n_users, n_movies)
```

Besides the rating matrix `data`, we also record the `movie ID` to column index mapping. The column index is from 0 to 3,705 as we have 3,706 movies.

It is always recommended to analyze the data distribution in order to identify if there is a class imbalance issue in the dataset. We do the following:

```
>>> values, counts = np.unique(data, return_counts=True)
... for value, count in zip(values, counts):
...      print(f'Number of rating {value}: {count}')
Number of rating 0: 21384031
Number of rating 1: 56174
Number of rating 2: 107557
Number of rating 3: 261197
Number of rating 4: 348971
Number of rating 5: 226310
```

As you can see, most ratings are unknown; for the known ones, 35% are of rating 4, followed by 26% of rating 3, 23% of rating 5, and then 11% and 6% of ratings 2 and 1, respectively.

Since most ratings are unknown, we take the movie with the most known ratings as our target movie for easier prediction validation. We look for rating counts for each movie as follows:

```
>>> print(df['movie_id'].value_counts())
2858    3428
260     2991
1196    2990
1210    2883
480     2672
        ...
3458       1
2226       1
1815       1
398        1
2909       1
Name: movie_id, Length: 3706, dtype: int64
```

So, the target movie is ID, and we will treat ratings of other movies as features. We only use rows with ratings available for the target movie so we can validate how good the prediction is. We construct the dataset accordingly as follows:

```
>>> target_movie_id = 2858
>>> X_raw = np.delete(data, movie_id_mapping[target_movie_id], axis=1)
>>> Y_raw = data[:, movie_id_mapping[target_movie_id]]
>>> X = X_raw[Y_raw > 0]
>>> Y = Y_raw[Y_raw > 0]
>>> print('Shape of X:', X.shape)
Shape of X: (3428, 3705)
>>> print('Shape of Y:', Y.shape)
Shape of Y: (3428,)
```

We can consider movies with ratings greater than 3 as being liked (being recommended):

```
>>> recommend = 3
>>> Y[Y <= recommend] = 0
>>> Y[Y > recommend] = 1
>>> n_pos = (Y == 1).sum()
>>> n_neg = (Y == 0).sum()
>>> print(f'{n_pos} positive samples and {n_neg} negative samples.')
2853 positive samples and 575 negative samples.
```

As a rule of thumb in solving classification problems, we need to always analyze the label distribution and see how balanced (or imbalanced) the dataset is.

Best practice

Dealing with imbalanced datasets in classification problems requires careful consideration and appropriate techniques to ensure that the model effectively learns from the data and produces reliable predictions. Here are several strategies to address class imbalance:

- **Oversampling:** We can increase the number of instances in the minority class by generating synthetic samples or duplicating existing ones.
- **Undersampling:** We can decrease the number of instances in the majority class by randomly removing samples. Note that we can even combine oversampling and undersampling for a more balanced dataset.
- **Class weighting:** We can also assign higher weights to minority class samples during model training. In this way, we penalize misclassifications of the minority class more heavily.

Next, to comprehensively evaluate our classifier's performance, we can randomly split the dataset into two sets, the training and testing sets, which simulate learning data and prediction data, respectively. Generally, the proportion of the original dataset to include in the testing split can be 20%, 25%, 30%, or 33.3%.

Best practice

Here are some guidelines for choosing the testing split:

- **Small datasets:** If you have a small dataset (e.g., less than a few thousand samples), a larger testing split (e.g., 25% to 30%) may be appropriate to ensure that you have enough data for training and testing.
- **Medium to large datasets:** For medium to large datasets (e.g., tens of thousands to millions of samples), a smaller testing split (e.g., 20%) may still provide enough data for evaluation while allowing more data to be used for training. A 20% testing split is a common choice in such cases.
- **Simple models:** Less complex models are generally less prone to overfitting, so using a smaller test set split may work.
- **Complex models:** Complex models like deep learning models can be more prone to overfitting. Hence, a larger test set split (e.g., 30%) is recommended.

We use the train_test_split function from scikit-learn to do the random splitting and to preserve the percentage of samples for each class:

```
>>> from sklearn.model_selection import train_test_split
>>> X_train, X_test, Y_train, Y_test = train_test_split(X, Y,
...         test_size=0.2, random_state=42)
```

It is a good practice to assign a fixed `random_state` (for example, 42) during experiments and exploration in order to guarantee that the same training and testing sets are generated every time the program runs. This allows us to make sure that the classifier functions and performs well on a fixed dataset before we incorporate randomness and proceed further.

We check the training and testing sizes as follows:

```
>>> print(len(Y_train), len(Y_test))
2742 686
```

Another good thing about the `train_test_split` function is that the resulting training and testing sets will have the same class ratio.

Training a Naïve Bayes model

Next, we train a Naïve Bayes model on the training set. You may notice that the values of the input features are from 0 to 5, as opposed to 0 or 1 in our toy example. Hence, we use the `MultinomialNB` module (https://scikit-learn.org/stable/modules/generated/sklearn.naive_bayes.MultinomialNB.html) from scikit-learn instead of the `BernoulliNB` module, as `MultinomialNB` can work with integer features as well as fractional counts. We import the module, initialize a model with a smoothing factor of 1.0 and prior learned from the training set, and train this model against the training set as follows:

```
>>> from sklearn.naive_bayes import MultinomialNB
>>> clf = MultinomialNB(alpha=1.0, fit_prior=True)
>>> clf.fit(X_train, Y_train)
```

Then, we use the trained model to make predictions on the testing set. We get the predicted probabilities as follows:

```
>>> prediction_prob = clf.predict_proba(X_test)
>>> print(prediction_prob[0:10])
[[7.50487439e-23 1.00000000e+00]
 [1.01806208e-01 8.98193792e-01]
 [3.57740570e-10 1.00000000e+00]
 [1.00000000e+00 2.94095407e-16]
 [1.00000000e+00 2.49760836e-25]
 [7.62630220e-01 2.37369780e-01]
 [3.47479627e-05 9.99965252e-01]
 [2.66075292e-11 1.00000000e+00]
 [5.88493563e-10 9.99999999e-01]
 [9.71326867e-09 9.99999990e-01]]
```

For each testing sample, we output the probability of class 0, followed by the probability of class 1.

We get the predicted class for the test set as follows:

```
>>> prediction = clf.predict(X_test)
>>> print(prediction[:10])
[[1. 1. 1. 0. 0. 0. 1. 1. 1. 1.]
```

Finally, we evaluate the model's performance with classification accuracy, which is the proportion of correct predictions:

```
>>> accuracy = clf.score(X_test, Y_test)
>>> print(f'The accuracy is: {accuracy*100:.1f}%')
The accuracy is: 71.6%
```

The classification accuracy is around 72%, which means that the Naïve Bayes classifier we've constructed accurately suggests movies to users about three quarters of the time. Ideally, we could also utilize movie genre information from the movies.dat file, and user demographics (gender, age, occupation, and ZIP code) information from the users.dat file. Obviously, movies in similar genres tend to attract similar users, and users of similar demographics likely have similar movie preferences. We will leave it as an exercise for you to explore further.

So far, we have covered in depth the first machine learning classifier and evaluated its performance by prediction accuracy. Are there any other classification metrics? Let's see in the next section.

Evaluating classification performance

Beyond accuracy, there are several metrics we can use to gain more insight and avoid class imbalance effects. These are as follows:

- Confusion matrix
- Precision
- Recall
- F1 score
- The area under the curve

A **confusion matrix** summarizes testing instances by their predicted values and true values, presented as a contingency table:

		Predicted	
		Negative	Positive
Actual	Negative	TN	FP
	Positive	FN	TP

TN = True Negative
FP = False Positive
FN = False Negative
TP =True Positive

Figure 2.8: Contingency table for a confusion matrix

To illustrate this, we can compute the confusion matrix of our Naïve Bayes classifier. We use the confusion_matrix function from scikit-learn to compute it, but it is very easy to code it ourselves:

```
>>> from sklearn.metrics import confusion_matrix
>>> print(confusion_matrix(Y_test, prediction, labels=[0, 1]))
[[ 60  47]
 [148 431]]
```

As you can see from the resulting confusion matrix, there are 47 false positive cases (where the model misinterprets a dislike as a like for a movie), and 148 false negative cases (where it fails to detect a like for a movie). Hence, classification accuracy is just the proportion of all true cases:

$$\frac{TN + TP}{TN + TP + FP + FN} = \frac{60 + 431}{60 + 431 + 47 + 148} = 71.6\%$$

Precision measures the fraction of positive calls that are correct, which are the following, in our case:

$$\frac{TP}{TP + FP} = \frac{431}{431 + 47} = 0.90$$

Recall, on the other hand, measures the fraction of true positives that are correctly identified, which are the following in our case:

$$\frac{TP}{TP + FN} = \frac{431}{431 + 148} = 0.74$$

Recall is also called the **true positive rate**.

The **f1 score** comprehensively includes both the precision and the recall and equates to their **harmonic mean**:

$$f_1 = 2 * \frac{precision * recall}{precision + recall}$$

We tend to value the **f1** score above precision or recall alone.

Let's compute these three measurements using corresponding functions from scikit-learn, as follows:

```
>>> from sklearn.metrics import precision_score, recall_score, f1_score
>>> precision_score(Y_test, prediction, pos_label=1)
0.9016736401673641
>>> recall_score(Y_test, prediction, pos_label=1)
0.7443868739205527
>>> f1_score(Y_test, prediction, pos_label=1)
0.815515610217597
```

On the other hand, the negative (dislike) class can also be viewed as positive, depending on the context. For example, assign the 0 class as pos_label and we have the following:

```
>>> f1_score(Y_test, prediction, pos_label=0)
0.38095238095238093
```

To obtain the precision, recall, and f1 score for each class, instead of exhausting all class labels in the three function calls as shown earlier, a quicker way is to call the classification_report function:

```
>>> from sklearn.metrics import classification_report
>>> report = classification_report(Y_test, prediction)
>>> print(report)
              precision    recall  f1-score   support
         0.0       0.29      0.56      0.38       107
         1.0       0.90      0.74      0.82       579
   micro avg       0.72      0.72      0.72       686
   macro avg       0.60      0.65      0.60       686
weighted avg       0.81      0.72      0.75       686
```

Here, weighted avg is the weighted average according to the proportions of the class.

The classification report provides a comprehensive view of how the classifier performs on each class. It is, as a result, useful in imbalanced classification, where we can easily obtain high accuracy by simply classifying every sample as the dominant class, while the precision, recall, and f1 score measurements for the minority class, however, will be significantly low.

Precision, recall, and the f1 score are also applicable to **multiclass** classification, where we can simply treat a class we are interested in as a positive case, and any other classes as negative cases.

During the process of tweaking a binary classifier (that is, trying out different combinations of hyperparameters, for example, the smoothing factor in our Naïve Bayes classifier), it would be perfect if there was a set of parameters in which the highest averaged and class individual f1 scores are achieved at the same time. It is, however, usually not the case. Sometimes, a model has a higher average f1 score than another model, but a significantly low f1 score for a particular class; sometimes, two models have the same average f1 scores, but one has a higher f1 score for one class and a lower score for another class. In situations such as these, how can we judge which model works better? The **Area Under the Curve (AUC)** of the **Receiver Operating Characteristic (ROC)** is a consolidated measurement frequently used in binary classification.

The ROC curve is a plot of the true positive rate versus the false positive rate at various probability thresholds, ranging from 0 to 1. For a testing sample, if the probability of a positive class is greater than the threshold, then a positive class is assigned; otherwise, we use a negative class. To recap, the true positive rate is equivalent to recall, and the false positive rate is the fraction of negatives that are incorrectly identified as positive. Let's code and exhibit the ROC curve (under thresholds of 0.0, 0.1, 0.2, ..., 1.0) of our model:

```
>>> pos_prob = prediction_prob[:, 1]
>>> thresholds = np.arange(0.0, 1.1, 0.05)
>>> true_pos, false_pos = [0]*len(thresholds), [0]*len(thresholds)
>>> for pred, y in zip(pos_prob, Y_test):
...         for i, threshold in enumerate(thresholds):
...             if pred >= threshold:
...                 # if truth and prediction are both 1
...                 if y == 1:
...                     true_pos[i] += 1
...                 # if truth is 0 while prediction is 1
...                 else:
...                     false_pos[i] += 1
...             else:
...                 break
```

Then, let's calculate the true and false positive rates for all threshold settings (remember, there are 516.0 positive testing samples and 1191 negative ones):

```
>>> n_pos_test = (Y_test == 1).sum()
>>> n_neg_test = (Y_test == 0).sum()
>>> true_pos_rate = [tp / n_pos_test for tp in true_pos]
>>> false_pos_rate = [fp / n_neg_test for fp in false_pos]
```

Now, we can plot the ROC curve with matplotlib:

```
>>> import matplotlib.pyplot as plt
>>> plt.figure()
>>> lw = 2
>>> plt.plot(false_pos_rate, true_pos_rate,
...          color='darkorange', lw=lw)
>>> plt.plot([0, 1], [0, 1], color='navy', lw=lw, linestyle='--')
>>> plt.xlim([0.0, 1.0])
>>> plt.ylim([0.0, 1.05])
>>> plt.xlabel('False Positive Rate')
>>> plt.ylabel('True Positive Rate')
>>> plt.title('Receiver Operating Characteristic')
>>> plt.legend(loc="lower right")
>>> plt.show()
```

Refer to *Figure 2.9* for the resulting ROC curve:

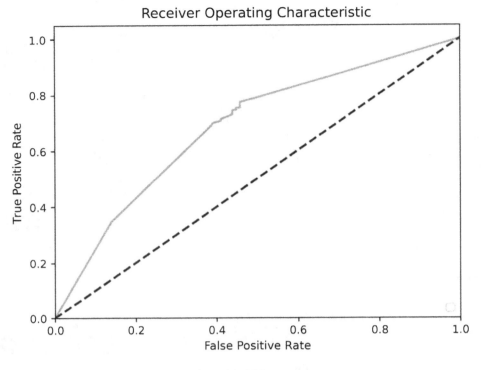

Figure 2.9: ROC curve

In the graph, the dashed line is the baseline representing random guessing, where the true positive rate increases linearly with the false positive rate; its AUC is 0.5. The solid line is the ROC plot of our model, and its AUC is somewhat less than 1. In a perfect case, the true positive samples have a probability of 1, so that the ROC starts at the point with 100% true positive and 0% false positive. The AUC of such a perfect curve is 1. To compute the exact AUC of our model, we can resort to the roc_auc_score function of scikit-learn:

```
>>> from sklearn.metrics import roc_auc_score
>>> roc_auc_score(Y_test, pos_prob)
0.6857375752586637
```

 What AUC value leads to the conclusion that a classifier is good? Unfortunately, there is no such "magic" number. We use the following rule of thumb as general guidelines: classification models achieving an AUC of 0.7 to 0.8 are considered acceptable, 0.8 to 0.9 are great, and anything above 0.9 are superb. Again, in our case, we are only using the very sparse movie rating data. Hence, an AUC of 0.69 is actually acceptable.

You have learned several classification metrics, and we will explore how to measure them properly and how to fine-tune our models in the next section.

Tuning models with cross-validation

Limiting the evaluation to a single fixed set may be misleading since it's highly dependent on the specific data points chosen for that set. We can simply avoid adopting the classification results from one fixed testing set, which we did in experiments previously. Instead, we usually apply the **k-fold cross-validation** technique to assess how a model will generally perform in practice.

In the k-fold cross-validation setting, the original data is first randomly divided into k equal-sized subsets, in which class proportion is often preserved. Each of these k subsets is then successively retained as the testing set for evaluating the model. During each trial, the rest of the k -1 subsets (excluding the one-fold holdout) form the training set for driving the model. Finally, the average performance across all k trials is calculated to generate an overall result:

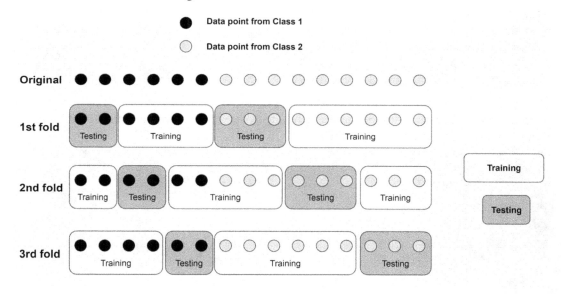

Figure 2.10: Diagram of 3-fold cross-validation

Statistically, the average performance of k-fold cross-validation is a better estimate of how a model performs in general. Given different sets of parameters pertaining to a machine learning model and/or data preprocessing algorithms, or even two or more different models, the goal of model tuning and/or model selection is to pick a set of parameters of a classifier so that the best average performance is achieved. With these concepts in mind, we can now start to tweak our Naïve Bayes classifier, incorporating cross-validation and the AUC of ROC measurements.

 In k-fold cross-validation, k is usually set at 3, 5, or 10. If the training size is small, a large k (5 or 10) is recommended to ensure sufficient training samples in each fold. If the training size is large, a small value (such as 3 or 4) works fine since a higher k will lead to an even higher computational cost of training on a large dataset.

We will use the `split()` method from the `StratifiedKFold` class of `scikit-learn` to divide the data into chunks with preserved class distribution:

```
>>> from sklearn.model_selection import StratifiedKFold
>>> k = 5
>>> k_fold = StratifiedKFold(n_splits=k, random_state=42)
```

After initializing a 5-fold generator, we choose to explore the following values for the following parameters:

- `alpha`: This represents the smoothing factor, the initial value for each feature
- `fit_prior`: This represents whether to use prior tailored to the training data

We start with the following options:

```
>>> smoothing_factor_option = [1, 2, 3, 4, 5, 6]
>>> fit_prior_option = [True, False]
>>> auc_record = {}
```

Then, for each fold generated by the `split()` method of the `k_fold` object, we repeat the process of classifier initialization, training, and prediction with one of the aforementioned combinations of parameters, and record the resulting AUCs:

```
>>> for train_indices, test_indices in k_fold.split(X, Y):
...     X_train_k, X_test _k= X[train_indices], X[test_indices]
...     Y_train_k, Y_test_k = Y[train_indices], Y[test_indices]
...     for alpha in smoothing_factor_option:
...         if alpha not in auc_record:
...             auc_record[alpha] = {}
...         for fit_prior in fit_prior_option:
...             clf = MultinomialNB(alpha=alpha,
                                    fit_prior=fit_prior)
...             clf.fit(X_train_k, Y_train_k)
...             prediction_prob = clf.predict_proba(X_test_k)
...             pos_prob = prediction_prob[:, 1]
...             auc = roc_auc_score(Y_test_k, pos_prob)
...             auc_record[alpha][fit_prior] = auc +
                                auc_record[alpha].get(fit_prior, 0.0)
```

Finally, we present the results as follows:

```
>>> for smoothing, smoothing_record in auc_record.items():
...     for fit_prior, auc in smoothing_record.items():
...         print(f'    {smoothing}        {fit_prior}
                    {auc/k:.5f}')
smoothing  fit prior  auc
```

```
1         True      0.65647
1         False     0.65708
2         True      0.65795
2         False     0.65823
3         True      0.65740
3         False     0.65801
4         True      0.65808
4         False     0.65795
5         True      0.65814
5         False     0.65694
6         True      0.65663
6         False     0.65719
```

The (2, False) set enables the best averaged AUC, at 0.65823.

Finally, we retrain the model with the best set of hyperparameters (2, False) and compute the AUC:

```
>>> clf = MultinomialNB(alpha=2.0, fit_prior=False)
>>> clf.fit(X_train, Y_train)
>>> pos_prob = clf.predict_proba(X_test)[:, 1]
>>> print('AUC with the best model:', roc_auc_score(Y_test,
...        pos_prob))
AUC with the best model:  0.6862056720417091
```

An AUC of 0.686 is achieved with the fine-tuned model. In general, tweaking model hyperparameters using cross-validation is one of the most effective ways to boost learning performance and reduce overfitting.

Summary

In this chapter, you learned about the fundamental concepts of machine learning classification, including types of classification, classification performance evaluation, cross-validation, and model tuning. You also learned about the simple, yet powerful, classifier, Naïve Bayes. We went in depth through the mechanics and implementations of Naïve Bayes with a couple of examples, the most important one being the movie recommendation project.

Binary classification using Naïve Bayes was the main talking point of this chapter. In the next chapter, we will solve ad click-through prediction using another binary classification algorithm: a **decision tree**.

Exercises

1. As mentioned earlier, we extracted user-movie relationships only from the movie rating data where most ratings are unknown. Can you also utilize data from the `movies.dat` and `users.dat` files?

2. Practice makes perfect—another great project to deepen your understanding could be heart disease classification. The dataset can be downloaded directly from `https://archive.ics.uci.edu/ml/datasets/Heart+Disease`.

3. Don't forget to fine-tune the model you obtained from Exercise 2 using the techniques you learned in this chapter. What is the best AUC it achieves?

References

To acknowledge the use of the MovieLens dataset in this chapter, I would like to cite the following paper:

F. Maxwell Harper and Joseph A. Konstan. 2015. *The MovieLens Datasets: History and Context*. ACM **Transactions on Interactive Intelligent Systems (TiiS)** 5, 4, Article 19 (December 2015), 19 pages. DOI: `http://dx.doi.org/10.1145/2827872`.

Unlock this book's exclusive benefits now

This book comes with additional benefits designed to elevate your learning experience.

Note: Have your purchase invoice ready before you begin.   `https://www.packtpub.com/unlock/9781835085622`

3

Predicting Online Ad Click-Through with Tree-Based Algorithms

In the previous chapter, we built a movie recommender. In this chapter and the next, we will be solving one of the most data-driven problems in digital advertising: ad click-through prediction—given a user and the page they are visiting, this predicts how likely it is that they will click on a given ad. We will focus on learning tree-based algorithms (including decision trees, random forest models, and boosted trees) and utilize them to tackle this billion-dollar problem.

We will be exploring decision trees from the root to the leaves, as well as the aggregated version, a forest of trees. This won't be a theory-only chapter, as there are a lot of hand calculations and implementations of tree models from scratch included. We will be using scikit-learn and XGBoost, a popular Python package for tree-based algorithms.

We will cover the following topics in this chapter:

- A brief overview of ad click-through prediction
- Exploring a decision tree from the root to the leaves
- Implementing a decision tree from scratch
- Implementing a decision tree with scikit-learn
- Predicting ad click-through with a decision tree
- Ensembling decision trees – random forests
- Ensembling decision trees – gradient-boosted trees

A brief overview of ad click-through prediction

Online display advertising is a multibillion-dollar industry. Online display ads come in different formats, including banner ads composed of text, images, and flash, and rich media such as audio and video. Advertisers, or their agencies, place ads on a variety of websites, and even mobile apps, across the internet in order to reach potential customers and deliver an advertising message.

Online display advertising has served as one of the greatest examples of machine learning utilization. Obviously, advertisers and consumers are keenly interested in well-targeted ads. In the last 20 years, the industry has relied heavily on the ability of machine learning models to predict the effectiveness of ad targeting: how likely it is that an audience of a certain age group will be interested in this product, that customers with a certain household income will purchase this product after seeing the ad, that frequent sports site visitors will spend more time reading this ad, and so on. The most common measurement of effectiveness is the **Click-Through Rate (CTR)**, which is the ratio of clicks on a specific ad to its total number of views. In general cases without clickbait or spammy content, a higher CTR indicates that an ad is targeted well and that an online advertising campaign is successful.

Click-through prediction entails both the promises and challenges of machine learning. It mainly involves the binary classification of whether a given ad on a given page (or app) will be clicked on by a given user, with predictive features from the following three aspects:

*   Ad content and information (category, position, text, format, and so on)
*   Page content and publisher information (category, context, domain, and so on)
*   User information (age, gender, location, income, interests, search history, browsing history, device, and so on)

Suppose we, as an agency, are operating ads on behalf of several advertisers, and our job is to place the right ads for the right audience. Let's say that we have an existing dataset in hand (the following small chunk is an example; the number of predictive features can easily go into the thousands in reality) taken from millions of records of campaigns run a month ago, and we need to develop a classification model to learn and predict future ad placement outcomes:

Ad category	Site category	Site domain	User age	User gender	User occupation	Interested in sports	Interested in tech	Click
Auto	News	cnn.com	25-34	M	Professional	True	True	1
Fashion	News	bbc.com	35-54	F	Professional	False	False	0
Auto	Education	onlinestudy.com	17-24	F	Student	True	True	0
Food	Entertainment	movie.com	25-34	M	Clerk	True	False	1
Fashion	Sports	football.com	55+	M	Retired	True	False	0
...	...	...	...	...	...	...	...	...
...	...	...	...	...	...	...	...	...

| Food | News | abc.com | 17-24 | M | Student | True | True | ? |
| Auto | Entertainment | movie.com | 35-54 | F | Professional | True | False | ? |

Figure 3.1: Ad samples for training and prediction

As you can see in *Figure 3.1*, the features are mostly categorical. In fact, data can be either numerical or categorical. Let's explore this in more detail in the next section.

Getting started with two types of data — numerical and categorical

At first glance, the features in the preceding dataset are **categorical** – for example, male or female, one of four age groups, one of the predefined site categories, and whether the user is interested in sports. Such data is different from the **numerical** feature data we have worked with until now.

Categorical features, also known as **qualitative features**, represent distinct characteristics or groups with a countable number of options. Categorical features may or may not have a logical order. For example, household income from low to medium to high is an **ordinal** feature, while the category of an ad is not ordinal.

Numerical (also called **quantitative**) features, on the other hand, have mathematical meaning as a measurement and, of course, are ordered. For instance, counts of items (e.g., number of children in a family, number of bedrooms in a house, and number of days until an event) are discrete numerical features; the height of individuals, temperature, and the weight of objects are continuous numerical. The cardiotocography dataset (`https://archive.ics.uci.edu/ml/datasets/Cardiotocography`) contains both discrete (such as the number of accelerations per second or the number of fetal movements per second) and continuous (such as the mean value of long-term variability) numerical features.

Categorical features can also take on numerical values. For example, 1 to 12 can represent months of the year, and 1 and 0 can indicate adult and minor. Still, these values do not have mathematical implications.

The Naïve Bayes classifier you learned about previously works for both numerical and categorical features as the likelihoods, $P(x \mid y)$ or $P(feature \mid class)$, are calculated in the same way.

Now, say we are thinking of predicting click-through using Naïve Bayes and trying to explain the model to our advertising clients. However, our clients may find it difficult to understand the prior and the likelihood of individual attributes and their multiplication. Is there a classifier that is easy to interpret and explain to clients, and that is able to directly handle categorical data? Decision trees are the answer!

Exploring a decision tree from the root to the leaves

A decision tree is a tree-like graph, that is, a sequential diagram illustrating all of the possible decision alternatives and their corresponding outcomes. Starting from the **root** of a tree, every internal **node** represents the basis on which a decision is made. Each branch of a node represents how a choice may lead to the next node. And, finally, each **terminal node**, the **leaf**, represents the outcome produced.

For example, we have just made a couple of decisions that brought us to the point of using a decision tree to solve our advertising problem:

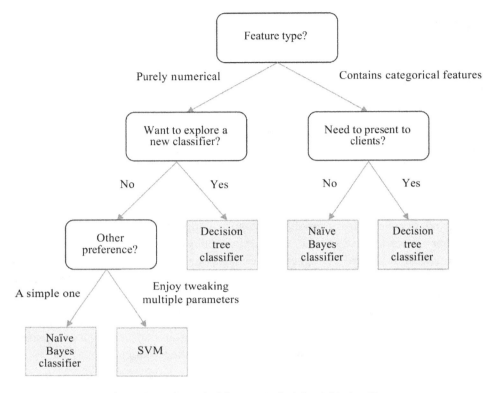

Figure 3.2: Using a decision tree to find the right algorithm

The first condition, or the root, is whether the feature type is numerical or categorical. Let's assume our ad clickstream data contains mostly categorical features, so it goes to the right branch. In the next node, our work needs to be interpretable by non-technical clients, so, it goes to the right branch and reaches the leaf for choosing the decision tree classifier.

You can also look at the paths and see what kinds of problems they can fit in. A decision tree classifier operates in the form of a decision tree. It maps observations to class assignments (symbolized as leaf nodes) through a series of tests (represented as internal nodes) based on feature values and corresponding conditions (represented as branches). In each node, a question regarding the values and characteristics of a feature is asked; depending on the answer to the question, the observations are split into subsets. Sequential tests are conducted until a conclusion about the observations' target label is reached. The paths from the root to the end leaves represent the decision-making process and the classification rules.

In a more simplified scenario, as shown in *Figure 3.3*, where we want to predict **Click** or **No click** on a self-driven car ad, we can manually construct a decision tree classifier that works for an available dataset. For example, if a user is interested in technology and has a car, they will tend to click on the ad; a person outside of this subset is unlikely to click on the ad, hypothetically. We then use the trained tree to predict two new inputs, whose results are **Click** and **No click**, respectively:

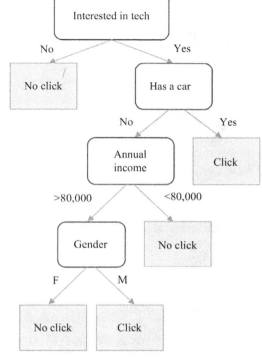

User gender	Annual income	Has a car	Interested in tech	Click
M	200,000	True	True	1
F	5,000	False	False	0
F	100,000	True	True	1
M	10,000	True	False	0
M	80,000	False	False	0
...	...	...	...	...
...	...	...	...	...

M	120,000	True	True	?
F	70,000	False	True	?

Figure 3.3: Predicting Click/No Click with a trained decision tree

After a decision tree has been constructed, classifying a new sample is straightforward, as you just saw: starting from the root, apply the test condition and follow the branch accordingly until a leaf node is reached, and the class label associated will be assigned to the new sample.

So, how can we build an appropriate decision tree?

Constructing a decision tree

A decision tree is constructed by partitioning the training samples into successive subsets. The partitioning process is repeated in a recursive fashion on each subset. For each partitioning at a node, a condition test is conducted based on the value of a feature of the subset. When the subset shares the same class label, or when no further splitting can improve the class purity of this subset, recursive partitioning on this node is finished.

Important note

Class purity refers to the homogeneity of the target variable (class labels) within a subset of data. A subset is considered to have high class purity if the majority of its instances belong to the same class. In other words, a subset with high class purity contains mostly instances of the same class label, while a subset with low class purity contains instances from multiple classes.

Theoretically, to partition a feature (numerical or categorical) with n different values, there are n different methods of binary splitting (**Yes** or **No** to the condition test, as illustrated in *Figure 3.4*), not to mention other ways of splitting (for example, three- and four-way splitting in *Figure 3.4*):

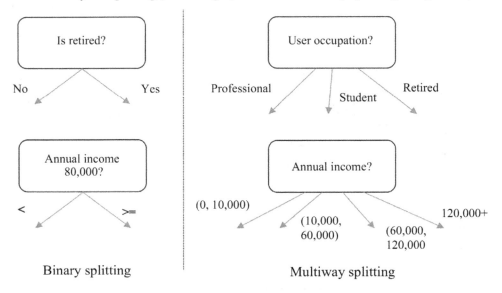

Figure 3.4: Examples of binary splitting and multiway splitting

Without considering the order of features that partitioning is taking place on, there are already n^m possible trees for an m-dimensional dataset.

Many algorithms have been developed to efficiently construct an accurate decision tree. Popular ones include the following:

- **Iterative Dichotomiser 3 (ID3):** This algorithm uses a greedy search in a top-down manner by selecting the best attribute to split the dataset on with each iteration without backtracking.
- **C4.5:** This is an improved version of ID3 that introduces backtracking. It traverses the constructed tree and replaces the branches with leaf nodes if purity is improved this way.
- **Classification and Regression Tree (CART):** This constructs the tree using binary splitting, which we will discuss in more detail shortly. CART's flexibility, efficiency, interpretability, and robustness make it a popular choice for various classification and regression tasks.
- **Chi-squared Automatic Interaction Detector (CHAID):** This algorithm is often used in direct marketing. It involves complicated statistical concepts, but basically, it determines the optimal way of merging predictive variables in order to best explain the outcome.

The basic idea of these algorithms is to grow the tree greedily by making a series of local optimizations when choosing the most significant feature to use to partition the data. The dataset is then split based on the optimal value of that feature. We will discuss the measurement of a significant feature and the optimal splitting value of a feature in the next section.

First, we will study the CART algorithm in more detail, and we will implement it as the most notable decision tree algorithm after that. It constructs the tree using binary splitting and grows each node into left and right children. In each partition, it greedily searches for the most significant combination of a feature and its value; all different possible combinations are tried and tested using a measurement function. With the selected feature and value as a splitting point, the algorithm then divides the dataset as follows:

- Samples with the feature of this value (for a categorical feature) or a greater value (for a numerical feature) become the right child
- The remaining samples become the left child

This partitioning process repeats and recursively divides up the input samples into two subgroups. The splitting process stops at a subgroup where either of the following two criteria is met:

- **The minimum number of samples for a new node:** When the number of samples is not greater than the minimum number of samples required for a further split, the partitioning stops in order to prevent the tree from excessively tailoring to the training set and, as a result, overfitting.
- **The maximum depth of the tree:** A node stops growing when its depth, which is defined as the number of partitions taking place from the top down, starting from the root node and ending in a terminal node, meets the maximum tree depth. Deeper trees are more specific to the training set and can lead to overfitting.

A node with no branches becomes a leaf, and the dominant class of samples at this node is the prediction. Once all the splitting processes have finished, the tree is constructed and is portrayed with the assigned labels at the terminal nodes and the splitting points (feature and value) at all the internal nodes above.

We will implement the CART decision tree algorithm from scratch after studying the metrics of selecting the optimal splitting feature and value, as promised.

The metrics for measuring a split

When selecting the best combination of a feature and a value as the splitting point, two criteria, such as **Gini Impurity** and **Information Gain**, can be used to measure the quality of separation.

Gini Impurity

Gini Impurity, as its name implies, measures the impurity rate of the class distribution of data points, or the class mixture rate. For a dataset with K classes, suppose that data from class $k(1 \le k \le K)$ takes up a fraction $f_k(0 \le f_k \le 1)$ of the entire dataset; then the *Gini Impurity* of this dataset is written as follows:

$$Gini\ impurity = 1 - \sum_{K=1}^{K} f_K{}^2$$

A lower Gini Impurity indicates a purer dataset. For example, when the dataset contains only one class, say, the fraction of this class is 1 and that of the others is 0, its Gini Impurity becomes $1 - (1^2 + 0^2) = 0$. In another example, a dataset records a large number of coin flips, and heads and tails each take up half of the samples. The Gini Impurity is $1 - (0.5^2 + 0.5^2) = 0.5$.

In binary cases, Gini Impurity, under different values of the positive class fraction, can be visualized with the following code blocks:

```
>>> import matplotlib.pyplot as plt
>>> import numpy as np
```

The fraction of the positive class varies from 0 to 1:

```
>>> pos_fraction = np.linspace(0.00, 1.00, 1000)
```

The Gini Impurity is calculated accordingly, followed by the plot of **Gini Impurity** versus **positive fraction**:

```
>>> gini = 1 - pos_fraction**2 - (1-pos_fraction)**2
```

Here, 1-pos_fraction is the negative fraction:

```
>>> plt.plot(pos_fraction, gini)
>>> plt.ylim(0, 1)
>>> plt.xlabel('Positive fraction')
>>> plt.ylabel('Gini impurity')
>>> plt.show()
```

Refer to *Figure 3.5* for the end result:

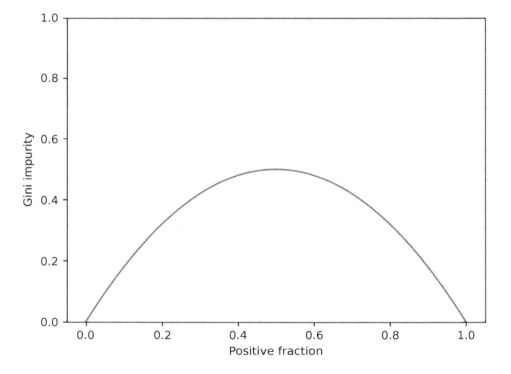

Figure 3.5: Gini Impurity versus positive fraction

As you can see, in binary cases, if the positive fraction is 50%, the impurity will be the highest at 0.5; if the positive fraction is 100% or 0%, it will reach 0 impurity.

Given the labels of a dataset, we can implement the Gini Impurity calculation function as follows:

```
>>> def gini_impurity(labels):
...     # When the set is empty, it is also pure
...     if len(labels) == 0:
...         return 0
...     # Count the occurrences of each label
...     counts = np.unique(labels, return_counts=True)[1]
...     fractions = counts / float(len(labels))
...     return 1 - np.sum(fractions ** 2)
```

Test it out with some examples:

```
>>> print(f'{gini_impurity([1, 1, 0, 1, 0]):.4f}')
0.4800
>>> print(f'{gini_impurity([1, 1, 0, 1, 0, 0]):.4f}')
0.5000
>>> print(f'{gini_impurity([1, 1, 1, 1]):.4f}')
0.0000
```

In order to evaluate the quality of a split, we simply add up the Gini Impurity of all resulting subgroups, combining the proportions of each subgroup as corresponding weight factors. And again, the smaller the weighted sum of the Gini Impurity, the better the split.

Take a look at the following self-driving car ad example. Here, we split the data based on a user's gender and interest in technology, respectively:

User gender	Interested in tech	Click	Group by gender
M	True	1	Group 1
F	False	0	Group 2
F	True	1	Group 2
M	False	0	Group 1
M	False	1	Group 1

User gender	Interested in tech	Click	Group by interest
M	True	1	Group 1
F	False	0	Group 2
F	True	1	Group 1
M	False	0	Group 2
M	False	1	Group 2

#1 split based on gender   #2 split based on interest in tech

Figure 3.6: Splitting the data based on gender or interest in tech

The weighted Gini Impurity of the first split can be calculated as follows:

$$\#1\ Gini\ Impurity = \frac{3}{5}[1 - (\frac{2^2}{3} + \frac{1^2}{3})] + \frac{2}{5}[1 - (\frac{1^2}{2} + \frac{1^2}{2})] = 0.467$$

The second split is as follows:

$$\#2\ Gini\ Impurity = \frac{2}{5}[1 - (1^2 + 0^2)] + \frac{3}{5}[1 - (\frac{1^2}{3} + \frac{2^2}{3})] = 0.267$$

Therefore, splitting data based on the user's interest in technology is a better strategy than gender.

Information Gain

Another metric, **Information Gain**, measures the improvement of purity after splitting or, in other words, the reduction of uncertainty due to a split. Higher Information Gain implies better splitting. We obtain the Information Gain of a split by comparing the **entropy** before and after the split.

Entropy is a probabilistic measure of uncertainty. Given a K-class dataset, and f_k $(0 \le f_k \le 1)$ denoted as the fraction of data from class k $(1 \le k \le K)$, the *entropy* of the dataset is defined as follows:

$$Entropy = -\sum_{K=1}^{K} f_k * log_2 f_k$$

Lower entropy implies a purer dataset with less ambiguity. In a perfect case, where the dataset contains only one class, the entropy is:

$$-(1 * log_2 1 + 0) = 0$$

In the coin flip example, the entropy becomes:

$$-(0.5 * log_2 0.5 + 0.5 * log_2 0.5) = 1$$

Similarly, we can visualize how entropy changes with different values of the positive class fraction in binary cases using the following lines of code:

```
>>> pos_fraction = np.linspace(0.001, 0.999, 1000)
>>> ent = - (pos_fraction * np.log2(pos_fraction) +
...          (1 - pos_fraction) * np.log2(1 - pos_fraction))
>>> plt.plot(pos_fraction, ent)
>>> plt.xlabel('Positive fraction')
>>> plt.ylabel('Entropy')
>>> plt.ylim(0, 1)
>>> plt.show()
```

This will give us the following output:

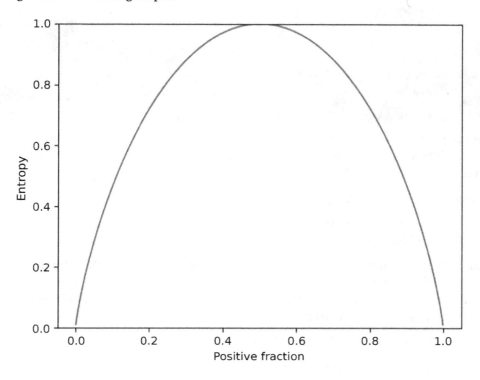

Figure 3.7: Entropy versus positive fraction

As you can see, in binary cases, if the positive fraction is 50%, the entropy will be the highest at 1; if the positive fraction is 100% or 0%, it will reach 0 entropy.

Given the labels of a dataset, the entropy calculation function can be implemented as follows:

```
>>> def entropy(labels):
...     if len(labels) == 0:
...         return 0
...     counts = np.unique(labels, return_counts=True)[1]
...     fractions = counts / float(len(labels))
...     return - np.sum(fractions * np.log2(fractions))
```

Test it out with some examples:

```
>>> print(f'{entropy([1, 1, 0, 1, 0]):.4f}')
0.9710
>>> print(f'{entropy([1, 1, 0, 1, 0, 0]):.4f}')
1.0000
>>> print(f'{entropy([1, 1, 1, 1]):.4f}')
-0.0000
```

Now that you have fully understood entropy, we can look into how Information Gain measures how much uncertainty was reduced after splitting, which is defined as the difference in entropy before a split (parent) and after a split (children):

Information Gain = Entropy(before) - Entropy(after) = Entropy(parent) - Entropy(children)

Entropy after a split is calculated as the weighted sum of the entropy of each child, which is similar to the weighted Gini Impurity.

During the process of constructing a node in a tree, our goal is to search for the splitting point where the maximum Information Gain is obtained. As the entropy of the parent node is unchanged, we just need to measure the entropy of the resulting children due to a split. The best split is the one with the lowest entropy of its resulting children.

To understand this better, let's look at the self-driving car ad example again.

For the first option, the entropy after the split can be calculated as follows:

$$\#1\ Entropy = \frac{3}{5}\left(-\left(\frac{2}{3} * log_2\frac{2}{3} + \frac{1}{3} * log_2\frac{1}{3}\right)\right) + \frac{2}{5}\left(-\left(\frac{1}{2} * log_2\frac{1}{2} + \frac{1}{2} * log_2\frac{1}{2}\right)\right) - 0.951$$

The second way of splitting is as follows:

$$\#2\ Entropy = \frac{2}{5}\left(-(1 * log_2 1 + 0)\right) + \frac{3}{5}\left(-\left(\frac{1}{3} * log_2\frac{1}{3} + \frac{2}{3} * log_2\frac{2}{3}\right)\right) = 0.551$$

For exploration purposes, we can also calculate the Information Gain with:

$$Entropy\ before = -\left(\frac{3}{5} * log_2\frac{3}{5} + \frac{2}{5} * log_2\frac{2}{5}\right) = 0.971$$

$$\#1\ Information\ Gain=0.971-0.951=0.020$$

$$\#2\ Information\ Gain=0.971-0.551=0.420$$

According to the **information Gain = entropy-based evaluation**, the second split is preferable, which is the conclusion of the Gini Impurity criterion.

In general, the choice between the two metrics, Gini Impurity and Information Gain, has little effect on the performance of the trained decision tree. They both measure the weighted impurity of the children after a split. We can combine them into one function to calculate the weighted impurity:

```
>>> criterion_function = {'gini': gini_impurity,
...                        'entropy': entropy}
>>> def weighted_impurity(groups, criterion='gini'):
...     """
...     Calculate weighted impurity of children after a split
...     @param groups: list of children, and a child consists a
...                    list of class labels
...     @param criterion: metric to measure the quality of a split,
...                       'gini' for Gini impurity or 'entropy' for
...                       information gain
...     @return: float, weighted impurity
...     """
...     total = sum(len(group) for group in groups)
...     weighted_sum = 0.0
...     for group in groups:
...         weighted_sum += len(group) / float(total) *
...                         criterion_function[criterion](group)
...     return weighted_sum
```

Test it with the example we just hand-calculated, as follows:

```
>>> children_1 = [[1, 0, 1], [0, 1]]
>>> children_2 = [[1, 1], [0, 0, 1]]
>>> print(f"Entropy of #1 split: {weighted_impurity(children_1,
...         'entropy'):.4f}")
Entropy of #1 split: 0.9510
>>> print(f"Entropy of #2 split: {weighted_impurity(children_2,
...         'entropy'):.4f}")
Entropy of #2 split: 0.5510
```

Now that you have a solid understanding of partitioning evaluation metrics, let's implement the CART tree algorithm from scratch in the next section.

Implementing a decision tree from scratch

We develop the CART tree algorithm by hand on a toy dataset as follows:

User interest	User occupation	Click
Tech	Professional	1
Fashion	Student	0
Fashion	Professional	0
Sports	Student	0
Tech	Student	1
Tech	Retired	0
Sports	Professional	1

Figure 3.8: An example of ad data

To begin with, we decide on the first splitting point, the root, by trying out all possible values for each of the two features. We utilize the `weighted_impurity` function we just defined to calculate the weighted Gini Impurity for each possible combination, as follows:

If we partition according to whether the user interest is tech, we have the 1st, 5th, and 6th samples for one group and the remaining samples for another group. Then the classes for the first group are [1, 1, 0], and the classes for the second group are [0, 0, 0, 1]:

```
Gini(interest, tech) = weighted_impurity([[1, 1, 0], [0, 0, 0, 1]])
                     = 0.405
```

If we partition according to whether the user's interest is fashion, we have the 2nd and 3rd samples for one group and the remaining samples for another group. Then the classes for the first group are [0, 0], and the classes for the second group are [1, 0, 1, 0, 1]:

```
Gini(interest, Fashion) = weighted_impurity([[0, 0], [1, 0, 1, 0, 1]])
                        = 0.343
```

Similarly, we have the following:

```
Gini(interest, Sports) = weighted_impurity([[0, 1], [1, 0, 0, 1, 0]])
                       = 0.486
Gini(occupation, professional) = weighted_impurity([[0, 0, 1, 0],
                                                    [1, 0, 1]]) = 0.405
Gini(occupation, student) = weighted_impurity([[0, 0, 1, 0],
                                              [1, 0, 1]]) = 0.405
Gini(occupation, retired) = weighted_impurity([[1, 0, 0, 0, 1, 1], [1]])
                          = 0.429
```

The root goes to the user interest feature with the fashion value, as this combination achieves the lowest weighted impurity or the highest Information Gain. We can now build the first level of the tree, as follows:

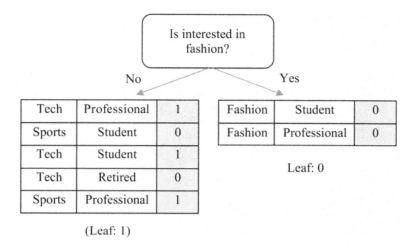

Figure 3.9: Partitioning the data according to "Is interested in fashion?"

If we are satisfied with a one-level-deep tree, we can stop here by assigning the right branch label 0 and the left branch label 1 as the majority class.

Alternatively, we can go further down the road, constructing the second level from the left branch (the right branch cannot be split further):

```
Gini(interest, tech) = weighted_impurity([[0, 1],
    [1, 1, 0]]) = 0.467
Gini(interest, Sports) = weighted_impurity([[1, 1, 0],
    [0, 1]]) = 0.467
Gini(occupation, professional) = weighted_impurity([[0, 1, 0],
    [1, 1]]) = 0.267
Gini(occupation, student) = weighted_impurity([[1, 0, 1],
    [0, 1]]) = 0.467
Gini(occupation, retired) = weighted_impurity([[1, 0, 1, 1],
    [0]]) = 0.300
```

With the second splitting point specified by (occupation, professional) with the lowest Gini Impurity, our tree becomes this:

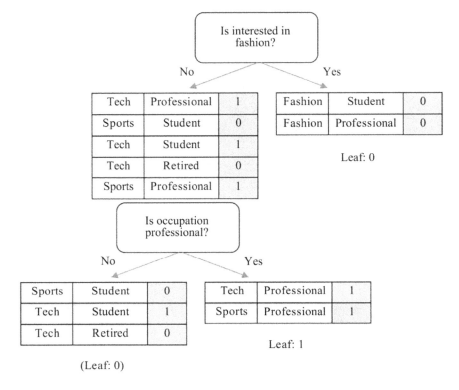

Figure 3.10: Further partitioning of the data according to "Is occupation professional?"

We can repeat the splitting process as long as the tree does not exceed the maximum depth and the node contains enough samples.

Now that the process of the tree construction has been made clear, it is time for coding.

We start with defining a utility function to split a node into left and right children based on a feature and a value:

```
>>> def split_node(X, y, index, value):
...     x_index = X[:, index]
...     # if this feature is numerical
...     if X[0, index].dtype.kind in ['i', 'f']:
...         mask = x_index >= value
...     # if this feature is categorical
...     else:
...         mask = x_index == value
...     # split into left and right child
```

```
...         left = [X[~mask, :], y[~mask]]
...         right = [X[mask, :], y[mask]]
...         return left, right
```

We check whether the feature is numerical or categorical and split the data accordingly.

With the splitting measurement and generation functions available, we now define the greedy search function, which tries out all possible splits and returns the best one given a selection criterion, along with the resulting children:

```
>>> def get_best_split(X, y, criterion):
...         best_index, best_value, best_score, children =
...                                     None, None, 1, None
...         for index in range(len(X[0])):
...             for value in np.sort(np.unique(X[:, index])):
...                 groups = split_node(X, y, index, value)
...                 impurity = weighted_impurity(
...                         [groups[0][1], groups[1][1]], criterion)
...                 if impurity < best_score:
...                     best_index, best_value, best_score, children =
...                                 index, value, impurity, groups
...         return {'index': best_index, 'value': best_value,
...                 'children': children}
```

The selection and splitting process occurs in a recursive manner on each of the subsequent children. When a stopping criterion is met, the process stops at a node, and the major label is assigned to this leaf node:

```
>>> def get_leaf(labels):
...         # Obtain the leaf as the majority of the labels
...         return np.bincount(labels).argmax()
```

And, finally, the recursive function links all of them together:

- It assigns a leaf node if one of two child nodes is empty
- It assigns a leaf node if the current branch depth exceeds the maximum depth allowed
- It assigns a leaf node if the node does not contain sufficient samples required for a further split
- Otherwise, it proceeds with a further split with the optimal splitting point

This can be done with the following function:

```
>>> def split(node, max_depth, min_size, depth, criterion):
...         left, right = node['children']
...         del (node['children'])
...         if left[1].size == 0:
...             node['right'] = get_leaf(right[1])
```

```
...         return
...     if right[1].size == 0:
...         node['left'] = get_leaf(left[1])
...         return
...     # Check if the current depth exceeds the maximal depth
...     if depth >= max_depth:
...         node['left'], node['right'] =
...                     get_leaf(left[1]), get_leaf(right[1])
...         return
...     # Check if the left child has enough samples
...     if left[1].size <= min_size:
...         node['left'] = get_leaf(left[1])
...     else:
...         # It has enough samples, we further split it
...         result = get_best_split(left[0], left[1], criterion)
...         result_left, result_right = result['children']
...         if result_left[1].size == 0:
...             node['left'] = get_leaf(result_right[1])
...         elif result_right[1].size == 0:
...             node['left'] = get_leaf(result_left[1])
...         else:
...             node['left'] = result
...             split(node['left'], max_depth, min_size,
...                                 depth + 1, criterion)
...     # Check if the right child has enough samples
...     if right[1].size <= min_size:
...         node['right'] = get_leaf(right[1])
...     else:
...         # It has enough samples, we further split it
...         result = get_best_split(right[0], right[1], criterion)
...         result_left, result_right = result['children']
...         if result_left[1].size == 0:
...             node['right'] = get_leaf(result_right[1])
...         elif result_right[1].size == 0:
...             node['right'] = get_leaf(result_left[1])
...         else:
...             node['right'] = result
...             split(node['right'], max_depth, min_size,
...                                 depth + 1, criterion)
```

The function first extracts the left and right children from the node dictionary. It then checks whether either the left or right child is empty. If so, it assigns a leaf node to the corresponding child. Next, it checks whether the current depth exceeds the maximum depth allowed for the tree. If so, it assigns leaf nodes to both children. If the left child has enough samples to split (greater than `min_size`), it computes the best split using the `get_best_split` function. If the resulting split produces empty children, it assigns a leaf node to the corresponding child; otherwise, it recursively calls the `split` function on the left child. Similar steps are repeated for the right child.

Finally, the entry point of the tree's construction is as follows:

```python
>>> def train_tree(X_train, y_train, max_depth, min_size,
...                 criterion='gini'):
...     X = np.array(X_train)
...     y = np.array(y_train)
...     root = get_best_split(X, y, criterion)
...     split(root, max_depth, min_size, 1, criterion)
...     return root
```

Now, let's test it with the preceding hand-calculated example:

```python
>>> X_train = [['tech', 'professional'],
...            ['fashion', 'student'],
...            ['fashion', 'professional'],
...            ['sports', 'student'],
...            ['tech', 'student'],
...            ['tech', 'retired'],
...            ['sports', 'professional']]
>>> y_train = [1, 0, 0, 0, 1, 0, 1]
>>> tree = train_tree(X_train, y_train, 2, 2)
```

To verify that the resulting tree from the model is identical to what we constructed by hand, we write a function displaying the tree:

```python
>>> CONDITION = {'numerical': {'yes': '>=', 'no': '<'},
...              'categorical': {'yes': 'is', 'no': 'is not'}}
>>> def visualize_tree(node, depth=0):
...     if isinstance(node, dict):
...         if node['value'].dtype.kind in ['i', 'f']:
...             condition = CONDITION['numerical']
...         else:
...             condition = CONDITION['categorical']
...         print('{}|- X{} {} {}'.format(depth * '  ',
...             node['index'] + 1, condition['no'], node['value']))
...         if 'left' in node:
```

```
...                visualize_tree(node['left'], depth + 1)
...            print('{}|- X{} {} {}'.format(depth * '   ',
...                node['index'] + 1, condition['yes'], node['value']))
...            if 'right' in node:
...                visualize_tree(node['right'], depth + 1)
...        else:
...            print(f"{depth * '   '}[{node}]")
>>> visualize_tree(tree)
|- X1 is not fashion
 |- X2 is not professional
   [0]
 |- X2 is professional
   [1]
|- X1 is fashion
 [0]
```

We can test it with a numerical example, as follows:

```
>>> X_train_n = [[6, 7],
...              [2, 4],
...              [7, 2],
...              [3, 6],
...              [4, 7],
...              [5, 2],
...              [1, 6],
...              [2, 0],
...              [6, 3],
...              [4, 1]]
>>> y_train_n = [0, 0, 0, 0, 0, 1, 1, 1, 1, 1]
>>> tree = train_tree(X_train_n, y_train_n, 2, 2)
>>> visualize_tree(tree)
|- X2 < 4
 |- X1 < 7
   [1]
 |- X1 >= 7
   [0]
|- X2 >= 4
 |- X1 < 2
   [1]
 |- X1 >= 2
   [0]
```

The resulting trees from our decision tree model are the same as those we hand-crafted.

Now that you have a more solid understanding of decision trees after implementing one from scratch, we can move on with implementing a decision tree with scikit-learn.

Implementing a decision tree with scikit-learn

Here, we'll use scikit-learn's decision tree module (https://scikit-learn.org/stable/modules/generated/sklearn.tree.DecisionTreeClassifier.html), which is already well developed and optimized:

```
>>> from sklearn.tree import DecisionTreeClassifier
>>> tree_sk = DecisionTreeClassifier(criterion='gini',
...                                  max_depth=2, min_samples_split=2)
>>> tree_sk.fit(X_train_n, y_train_n)
```

To visualize the tree we just built, we utilize the built-in export_graphviz function, as follows:

```
>>> from sklearn.tree import export_graphviz
>>> export_graphviz(tree_sk, out_file='tree.dot',
...                 feature_names=['X1', 'X2'], impurity=False,
...                 filled=True, class_names=['0', '1'])
```

Running this will generate a file called tree.dot, which can be converted into a PNG image file using **Graphviz** (the introduction and installation instructions can be found at http://www.graphviz.org) by running the following command in the terminal:

```
dot -Tpng tree.dot -o tree.png
```

Refer to *Figure 3.11* for the result:

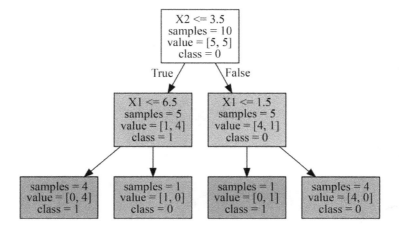

Figure 3.11: Tree visualization

The generated tree is essentially the same as the one we had before.

I know you can't wait to employ a decision tree to predict ad click-through. Let's move on to the next section.

Predicting ad click-through with a decision tree

After several examples, it is now time to predict ad click-through using the decision tree algorithm you have just thoroughly learned about and practiced with. We will use the dataset from a Kaggle machine learning competition, *Click-Through Rate Prediction* (`https://www.kaggle.com/c/avazu-ctr-prediction`). The dataset can be downloaded from `https://www.kaggle.com/c/avazu-ctr-prediction/data`.

Only the `train.gz` file contains labeled samples, so we only need to download this and unzip it (it will take a while). In this chapter, we will focus on only the first 300,000 samples from the `train.csv` file unzipped from `train.gz`.

The fields in the raw file are as follows:

Field	Description	Example values
id	ad identifier	such as '1000009418151094273', '10000169349117863715'
click	'0' for non-click, '1' for click	0, 1
hour	in the format of YYMMDDHH	'14102100'
C1	anonymized categorical variable	'1005', '1002'
banner_pos	where banner is located	1, 0
site_id	site identifier	'1fbe01fe', 'fe8cc448', 'd6137915'
site_domain	hashed site domain	'bb1ef334', 'f3845767'
site_category	hashed site category	'28905ebd', '28905ebd'
app_id	mobile app identifier	'ecad2386'
app_domain	mobile app domain	'7801e8d9'
app_category	category of app	'07d7df22'
device_id	mobile device identifier	'a99f214a'
device_ip	IP address	'ddd2926e'
device_model	such as iphone 6, Samsung, hashed	'44956a24'
device_type	such as tablet, smartphone, hashed	1
device_conn_type	Wi-Fi or 3G for example, again hashed in the data	0, 2
C14-C21	anonymized categorical variables	

Figure 3.12: Description and example values of the dataset

We take a glance at the head of the file by running the following command:

```
head train | sed 's/,,/, ,/g;s/,,/, ,/g' | column -s, -t
```

Rather than a simple `head train`, the output is cleaner as all the columns are aligned:

```
id                      click  hour      C1    banner_pos  site_id   site_domain  site_category  app_id
app_domain  app_category  device_id  device_ip  device_model  device_type  device_conn_type  C14    C15
C16  C17   C18  C19  C20      C21
1000009418151094273   0         14102100  1005  0                     1fbe01fe  f3845767     28905ebd       ecad2386
7801e8d9    07d7df22      a99f214a   ddd2926e   44956a24      1            2                 15706  320
50   1722  0    35   -1       79
10000169349117863715  0         14102100  1005  0                     1fbe01fe  f3845767     28905ebd       ecad2386
7801e8d9    07d7df22      a99f214a   96809ac8   711ee120      1            0                 15704  320
50   1722  0    35   100084   79
10000371904215119486  0         14102100  1005  0                     1fbe01fe  f3845767     28905ebd       ecad2386
7801e8d9    07d7df22      a99f214a   b3cf8def   8a4875bd      1            0                 15704  320
50   1722  0    35   100084   79
10000640724480838376  0         14102100  1005  0                     1fbe01fe  f3845767     28905ebd       ecad2386
7801e8d9    07d7df22      a99f214a   e8275b8f   6332421a      1            0                 15706  320
50   1722  0    35   100084   79
10000679056417042096  0         14102100  1005  1                     fe8cc448  9166c161     0569f928       ecad2386
7801e8d9    07d7df22      a99f214a   9644d0bf   779d90c2      1            0                 18993  320
50   2161  0    35   -1       157
10000720757801103869  0         14102100  1005  0                     d6137915  bb1ef334     f028772b       ecad2386
7801e8d9    07d7df22      a99f214a   05241af0   8a4875bd      1            0                 16920  320
50   1899  0    431  100077   117
10000724729988544911  0         14102100  1005  0                     8fda644b  25d4cfcd     f028772b       ecad2386
7801e8d9    07d7df22      a99f214a   b264c159   be6db1d7      1            0                 20362  320
50   2333  0    39   -1       157
10000918755742328737  0         14102100  1005  1                     e151e245  7e091613     f028772b       ecad2386
7801e8d9    07d7df22      a99f214a   e6f67278   be74e6fe      1            0                 20632  320
50   2374  3    39   -1       23
10000949271186029916  1         14102100  1005  0                     1fbe01fe  f3845767     28905ebd       ecad2386
7801e8d9    07d7df22      a99f214a   37e8da74   5db079b5      1            2                 15707  320
50   1722  0    35   -1       79
```

Figure 3.13: The first few rows of the data

Don't be scared by the anonymized and hashed values. They are categorical features, and each of their possible values corresponds to a real and meaningful value, but it is presented this way due to the privacy policy. Possibly, `C1` means user gender, and `1005` and `1002` represent male and female, respectively.

Now, let's start by reading the dataset using `pandas`. That's right, `pandas` is extremely good at handling data in a tabular format:

```
>>> import pandas as pd
>>> n_rows = 300000
>>> df = pd.read_csv("train.csv", nrows=n_rows)
```

The first 300,000 lines of the file are loaded and stored in a DataFrame. Take a quick look at the first five rows of the DataFrame:

```
>>> print(df.head(5))
   id        click      hour C1 banner_pos    site_id ... C16 C17 C18 C19    C20 C21
0  1.000009e+18        0 14102100 1005         0 1fbe01fe ... 50 1722 0   35 -1
79
```

```
1  1.000017e+19        0 14102100 1005        0 1fbe01fe ... 50 1722 0  35
100084 79
2  1.000037e+19        0 14102100 1005        0 1fbe01fe ... 50 1722 0  35
100084 79
3  1.000064e+19        0 14102100 1005        0 1fbe01fe ... 50 1722 0  35
100084 79
4  1.000068e+19        0 14102100 1005        1 fe8cc448 ... 50 2161 0  35 -1
157
```

The target variable is the `click` column:

```
>>> Y = df['click'].values
```

For the remaining columns, there are several columns that should be removed from the features (`id`, `hour`, `device_id`, and `device_ip`) as they do not contain much useful information:

```
>>> X = df.drop(['click', 'id', 'hour', 'device_id', 'device_ip'],
            axis=1).values
>>> print(X.shape)
(300000, 19)
```

Each sample has 19 predictive attributes.

Next, we need to split the data into training and testing sets. Normally, we do this by randomly picking samples. However, in our case, the samples are in chronological order, as indicated in the `hour` field. Obviously, we cannot use future samples to predict past ones. Hence, we take the first 90% as training samples and the rest as testing samples:

```
>>> n_train = int(n_rows * 0.9)
>>> X_train = X[:n_train]
>>> Y_train = Y[:n_train]
>>> X_test = X[n_train:]
>>> Y_test = Y[n_train:]
```

As mentioned earlier, decision tree models can take in categorical features. However, because the tree-based algorithms in scikit-learn (the current version is 1.4.1 as of early 2024) only allow numeric input, we need to transform the categorical features into numerical ones. But note that, in general, we do not need to do this; for example, the decision tree classifier we developed from scratch earlier can directly take in categorical features.

We will now transform string-based categorical features into one-hot encoded vectors using the `OneHotEncoder` module from `scikit-learn`. One-hot encoding was briefly mentioned in *Chapter 1, Getting Started with Machine Learning and Python*. To recap, it basically converts a categorical feature with k possible values into k binary features. For example, the site category feature with three possible values, `news`, `education`, and `sports`, will be encoded into three binary features, such as `is_news`, `is_education`, and `is_sports`, whose values are either 1 or 0.

We initialize a `OneHotEncoder` object as follows:

```
>>> from sklearn.preprocessing import OneHotEncoder
>>> enc = OneHotEncoder(handle_unknown='ignore')
```

We fit it on the training set as follows:

```
>>> X_train_enc = enc.fit_transform(X_train)
>>> X_train_enc[0]
<1x8385 sparse matrix of type '<class 'numpy.float64'>'
with 19 stored elements in Compressed Sparse Row format>
>>> print(X_train_enc[0])
  (0, 2)          1.0
  (0, 6)          1.0
  (0, 188)        1.0
  (0, 2608)       1.0
  (0, 2679)       1.0
  (0, 3771)       1.0
  (0, 3885)       1.0
  (0, 3929)       1.0
  (0, 4879)       1.0
  (0, 7315)       1.0
  (0, 7319)       1.0
  (0, 7475)       1.0
  (0, 7824)       1.0
  (0, 7828)       1.0
  (0, 7869)       1.0
  (0, 7977)       1.0
  (0, 7982)       1.0
  (0, 8021)       1.0
  (0, 8189)       1.0
```

Each converted sample is a sparse vector.

We transform the testing set using the trained one-hot encoder as follows:

```
>>> X_test_enc = enc.transform(X_test)
```

Remember, we specified the `handle_unknown='ignore'` parameter in the one-hot encoder earlier. This is to prevent errors due to any unseen categorical values. To use the previous site category example, if there is a sample with the value `movie`, all of the three converted binary features (`is_news`, `is_education`, and `is_sports`) become `0`. If we do not specify `ignore`, an error will be raised.

The way we have conducted cross-validation so far is to explicitly split data into folds and repetitively write a `for` loop to consecutively examine each hyperparameter. To make this less redundant, we'll introduce a more elegant approach utilizing the `GridSearchCV` module from scikit-learn. `GridSearchCV` handles the entire process implicitly, including data splitting, fold generation, cross-training and validation, and finally, an exhaustive search over the best set of parameters. What is left for us is just to specify the hyperparameter(s) to tune and the values to explore for each individual hyperparameter. For demonstration purposes, we will only tweak the `max_depth` hyperparameter (other hyperparameters, such as `min_samples_split` and `class_weight`, are also highly recommended):

```
>>> from sklearn.tree import DecisionTreeClassifier
>>> parameters = {'max_depth': [3, 10, None]}
```

We pick three options for the maximal depth – 3, 10, and unbounded. We initialize a decision tree model with Gini Impurity as the metric and 30 as the minimum number of samples required to split further:

```
>>> decision_tree = DecisionTreeClassifier(criterion='gini',
...                                         min_samples_split=30)
```

The classification metric should be the AUC of the ROC curve, as it is an imbalanced binary case (only 51,211 out of 300,000 training samples are clicks, which is a 17% positive CTR; I encourage you to figure out the class distribution yourself). As for grid search, we use three-fold (as the training set is relatively small) cross-validation and select the best-performing hyperparameter measured by the AUC:

```
>>> grid_search = GridSearchCV(decision_tree, parameters,
...                             n_jobs=-1, cv=3, scoring='roc_auc')
```

Note, `n_jobs=-1` means that we use all of the available CPU processors:

```
>>> grid_search.fit(X_train, y_train)
>>> print(grid_search.best_params_)
{'max_depth': 10}
```

We use the model with the optimal parameter to predict any future test cases as follows:

```
>>> decision_tree_best = grid_search.best_estimator_
>>> pos_prob = decision_tree_best.predict_proba(X_test)[:, 1]
>>> from sklearn.metrics import roc_auc_score
>>> print(f'The ROC AUC on testing set is: {roc_auc_score(Y_test,
...          pos_prob):.3f}')
The ROC AUC on testing set is: 0.719
```

The AUC we can achieve with the optimal decision tree model is 0.72. This does not seem to be very high, but click-through involves many intricate human factors, which is why predicting it is not an easy task. Although we can further optimize the hyperparameters, an AUC of 0.72 is actually pretty good. As a comparison, randomly selecting 17% of the samples to be clicked on will generate an AUC of 0.499:

```
>>> pos_prob = np.zeros(len(Y_test))
```

```
>>> click_index = np.random.choice(len(Y_test),
...                    int(len(Y_test) * 51211.0/300000),
...                    replace=False)
>>> pos_prob[click_index] = 1
>>> print(f'The ROC AUC on testing set using random selection is: {roc_auc_
score(Y_test, pos_prob):.3f}')
The ROC AUC on testing set using random selection is: 0.499
```

Our decision tree model significantly outperforms the random predictor. Looking back, we can see that a decision tree is a sequence of greedy searches for the best splitting point at each step, based on the training dataset. However, this tends to cause overfitting as it is likely that the optimal points only work well for the training samples. Fortunately, ensembling is the technique to correct this, and random forest is an ensemble tree model that usually outperforms a simple decision tree.

Best practice

Here are two best practices for getting data ready for tree-based algorithms:

- **Encode categorical features:** As mentioned earlier, we need to encode categorical features before feeding them into the models. One-hot encoding and label encoding are popular choices.
- **Scale numerical features:** We need to pay attention to the scales of numerical features to prevent features with larger scales from dominating the splitting decisions in the tree. Normalization or standardization are commonly used for this purpose.

Ensembling decision trees — random forests

The **ensemble** technique of **bagging** (which stands for **bootstrap aggregating**), which I briefly mentioned in *Chapter 1, Getting Started with Machine Learning and Python*, can effectively overcome overfitting. To recap, different sets of training samples are randomly drawn with replacements from the original training data; each resulting set is used to fit an individual classification model. The results of these separately trained models are then combined together through a **majority vote** to make the final decision.

Tree bagging, as described in the preceding paragraph, reduces the high variance that a decision tree model suffers from and, hence, in general, performs better than a single tree. However, in some cases, where one or more features are strong indicators, individual trees are constructed largely based on these features and, as a result, become highly correlated. Aggregating multiple correlated trees will not make much difference. To force each tree to become uncorrelated, random forest only considers a random subset of the features when searching for the best splitting point at each node. Individual trees are now trained based on different sequential sets of features, which guarantees more diversity and better performance. Random forest is a variant of the tree bagging model with additional **feature-based bagging**.

To employ random forest in our click-through prediction project, we can use the package from scikit-learn. Similarly to the way we implemented the decision tree in the preceding section, we only tweak the `max_depth` parameter:

```
>>> from sklearn.ensemble import RandomForestClassifier
>>> random_forest = RandomForestClassifier(n_estimators=100,
...                       criterion='gini', min_samples_split=30,
...                       n_jobs=-1)
```

Besides `max_depth`, `min_samples_split`, and `class_weight`, which are important hyperparameters related to a single decision tree, hyperparameters that are related to a random forest (a set of trees) such as `n_estimators` are also highly recommended. We fine-tune `max_depth` as follows:

```
>>> grid_search = GridSearchCV(random_forest, parameters,
...                       n_jobs=-1, cv=3, scoring='roc_auc')
>>> grid_search.fit(X_train, y_train)
>>> print(grid_search.best_params_)
{'max_depth': None}
```

We use the model with the optimal parameter `None` for `max_depth` (the nodes are expanded until another stopping criterion is met) to predict any future unseen cases:

```
>>> random_forest_best = grid_search.best_estimator_
>>> pos_prob = random_forest_best.predict_proba(X_test)[:, 1]
>>> print(f'The ROC AUC on testing set using random forest is: {roc_auc_
...        score(Y_test, pos_prob):.3f}')
The ROC AUC on testing set using random forest is: 0.759
```

It turns out that the random forest model gives a substantial lift to the performance.

Let's summarize several critical hyperparameters to tune:

- `max_depth`: This is the deepest individual tree. It tends to overfit if it is too deep or underfit if it is too shallow.

- `min_samples_split`: This hyperparameter represents the minimum number of samples required for further splitting at a node. Too small a value tends to cause overfitting, while too large a value is likely to introduce underfitting. `10`, `30`, and `50` might be good options to start with.

The preceding two hyperparameters are generally related to individual decision trees. The following two parameters are more related to a random forest or collection of trees:

- `max_features`: This parameter represents the number of features to consider for each best splitting point search. Typically, for an m-dimensional dataset, $\sqrt{m}$ (rounded) is a recommended value for `max_features`. This can be specified as `max_features="sqrt"` in scikit-learn. Other options include `log2`, 20%, and 50% of the original features.

- n_estimators: This parameter represents the number of trees considered for majority voting. Generally speaking, the more trees, the better the performance but the longer the computation time. It is usually set as 100, 200, 500, and so on.

Next, we'll discuss gradient-boosted trees.

Ensembling decision trees — gradient-boosted trees

Boosting, which is another ensemble technique, takes an iterative approach instead of combining multiple learners in parallel. In boosted trees, individual trees are no longer trained separately. Specifically, in **Gradient-Boosted Trees (GBT)** (also called **Gradient-Boosting Machines**), individual trees are trained in succession where a tree aims to correct the errors made by the previous tree. The following two diagrams illustrate the difference between random forest and GBT.

The random forest model builds each tree independently using a different subset of the dataset, and then combines the results at the end by majority votes or averaging:

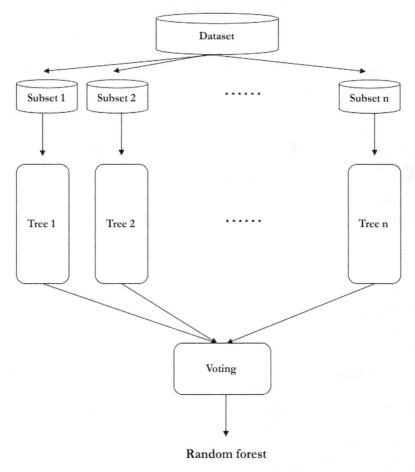

Figure 3.14: The random forest workflow

The GBT model builds one tree at a time and combines the results along the way:

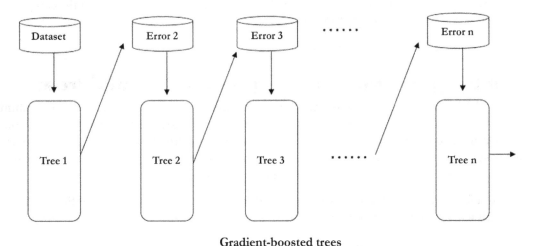

Gradient-boosted trees

Figure 3.15: The GBT workflow

GBT works by iteratively improving the ensemble's predictions through the addition of sequentially trained decision trees, with each tree focusing on the residuals of the previous ones. Here's how it works:

- **Initialization:** The process starts with an initial simple model, often a single decision tree, which serves as the starting point for the ensemble.
- **Sequential training:** Subsequent decision trees are trained sequentially, with each tree attempting to correct the errors of the previous ones. Each new tree is trained on the residuals (the differences between the actual and predicted values) of the ensemble's predictions from the previous trees.
- **Additive modeling:** Each new decision tree is added to the ensemble in a way that minimizes the overall error. The trees are typically shallow, with a limited number of nodes, to avoid overfitting and improve generalization.
- **Learning rate:** GBT introduces a learning rate parameter, which controls the contribution of each tree to the ensemble. A lower learning rate leads to slower learning but can enhance the overall performance and stability of the ensemble.
- **Ensemble prediction:** The final prediction is made by combining the predictions of all the trees in the ensemble.

We will use the XGBoost package (https://xgboost.readthedocs.io/en/latest/) to implement GBT. We first install the XGBoost Python API via the following command with conda:

```
conda install -c conda-forge xgboost
```

We can also use pip, as follows:

```
pip install xgboost
```

If you run into a problem, please install or upgrade `CMake` (a cross-platform build system generator), as follows:

```
pip install CMake
```

Let's now take a look at the following steps. You will see how we predict clicks using GBT:

1. We import XGBoost and initialize a GBT model:

    ```
    >>> import xgboost as xgb
    >>> model = xgb.XGBClassifier(learning_rate=0.1, max_depth=10,
    ...                           n_estimators=1000)
    ```

    We set the learning rate to `0.1`, which determines how fast or slow we want to proceed with learning in each step (in each tree, in GBT). We will discuss the learning rate in more detail in *Chapter 4, Predicting Online Ad Click-Through with Logistic Regression*. `max_depth` for individual trees is set to 10. Additionally, 1,000 trees will be trained in sequence in our GBT model.

2. Next, we train the GBT model on the training set we prepared previously:

    ```
    >>> model.fit(X_train_enc, Y_train)
    ```

3. We use the trained model to make predictions on the testing set and calculate the ROC AUC accordingly:

    ```
    >>> pos_prob = model.predict_proba(X_test_enc)[:, 1]
    >>> print(f'The ROC AUC on testing set using GBT is: {roc_auc_score(Y_
    test, pos_prob):.3f}')
    The ROC AUC on testing set using GBT is: 0.771
    ```

We are able to achieve `0.77` AUC using the XGBoost GBT model.

In this section, you learned about another type of tree ensembling, GBT, and applied it to our ad click-through prediction.

Best practice

So, you've learned about several tree-based algorithms in the chapter – awesome! But picking the right one can be tricky. Here is a practical guide:

- **Decision tree (CART):** This is the most simple and interpretable algorithm. We usually use it for smaller datasets.
- **Random forest:** This is more robust to overfitting, and can handle larger or complex datasets well.
- **GBT:** This is considered the most powerful algorithm for complex problems, and the most popular tree-based algorithm in the industry. At the same time, however, it can be prone to overfitting. Hence, using hyperparameter tuning and regularization techniques to avoid overfitting is recommended.

Summary

In this chapter, we started with an introduction to a typical machine learning problem, online ad click-through prediction, and its inherent challenges, including categorical features. We then looked at tree-based algorithms that can take in both numerical and categorical features.

Next, we had an in-depth discussion about the decision tree algorithm: its mechanics, its different types, how to construct a tree, and two metrics (Gini Impurity and entropy) that measure the effectiveness of a split at a node. After constructing a tree by hand, we implemented the algorithm from scratch.

You also learned how to use the decision tree package from scikit-learn and applied it to predict the CTR. We continued to improve performance by adopting the feature-based random forest bagging algorithm. Finally, the chapter ended with several ways in which to tune a random forest model, along with two different ways of ensembling decision trees, random forest and GBT modeling. Bagging and boosting are two approaches to model ensembling that can improve learning performance.

More practice is always good for honing your skills. I recommend that you complete the following exercises before moving on to the next chapter, where we will solve ad click-through prediction using another algorithm: **logistic regression**.

Exercises

1. In the decision tree click-through prediction project, can you also tweak other hyperparameters, such as `min_samples_split` and `class_weight`? What is the highest AUC you are able to achieve?

2. In the random forest-based click-through prediction project, can you also tweak other hyperparameters, such as `min_samples_split`, `max_features`, and `n_estimators`, in scikit-learn? What is the highest AUC you are able to achieve?

3. In the GBT-based click-through prediction project, what hyperparameters can you tweak? What is the highest AUC you are able to achieve? You can read `https://xgboost.readthedocs.io/en/latest/python/python_api.html#module-xgboost.sklearn` to figure it out.

Join our book's Discord space

Join our community's Discord space for discussions with the authors and other readers:

`https://packt.link/yuxi`

4

Predicting Online Ad Click-Through with Logistic Regression

In the previous chapter, we predicted ad click-through using tree algorithms. In this chapter, we will continue our journey of tackling the billion-dollar problem. We will focus on learning a very (probably the most) scalable classification model – logistic regression. We will explore what the logistic function is, how to train a logistic regression model, adding regularization to the model, and variants of logistic regression that are applicable to very large datasets. Besides its application in classification, we will also discuss how logistic regression and random forest models are used to pick significant features. You won't get bored as there will be lots of implementations from scratch with scikit-learn and TensorFlow.

In this chapter, we will cover the following topics:

- Converting categorical features to numerical – one-hot encoding and original encoding
- Classifying data with logistic regression
- Training a logistic regression model
- Training on large datasets with online learning
- Handling multiclass classification
- Implementing logistic regression using TensorFlow

Converting categorical features to numerical — one-hot encoding and ordinal encoding

In *Chapter 3, Predicting Online Ad Click-Through with Tree-Based Algorithms*, I mentioned how **one-hot encoding** transforms categorical features to numerical features in order to use them in the tree algorithms in scikit-learn and TensorFlow. If we transform categorical features into numerical ones using one-hot encoding, we don't limit our choice of algorithms to the tree-based ones that can work with categorical features.

The simplest solution we can think of in terms of transforming a categorical feature with k possible values is to map it to a numerical feature with values from 1 to k. For example, [Tech, Fashion, Fashion, Sports, Tech, Tech, Sports] becomes [1, 2, 2, 3, 1, 1, 3]. However, this will impose an ordinal characteristic, such as Sports being greater than Tech, and a distance property, such as Sports being closer to Fashion than to Tech.

Instead, one-hot encoding converts the categorical feature to k binary features. Each binary feature indicates the presence or absence of a corresponding possible value. Hence, the preceding example becomes the following:

User interest		Interest: tech	Interest: fashion	Interest: sports
Tech		1	0	0
Fashion		0	1	0
Fashion	⟹	0	1	0
Sports		0	0	1
Tech		1	0	0
Tech		1	0	0
Sports		0	0	1

Figure 4.1: Transforming user interest into numerical features with one-hot encoding

Previously, we used OneHotEncoder from scikit-learn to convert a matrix of strings into a binary matrix, but here, let's take a look at another module, DictVectorizer, which also provides an efficient conversion. It transforms dictionary objects (categorical feature: value) into one-hot encoded vectors.

For example, take a look at the following code, which performs one-hot encoding on a list of dictionaries containing categorical features:

```
>>> from sklearn.feature_extraction import DictVectorizer
>>> X_dict = [{'interest': 'tech', 'occupation': 'professional'},
...          {'interest': 'fashion', 'occupation': 'student'},
...          {'interest': 'fashion','occupation':'professional'},
...          {'interest': 'sports', 'occupation': 'student'},
...          {'interest': 'tech', 'occupation': 'student'},
...          {'interest': 'tech', 'occupation': 'retired'},
...          {'interest': 'sports','occupation': 'professional'}]
>>> dict_one_hot_encoder = DictVectorizer(sparse=False)
>>> X_encoded = dict_one_hot_encoder.fit_transform(X_dict)
>>> print(X_encoded)
[[ 0.  0.  1.  1.  0.  0.]
 [ 1.  0.  0.  0.  0.  1.]
 [ 1.  0.  0.  1.  0.  0.]
```

```
[ 0.  1.  0.  0.  0.  1.]
[ 0.  0.  1.  0.  0.  1.]
[ 0.  0.  1.  0.  1.  0.]
[ 0.  1.  0.  1.  0.  0.]]
```

We can also see the mapping by executing the following:

```
>>> print(dict_one_hot_encoder.vocabulary_)
{'interest=fashion': 0, 'interest=sports': 1,
 'occupation=professional': 3, 'interest=tech': 2,
 'occupation=retired': 4, 'occupation=student': 5}
```

When it comes to new data, we can transform it with the following:

```
>>> new_dict = [{'interest': 'sports', 'occupation': 'retired'}]
>>> new_encoded = dict_one_hot_encoder.transform(new_dict)
>>> print(new_encoded)
[[ 0.  1.  0.  0.  1.  0.]]
```

We can inversely transform the encoded features back to the original features like this:

```
>>> print(dict_one_hot_encoder.inverse_transform(new_encoded))
[{'interest=sports': 1.0, 'occupation=retired': 1.0}]
```

One important thing to note is that if a new (not seen in training data) category is encountered in new data, it should be ignored (otherwise, the encoder will complain about the unseen categorical value). DictVectorizer handles this implicitly (while OneHotEncoder needs to specify the ignore parameter):

```
>>> new_dict = [{'interest': 'unknown_interest',
                 'occupation': 'retired'},
...              {'interest': 'tech', 'occupation':
                 'unseen_occupation'}]
>>> new_encoded = dict_one_hot_encoder.transform(new_dict)
>>> print(new_encoded)
[[ 0.  0.  0.  0.  1.  0.]
 [ 0.  0.  1.  0.  0.  0.]]
```

Sometimes, we prefer transforming a categorical feature with k possible values into a numerical feature with values ranging from 1 to k. This is **ordinal encoding** and we conduct it in order to employ ordinal or ranking knowledge in our learning; for example, large, medium, and small become 3, 2, and 1, respectively; good and bad become 1 and 0, while one-hot encoding fails to preserve such useful information. We can realize ordinal encoding easily through the use of pandas, for example:

```
>>> import pandas as pd
>>> df = pd.DataFrame({'score': ['low',
...                              'high',
```

```
...                              'medium',
...                              'medium',
...                              'low']})
>>> print(df)
     score
0      low
1     high
2   medium
3   medium
4      low
>>> mapping = {'low':1, 'medium':2, 'high':3}
>>> df['score'] = df['score'].replace(mapping)
>>> print(df)
     score
0        1
1        3
2        2
3        2
4        1
```

We convert the string feature into ordinal values based on the mapping we define.

Best practice

Handling high dimensionality resulting from one-hot encoding can be challenging. It may increase computational complexity or lead to overfitting. Here are some strategies to handle high dimensionality when using one-hot encoding:

- **Feature selection:** This can reduce the number of one-hot encoded features while retaining the most informative ones.
- **Dimensionality reduction:** It transforms the high-dimensional feature space into a lower-dimensional representation.
- **Feature aggregation:** Instead of one-hot encoding every category individually, consider aggregating categories that share similar characteristics. For example, group rare categories into an "Other" category.

We've covered transforming categorical features into numerical ones. Next, we will talk about logistic regression, a classifier that only takes in numerical features.

Classifying data with logistic regression

In the last chapter, we trained tree-based models only based on the first 300,000 samples out of 40 million. We did so simply because training a tree on a large dataset is extremely computationally expensive and time consuming. Since we are not limited to algorithms directly taking in categorical features thanks to one-hot encoding, we should turn to a new algorithm with high scalability for large datasets. As mentioned, logistic regression is one of the most, or perhaps the most, scalable classification algorithms.

Getting started with the logistic function

Let's start with an introduction to the **logistic function** (which is more commonly referred to as the **sigmoid function**) as the algorithm's core before we dive into the algorithm itself. It basically maps an input to an output of a value between *0* and *1*, and is defined as follows:

$$y(z) = \frac{1}{1 + \exp(-z)}$$

We define the logistic function as follows:

```
>>> import numpy as np
>>> import matplotlib.pyplot as plt
>>> def sigmoid(input):
...     return 1.0 / (1 + np.exp(-input))
```

Next, we visualize what it looks like with input variables from -8 to 8, as follows:

```
>>> z = np.linspace(-8, 8, 1000)
>>> y = sigmoid(z)
>>> plt.plot(z, y)
>>> plt.axhline(y=0, ls='dotted', color='k')
>>> plt.axhline(y=0.5, ls='dotted', color='k')
>>> plt.axhline(y=1, ls='dotted', color='k')
>>> plt.yticks([0.0, 0.25, 0.5, 0.75, 1.0])
>>> plt.xlabel('z')
>>> plt.ylabel('y(z)')
>>> plt.show()
```

Refer to the following screenshot for the result:

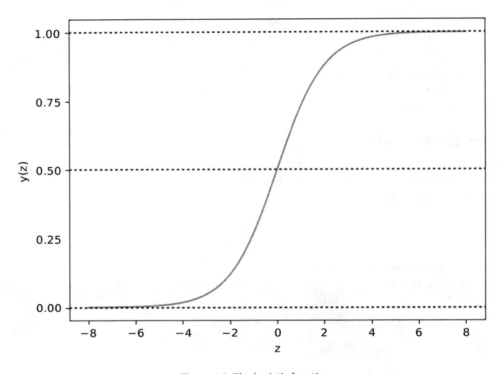

Figure 4.2: The logistic function

In the S-shaped curve, all inputs are transformed into the range from 0 to 1. For positive inputs, a greater value results in an output closer to 1; for negative inputs, a smaller value generates an output closer to 0; when the input is 0, the output is the midpoint, 0.5.

Jumping from the logistic function to logistic regression

Now that you have some knowledge of the logistic function, it is easy to map it to the algorithm that stems from it. In logistic regression, the function input z becomes the weighted sum of features. Given a data sample x with n features, $x_1, x_2, ..., x_n$ (x represents a feature vector and $x = (x_1, x_2, ..., x_n)$), and **weights** (also called **coefficients**) of the model w (w represents a vector $(w_1, w_2, ..., w_n)$), z is expressed as follows:

$$z = w_1 x_1 + w_2 x_2 + \cdots + w_n x_n = w^T x$$

Here, T is the transpose operator.

Occasionally, the model comes with an **intercept** (also called **bias**), w_0, which accounts for the inherent bias or baseline probability. In this instance, the preceding linear relationship becomes:

$$z = w_0 + w_1 x_1 + w_2 x_2 + \cdots + w_n x_n = w^T x$$

As for the output $y(z)$ in the range of 0 to 1, in the algorithm, it becomes the probability of the target being 1 or the positive class:

$$\hat{y} = P(y = 1|x) = \frac{1}{1 + \exp{(-w^T x)}}$$

Hence, logistic regression is a probabilistic classifier, similar to the Naïve Bayes classifier.

A logistic regression model, or, more specifically, its weight vector w, is learned from the training data, with the goal of predicting a positive sample as close to 1 as possible and predicting a negative sample as close to 0 as possible. In mathematical language, the weights are trained to minimize the cost defined as the **Mean Squared Error (MSE)**, which measures the average of squares of the difference between the truth and the prediction. Given m training samples:

$$\left(x^{(1)}, y^{(1)}\right), \left(x^{(2)}, y^{(2)}\right), \dots \left(x^{(i)}, y^{(i)}\right), \dots \left(x^{(m)}, y^{(m)}\right)$$

Here, $y^{(i)}$ is either 1 (positive class) or 0 (negative class), and the cost function $J(w)$ regarding the weights to be optimized is expressed as follows:

$$J(w) = \frac{1}{m}\sum_{i=1}^{m}\frac{1}{2}\left(\hat{y}\left(x^{(i)}\right) - y^{(i)}\right)^2$$

However, the preceding cost function is **non-convex**, which means that, when searching for the optimal w, many local (suboptimal) optimums are found and the function does not converge to a global optimum.

Examples of the **convex** and **non-convex** functions are plotted respectively in the following figure:

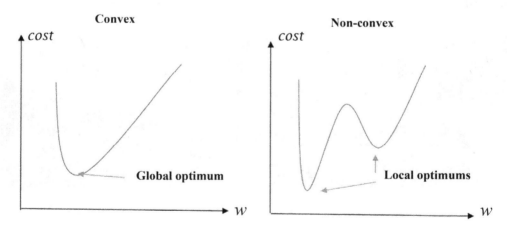

Figure 4.3: Examples of convex and non-convex functions

In the convex example, there is only one global optimum, while there are two optimums in the non-convex example.

 For more about convex and non-convex functions, check out `https://web.stanford.edu/class/ee364a/lectures/functions.pdf`.

To overcome this, in practice, we use the cost function that results in a convex optimization problem, which is defined as follows:

$$J(w) = \frac{1}{m} \sum_{i=1}^{m} -\left[y^{(i)} log(\hat{y}(x^{(i)})) + (1 - y^{(i)}) log\left(1 - \hat{y}(x^{(i)})\right) \right]$$

We can take a closer look at the cost of a single training sample:

$$j(w) = -y^{(i)} log\left(\hat{y}(x^{(i)})\right) - (1 - y^{(i)}) log\left(1 - \hat{y}(x^{(i)})\right)$$

$$= \begin{cases} -log\left(\hat{y}(x^{(i)})\right), if\ y^{(i)} = 1 \\ -log\left(1 - \hat{y}(x^{(i)})\right), if\ y^{(i)} = 0 \end{cases}$$

When the ground truth $y^{(i)}$ = 1, if the model predicts correctly with full confidence (the positive class with 100% probability), the sample cost j is 0; the cost j increases when the predicted probability $\hat{y}$ decreases. If the model incorrectly predicts that there is no chance of the positive class, the cost is infinitely high. We can visualize it as follows:

```
>>> y_hat = np.linspace(0.001, 0.999, 1000)
>>> cost = -np.log(y_hat)
>>> plt.plot(y_hat, cost)
>>> plt.xlabel('Prediction')
>>> plt.ylabel('Cost')
>>> plt.xlim(0, 1)
>>> plt.ylim(0, 7)
>>> plt.show()
```

Refer to the following graph for the end result:

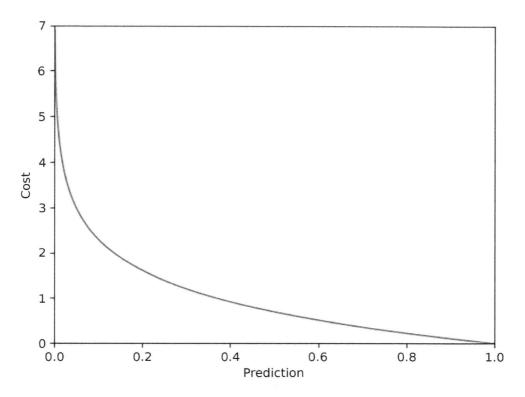

Figure 4.4: Cost function of logistic regression when y=1

On the contrary, when the ground truth $y^{(i)} = 0$, if the model predicts correctly with full confidence (the positive class with 0 probability, or the negative class with 100% probability), the sample cost *j* is *0*; the cost j increases when the predicted probability $\hat{y}$ increases. When it incorrectly predicts that there is no chance of the negative class, the cost becomes infinitely high. We can visualize it using the following code:

```
>>> y_hat = np.linspace(0.001, 0.999, 1000)
>>> cost = -np.log(1 - y_hat)
>>> plt.plot(y_hat, cost)
>>> plt.xlabel('Prediction')
>>> plt.ylabel('Cost')
>>> plt.xlim(0, 1)
>>> plt.ylim(0, 7)
>>> plt.show()
```

The following graph is the resultant output:

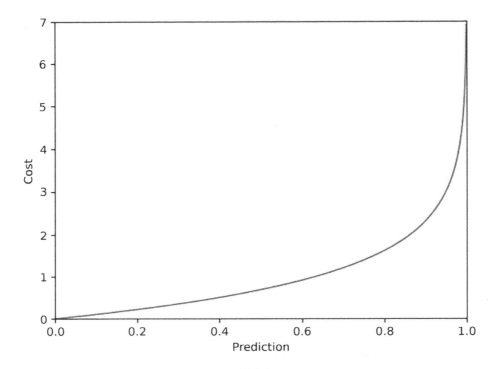

Figure 4.5: Cost function of logistic regression when y=0

Minimizing this alternative cost function is actually equivalent to minimizing the MSE-based cost function. The advantages of choosing it over the MSE version include the following:

- It is convex, so the optimal model weights can be found
- A summation of the logarithms of prediction, which are as follows, simplifies the calculation of its derivative with respect to the weights, which we will talk about later:

$$\hat{y}\left(x^{(i)}\right)$$

Or:

$$1 - \hat{y}(x^{(i)})$$

- Due to the logarithmic function, the cost function, which is as follows, is also called **logarithmic loss**, or simply **log loss**:

$$J(w) = \frac{1}{m}\sum_{i=1}^{m} -\left[y^{(i)}log\left(\hat{y}(x^{(i)})\right) + (1 - y^{(i)})log\left(1 - \hat{y}(x^{(i)})\right)\right]$$

Now that the cost function is ready, how can we train the logistic regression model to minimize the cost function? Let's see this in the next section.

Training a logistic regression model

Now, the question is as follows: how can we obtain the optimal w such that $J(w)$ is minimized? We can do so using gradient descent.

Training a logistic regression model using gradient descent

Gradient descent (also called **steepest descent**) is a procedure for minimizing a loss function by first-order iterative optimization. In each iteration, the model parameters move a small step that is proportional to the negative derivative of the objective function at the current point. This means the to-be-optimal point iteratively moves downhill toward the minimal value of the objective function. The proportion we just mentioned is called the **learning rate**, or **step size**. It can be summarized in a mathematical equation as follows:

$$w := w - \eta \Delta w$$

Here, the left w is the weight vector after a learning step, and the right w is the one before moving, η is the learning rate, and Δw is the first-order derivative, the gradient.

To train a logistic regression model using gradient descent, let's start with the derivative of the cost function $J(w)$ with respect to w. It might require some knowledge of calculus but don't worry, we will walk through it step by step:

1. We first calculate the derivative of $\hat{y}(x)$ with respect to w. We herein take the j-th weight, w_j, as an example (note $z = w^T x$, and we omit the $^{(i)}$ for simplicity):

$$\frac{\partial}{\partial w_j} \hat{y}(z) = \frac{\partial}{\partial w_j} \frac{1}{1 + exp(-z)} = \frac{\partial}{\partial z} \frac{1}{1 + exp(-z)} \frac{\partial}{\partial w_j} z$$

$$= \frac{1}{[1 + exp(-z)]^2} exp(-z) \frac{\partial}{\partial w_j} z$$

$$= \frac{1}{1 + exp(-z)} \left[1 - \frac{1}{1 + exp(-z)} \right] \frac{\partial}{\partial w_j} z = \hat{y}(z)(1 - \hat{y}(z)) \frac{\partial}{\partial w_j} z$$

2. Then, we calculate the derivative of the sample cost $J(w)$ as follows:

$$\frac{\partial}{\partial w_j} J(w) = -y \frac{\partial}{\partial w_j} log(\hat{y}(z)) + (1 - y) \frac{\partial}{\partial w_j} log(1 - \hat{y}(z))$$

$$= \left[-y \frac{1}{\hat{y}(z)} + (1 - y) \frac{1}{1 - \hat{y}(z)} \right] \frac{\partial}{\partial w_j} \hat{y}(z)$$

$$= \left[-y \frac{1}{\hat{y}(z)} + (1 - y) \frac{1}{1 - \hat{y}(z)} \right] \hat{y}(z)(1 - \hat{y}(z)) \frac{\partial}{\partial w_j} z$$

$$= (-y + \hat{y}(z)) x_j$$

3. Finally, we calculate the entire cost over m samples as follows:

$$\Delta w_j = \frac{\delta}{\delta w_j} J(w) = \frac{1}{m}\sum_{i=1}^{m} -(y^{(i)} - \hat{y}(z^{(i)}))x_j^{(i)}$$

4. We then generalize it to Δw:

$$\Delta w = \frac{1}{m}\sum_{i=1}^{m} -\left(y^{(i)} - \hat{y}(z^{(i)})\right)x^{(i)}$$

5. Combined with the preceding derivations, the weights can be updated as follows:

$$w := w + \eta\frac{1}{m}\sum_{i-1}^{m} \left(y^{(i)} - \hat{y}(z^{(i)})\right)x^{(i)}$$

Here, w gets updated in each iteration.

6. After a substantial number of iterations, the learned parameter w is then used to classify a new sample x' by means of the following equation:

$$y' = \frac{1}{1 + \exp(-w^T x')}$$

$$\begin{cases}1, if\ y' \geq 0.5 \\ 0, if\ y' < 0.5\end{cases}$$

The decision threshold is 0.5 by default, but it definitely can be other values. In cases where a false negative is supposed to be avoided, for example, when predicting fire occurrence (the positive class) for alerts, the decision threshold can be lower than 0.5, such as 0.3, depending on how paranoid we are and how proactively we want to prevent the positive event from happening. On the other hand, when the false positive class is the one that should be evaded, for instance, when predicting the product success (the positive class) rate for quality assurance, the decision threshold can be greater than 0.5, such as 0.7, or lower than 0.5, depending on how high a standard you set.

With a thorough understanding of the gradient descent-based training and predicting process, we will now implement the logistic regression algorithm from scratch:

1. We begin by defining the function that computes the prediction $\hat{y}(x)$ with the current weights:

```
>>> def compute_prediction(X, weights):
...     """
...     Compute the prediction y_hat based on current weights
...     """
...     z = np.dot(X, weights)
...     return sigmoid(z)
```

2. With this, we are able to continue with the function updating the weights, which is as follows, by one step in a gradient descent manner:

$$w := w + \eta \frac{1}{m} \sum_{i=1}^{m} \left(y^{(i)} - \hat{y}(z^{(i)}) \right) x^{(i)}$$

Take a look at the following code:

```
>>> def update_weights_gd(X_train, y_train, weights,
                                       learning_rate):
...     """
...     Update weights by one step
...     """
...     predictions = compute_prediction(X_train, weights)
...     weights_delta = np.dot(X_train.T, y_train - predictions)
...     m = y_train.shape[0]
...     weights += learning_rate / float(m) * weights_delta
...     return weights
```

3. Then, the function calculating the cost $J(w)$ is implemented as well:

```
>>> def compute_cost(X, y, weights):
...     """
...     Compute the cost J(w)
...     """
...     predictions = compute_prediction(X, weights)
...     cost = np.mean(-y * np.log(predictions)
                    - (1 - y) * np.log(1 - predictions))
...     return cost
```

4. Now, we connect all these functions to the model training function by executing the following:

    *   Updating the `weights` vector in each iteration
    *   Printing out the current cost for every `100` (this can be another value) iterations to ensure cost is decreasing and that things are on the right track

They are implemented in the following function:

```
>>> def train_logistic_regression(X_train, y_train, max_iter,
                              learning_rate, fit_intercept=False):
...     """ Train a logistic regression model
...     Args:
...         X_train, y_train (numpy.ndarray, training data set)
...         max_iter (int, number of iterations)
...         learning_rate (float)
```

```
...             fit_intercept (bool, with an intercept w0 or not)
...         Returns:
...             numpy.ndarray, learned weights
...         """
...         if fit_intercept:
...             intercept = np.ones((X_train.shape[0], 1))
...             X_train = np.hstack((intercept, X_train))
...         weights = np.zeros(X_train.shape[1])
...         for iteration in range(max_iter):
...             weights = update_weights_gd(X_train, y_train,
...                                         weights, learning_rate)
...             # Check the cost for every 100 (for example)
...              iterations
...             if iteration % 100 == 0:
...                 print(compute_cost(X_train, y_train, weights))
...         return weights
```

5. Finally, we predict the results of new inputs using the trained model as follows:

```
>>> def predict(X, weights):
...         if X.shape[1] == weights.shape[0] - 1:
...             intercept = np.ones((X.shape[0], 1))
...             X = np.hstack((intercept, X))
...         return compute_prediction(X, weights)
```

Implementing logistic regression is very simple, as you just saw. Let's now examine it using a toy example:

```
>>> X_train = np.array([[6, 7],
...                      [2, 4],
...                      [3, 6],
...                      [4, 7],
...                      [1, 6],
...                      [5, 2],
...                      [2, 0],
...                      [6, 3],
...                      [4, 1],
...                      [7, 2]])
>>> y_train = np.array([0,
...                      0,
...                      0,
...                      0,
...                      0,
```

```
...                    1,
...                    1,
...                    1,
...                    1,
...                    1])
```

We train a logistic regression model for `1000` iterations, at a learning rate of `0.1` based on intercept-included weights:

```
>>> weights = train_logistic_regression(X_train, y_train,
           max_iter=1000, learning_rate=0.1, fit_intercept=True)
0.574404237166
0.0344602233925
0.0182655727085
0.012493458388
0.00951532913855
0.00769338806065
0.00646209433351
0.00557351184683
0.00490163225453
0.00437556774067
```

> 💡 **Quick tip:** Enhance your coding experience with the **AI Code Explainer** and **Quick Copy** features. Open this book in the next-gen Packt Reader. Click the **Copy** button (1) to quickly copy code into your coding environment, or click the **Explain** button (2) to get the AI assistant to explain a block of code to you.

```
                                              Copy      Explain
function calculate(a, b) {
  return {sum: a + b};                          1          2
};
```

🔒 **The next-gen Packt Reader** is included for free with the purchase of this book. Unlock it by scanning the QR code below or visiting
`https://www.packtpub.com/unlock/9781835085622`.

The decreasing cost means that the model is being optimized over time. We can check the model's performance on new samples as follows:

```
>>> X_test = np.array([[6, 1],
...                     [1, 3],
...                     [3, 1],
...                     [4, 5]])
>>> predictions = predict(X_test, weights)
>>> print(predictions)
array([ 0.9999478 , 0.00743991, 0.9808652 , 0.02080847])
```

To visualize this, execute the following code using 0.5 as the classification decision threshold:

```
>>> plt.scatter(X_train[:5,0], X_train[:5,1], c='b', marker='x')
>>> plt.scatter(X_train[5:,0], X_train[5:,1], c='k', marker='.')
>>> for i, prediction in enumerate(predictions):
        marker = 'X' if prediction < 0.5 else 'o'
        c = 'b' if prediction < 0.5 else 'k'
        plt.scatter(X_test[i,0], X_test[i,1], c=c, marker=marker)
```

Blue-filled crosses are testing samples predicted from class 0, while black-filled dots are those predicted from class 1:

```
>>> plt.show()
```

Refer to the following screenshot for the result:

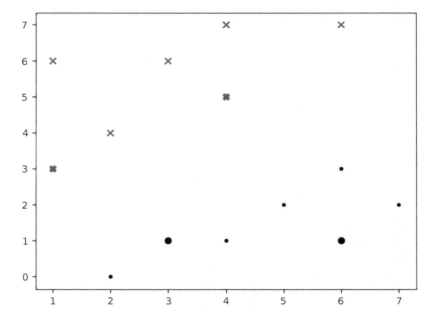

Figure 4.6: Training and testing sets of the toy example

The model we trained correctly predicts classes of new samples (filled crosses and filled dots).

Predicting ad click-through with logistic regression using gradient descent

We will now deploy the algorithm we just developed in our click-through prediction project.

We will start with only 10,000 training samples (you will soon see why we don't start with 270,000, as we did in the previous chapter):

```
>>> import pandas as pd
>>> n_rows = 300000
>>> df = pd.read_csv("train.csv", nrows=n_rows)
>>> X = df.drop(['click', 'id', 'hour', 'device_id', 'device_ip'],
                                              axis=1).values
>>> Y = df['click'].values
>>> n_train = 10000
>>> X_train = X[:n_train]
>>> Y_train = Y[:n_train]
>>> X_test = X[n_train:]
>>> Y_test = Y[n_train:]
>>> from sklearn.preprocessing import OneHotEncoder
>>> enc = OneHotEncoder(handle_unknown='ignore')
>>> X_train_enc = enc.fit_transform(X_train)
>>> X_test_enc = enc.transform(X_test)
```

We train a logistic regression model over `10000` iterations, at a learning rate of `0.01` with bias:

```
>>> import timeit
>>> start_time = timeit.default_timer()
>>> weights = train_logistic_regression(X_train_enc.toarray(),
            Y_train, max_iter=10000, learning_rate=0.01,
            fit_intercept=True)
0.6820019456743648
0.4608619713011896
0.4503715555130051
…
…
…
0.41485094023829017
0.41477416506724385
0.41469802145452467
>>> print(f"--- {(timeit.default_timer() - start_time :.3f} seconds ---")
--- 183.840 seconds ---
```

It takes 184 seconds to optimize the model. The trained model performs on the testing set as follows:

```
>>> pred = predict(X_test_enc.toarray(), weights)
>>> from sklearn.metrics import roc_auc_score
>>> print(f'Training samples: {n_train}, AUC on testing set: {roc_auc_score(Y_
test, pred):.3f}')
Training samples: 10000, AUC on testing set: 0.703
```

Now, let's use 100,000 training samples (n_train = 100000) and repeat the same process. It will take more than an hour – 22 times longer to fit data of 10 times the size. As I mentioned at the beginning of the chapter, the logistic regression classifier can be good at training on large datasets. But our testing results seem to contradict this. How could we handle even larger training datasets efficiently, not just 100,000 samples, but millions? Let's look at a more efficient way to train a logistic regression model in the next section.

Training a logistic regression model using stochastic gradient descent (SGD)

In gradient descent-based logistic regression models, **all** training samples are used to update the weights in every single iteration. Hence, if the number of training samples is large, the whole training process will become very time consuming and computationally expensive, as you just witnessed in our last example.

Fortunately, a small tweak will make logistic regression suitable for large-sized datasets. For each weight update, **only one** training sample is consumed, instead of the **complete** training set. The model moves a step based on the error calculated by a single training sample. Once all samples are used, one iteration finishes. This advanced version of gradient descent is called **SGD**. Expressed in a formula, for each iteration, we do the following:

$$for\ i\ in\ 1\ to\ m:$$

$$w := w + \eta \left(y^{(i)} - \hat{y}(z^{(i)}) \right) x^{(i)}$$

SGD generally converges much faster than gradient descent where a large number of iterations is usually needed.

To implement SGD-based logistic regression, we just need to slightly modify the update_weights_gd function:

```
>>> def update_weights_sgd(X_train, y_train, weights,
                                         learning_rate):
...     """ One weight update iteration: moving weights by one
            step based on each individual sample
...     Args:
...     X_train, y_train (numpy.ndarray, training data set)
...     weights (numpy.ndarray)
```

```
...         learning_rate (float)
...         Returns:
...         numpy.ndarray, updated weights
...         """
...         for X_each, y_each in zip(X_train, y_train):
...             prediction = compute_prediction(X_each, weights)
...             weights_delta = X_each.T * (y_each - prediction)
...             weights += learning_rate * weights_delta
...         return weights
```

In the `train_logistic_regression` function, SGD is applied:

```
>>> def train_logistic_regression_sgd(X_train, y_train, max_iter,
...                           learning_rate, fit_intercept=False):
...     """ Train a logistic regression model via SGD
...     Args:
...     X_train, y_train (numpy.ndarray, training data set)
...     max_iter (int, number of iterations)
...     learning_rate (float)
...     fit_intercept (bool, with an intercept w0 or not)
...     Returns:
...     numpy.ndarray, learned weights
...     """
...     if fit_intercept:
...         intercept = np.ones((X_train.shape[0], 1))
...         X_train = np.hstack((intercept, X_train))
...     weights = np.zeros(X_train.shape[1])
...     for iteration in range(max_iter):
...         weights = update_weights_sgd(X_train, y_train, weights,
...                                              learning_rate)
...         # Check the cost for every 2 (for example) iterations
...         if iteration % 2 == 0:
...             print(compute_cost(X_train, y_train, weights))
...     return weights
```

Now, let's see how powerful SGD is. We will work with 100,000 training samples and choose 10 as the number of iterations, 0.01 as the learning rate, and print out the current costs for every other iteration:

```
>>> start_time = timeit.default_timer()
>>> weights = train_logistic_regression_sgd(X_train_enc.toarray(),
...         Y_train, max_iter=10, learning_rate=0.01, fit_intercept=True)
0.4127864859625796
0.4078504597223988
```

```
0.40545733114863264
0.403811787845451
0.4025431351250833
>>> print(f"--- {(timeit.default_timer() - start_time)}.3fs seconds ---")
--- 25.122 seconds ---
>>> pred = predict(X_test_enc.toarray(), weights)
>>> print(f'Training samples: {n_train}, AUC on testing set: {roc_auc_score(Y_
test, pred):.3f}')
Training samples: 100000, AUC on testing set: 0.732
```

The training process finishes in just 25 seconds!

After successfully implementing the SGD-based logistic regression algorithm from scratch, we implement it using the `SGDClassifier` module of scikit-learn:

```
>>> from sklearn.linear_model import SGDClassifier
>>> sgd_lr = SGDClassifier(loss='log_loss', penalty=None,
            fit_intercept=True, max_iter=20,
            learning_rate='constant', eta0=0.01)
```

Here, `'log_loss'` for the `loss` parameter indicates that the cost function is log loss, `penalty` is the regularization term to reduce overfitting, which we will discuss further in the next section, `max_iter` is the number of iterations, and the remaining two parameters mean the learning rate is `0.01` and unchanged during the course of training. It should be noted that the default `learning_rate` is `'optimal'`, where the learning rate slightly decreases as more and more updates are made. This can be beneficial for finding the optimal solution on large datasets.

Now, train the model and test it:

```
>>> sgd_lr.fit(X_train_enc.toarray(), Y_train)
>>> pred = sgd_lr.predict_proba(X_test_enc.toarray())[:, 1]
>>> print(f'Training samples: {n_train}, AUC on testing set: {roc_auc_score(Y_
test, pred):.3f}')
Training samples: 100000, AUC on testing set: 0.732
```

Quick and easy!

Training a logistic regression model with regularization

As I briefly mentioned in the previous section, the `penalty` parameter in the logistic regression `SGDClassifier` is related to model **regularization**. There are two basic forms of regularization, **L1** (also called **Lasso**) and **L2** (also called **Ridge**). In either way, the regularization is an additional term on top of the original cost function:

$$J(w) = \frac{1}{m} \sum_{i=1}^{m} -\left[y^{(i)} log\left(\hat{y}(x^{(i)}) \right) + (1 - y^{(i)}) log\left(1 - \hat{y}(x^{(i)}) \right) \right] + \alpha \parallel w \parallel^q$$

Here, α is the constant that multiplies the regularization term, and q is either *1* or *2* representing L1 or L2 regularization where the following applies:

$$\|W\|^1 = \sum_{j=1}^{n} |w_j|$$

Training a logistic regression model is the process of reducing the cost as a function of weights w. If it gets to a point where some weights, such as w_i, w_j, and w_k are considerably large, the whole cost will be determined by these large weights. In this case, the learned model may just memorize the training set and fail to generalize to unseen data. The regularization term is introduced in order to penalize large weights, as the weights now become part of the cost to minimize.

Regularization as a result eliminates overfitting. Finally, parameter α provides a trade-off between log loss and generalization. If α is too small, it is not able to compress large weights and the model may suffer from high variance or overfitting; on the other hand, if α is too large, the model may become over-generalized and perform poorly in terms of fitting the dataset, which is the syndrome of underfitting. α is an important parameter to tune in order to obtain the best logistic regression model with regularization.

As for choosing between the L1 and L2 forms, the rule of thumb is based on whether **feature selection** is expected. In **Machine Learning** (**ML**) classification, feature selection is the process of picking a subset of significant features for use in better model construction. In practice, not every feature in a dataset carries information that is useful for discriminating samples; some features are either redundant or irrelevant and hence can be discarded with little loss.

In a logistic regression classifier, feature selection can only be achieved with L1 regularization. To understand this, let's consider two weight vectors, $w_1 = (1, 0)$ and $w_2 = (0.5, 0.5)$; supposing they produce the same amount of log loss, the L1 and L2 regularization terms of each weight vector are as follows:

$$|w_1|^1 = |1| + |0| = 1, |w_2|^1 = |0.5| + |0.5| = 1$$

$$|w_1|^2 = 1^2 + 0^2 = 1, |w_2|^2 = 0.5^2 + 0.5^2 = 0.5$$

The L1 term of both vectors is equivalent, while the L2 term of w_2 is less than that of w_1. This indicates that L2 regularization penalizes weights composed of significantly large and small weights more than L1 regularization does. In other words, L2 regularization favors relatively small values for all weights, and avoids significantly large and small values for any weight, while L1 regularization allows some weights with a significantly small value and some with a significantly large value. Only with L1 regularization can some weights be compressed to close to or exactly *0*, which enables feature selection.

In scikit-learn, the regularization type can be specified by the `penalty` parameter with the none (without regularization), `"l1"`, `"l2"`, and `"elasticnet"` (a mixture of L1 and L2) options, and the multiplier α can be specified by the `alpha` parameter.

Feature selection using L1 regularization

We herein examine L1 regularization for feature selection.

Initialize an SGD logistic regression model with L1 regularization, and train the model based on 10,000 samples:

```
>>> sgd_lr_l1 = SGDClassifier(loss='log_loss',
                              penalty='l1',
                              alpha=0.0001,
                              fit_intercept=True,
                              max_iter=10,
                              learning_rate='constant',
                              eta0=0.01,
                              random_state=42)
>>> sgd_lr_l1.fit(X_train_enc.toarray(), Y_train)
```

With the trained model, we obtain the absolute values of its coefficients:

```
>>> coef_abs = np.abs(sgd_lr_l1.coef_)
>>> print(coef_abs)
[[0. 0.16654682 0. ... 0. 0. 0.12803394]]
```

The bottom `10` coefficients and their values are printed as follows:

```
>>> print(np.sort(coef_abs)[0][:10])
[0. 0. 0. 0. 0. 0. 0. 0. 0. 0.]
>>> bottom_10 = np.argsort(coef_abs)[0][:10]
```

We can see what these 10 features are using the following code:

```
>>> feature_names = enc.get_feature_names_out()
>>> print('10 least important features are:\n', feature_names[bottom_10])
10 least important features are:
 ['x0_1001' 'x8_84c2f017' 'x8_84ace234'  'x8_84a9d4ba' 'x8_84915a27'
'x8_8441e1f3' 'x8_840161a0' 'x8_83fbdb80' 'x8_83fb63cd' 'x8_83ed0b87']
```

They are `1001` from the `0` column (that is the `C1` column) in `X_train`, `84c2f017` from the `8` column (that is the `device_model` column), and so on and so forth.

Similarly, the top 10 coefficients and their values can be obtained as follows:

```
>>> print(np.sort(coef_abs)[0][-10:])
[0.67912376 0.70885933 0.75157162 0.81783177 0.94672827 1.00864062
 1.08152137 1.130848   1.14859459 1.37750805]
>>> top_10 = np.argsort(coef_abs)[0][-10:]
```

```
>>> print('10 most important features are:\n', feature_names[top_10])
10 most important features are:
 ['x4_28905ebd' 'x3_7687a86e' 'x18_61' 'x18_15' 'x5_5e3f096f' 'x5_9c13b419'
 'x2_763a42b5' 'x3_27e3c518' 'x2_d9750ee7' 'x5_1779deee']
```

They are 28905ebd from the 4 column (that is `site_category`) in X_train, 7687a86e from the 3 column (that is site_domain), and so on and so forth.

You have seen how feature selection works with L1-regularized logistic regression in this section, where weights of unimportant features are compressed to close to, or exactly, 0. Besides L1-regularized logistic regression, random forest is another frequently used feature selection technique. Let's see more in the next section.

Feature selection using random forest

To recap, random forest is bagging over a set of individual decision trees. Each tree considers a random subset of the features when searching for the best splitting point at each node. In a decision tree, only those significant features (along with their splitting values) are used to constitute tree nodes. Consider the forest as a whole: the more frequently a feature is used in a tree node, the more important it is.

In other words, we can rank the importance of features based on their occurrences in nodes among all trees, and select the top most important ones.

The trained `RandomForestClassifier` module in scikit-learn comes with an attribute, `feature_importances_`, indicating the feature importance, which is calculated as the proportion of occurrences in tree nodes. Again, we will examine feature selection with random forest on the dataset with 100,000 ad click samples:

```
>>> from sklearn.ensemble import RandomForestClassifier
>>> random_forest = RandomForestClassifier(n_estimators=100,
                    criterion='gini', min_samples_split=30, n_jobs=-1)
>>> random_forest.fit(X_train_enc.toarray(), Y_train)
```

After fitting the random forest model, we obtain the feature importance scores with the following:

```
>>> feature_imp = random_forest.feature_importances_
>>> print(feature_imp)
[1.22776093e-05 1.42544940e-03 8.11601536e-04 ... 7.51812083e-04 8.79340746e-04
 8.49537255e-03]
```

Take a look at the bottom 10 feature scores and the corresponding 10 least important features:

```
>>> feature_names = enc.get_feature_names()
>>> print(np.sort(feature_imp)[:10])
[0. 0. 0. 0. 0. 0. 0. 0. 0. 0.]
>>> bottom_10 = np.argsort(feature_imp)[:10]
```

```
>>> print('10 least important features are:\n', feature_names[bottom_10])
10 least important features are:
 ['x5_f0222e42' 'x8_7d196936' 'x2_ba8f6070' 'x2_300ede9d' 'x5_72c55d0b'
'x2_4390d4c5' 'x5_69e5a5ec' 'x8_023a5294' 'x11_15541' 'x6_2022d54e']
```

Now, take a look at the top 10 feature scores and the corresponding 10 most important features:

```
>>> print(np.sort(feature_imp)[-10:])
[0.00849437 0.00849537 0.00872154 0.01010324 0.0109653  0.01099363 0.01319093
0.01471638 0.01802233 0.01889752]
>>> top_10 = np.argsort(feature_imp)[-10:]
>>> print('10 most important features are:\n', feature_names[top_10])
10 most important features are:
 ['x3_7687a86e' 'x18_157' 'x17_-1' 'x14_1993' 'x8_8a4875bd' 'x2_d9750ee7'
'x3_98572c79' 'x16_1063' 'x15_2' 'x18_33']
```

In this section, we covered how random forest is used for feature selection. We ranked the features of the ad click data using a random forest. Can you use the top 10 or 20 features to build another logistic regression model for ad click prediction?

Training on large datasets with online learning

So far, we have trained our model on no more than 300,000 samples. If we go beyond this figure, memory might be overloaded since it holds too much data, and the program will crash. In this section, we will explore how to train on a large-scale dataset with **online learning**.

SGD evolves from gradient descent by sequentially updating the model with individual training samples one at a time, instead of the complete training set at once. We can scale up SGD further with online learning techniques. In online learning, new data for training is available in sequential order or in real time, as opposed to all at once in an offline learning environment. A relatively small chunk of data is loaded and preprocessed for training at a time, which releases the memory used to hold the entire large dataset. Besides better computational feasibility, online learning is also used because of its adaptability to cases where new data is generated in real time and is needed for modernizing the model. For instance, stock price prediction models are updated in an online learning manner with timely market data; click-through prediction models need to include the most recent data reflecting users' latest behaviors and tastes; spam email detectors have to be reactive to ever-changing spammers by considering new features that are dynamically generated.

The existing model trained by previous datasets can now be updated based on the most recently available dataset only, instead of rebuilding it from scratch based on previous and recent datasets together, as is the case in offline learning:

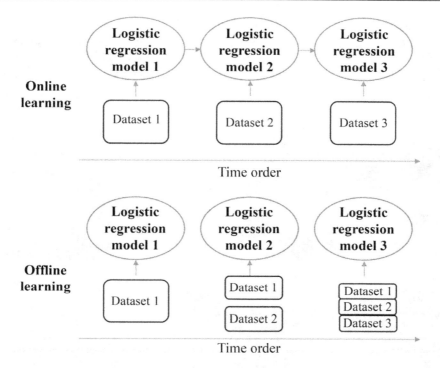

Figure 4.7: Online versus offline learning

In the preceding example, online learning allows the model to continue training with new arriving data. However, in offline learning, we have to retrain the whole model with the new arriving data along with the old data.

The SGDClassifier module in scikit-learn implements online learning with the partial_fit method (while the fit method is applied in offline learning, as you have seen). We will train the model with 1,000,000 samples, where we feed in 100,000 samples at one time to simulate an online learning environment. Also, we will test the trained model on another 100,000 samples as follows:

```
>>> n_rows = 100000 * 11
>>> df = pd.read_csv("train.csv", nrows=n_rows)
>>> X = df.drop(['click', 'id', 'hour', 'device_id', 'device_ip'],
                                            axis=1).values
>>> Y = df['click'].values
>>> n_train = 100000 * 10
>>> X_train = X[:n_train]
>>> Y_train = Y[:n_train]
>>> X_test = X[n_train:]
>>> Y_test = Y[n_train:]
```

Fit the encoder on the whole training set as follows:

```
>>> enc = OneHotEncoder(handle_unknown='ignore')
>>> enc.fit(X_train)
```

Initialize an SGD logistic regression model where we set the number of iterations to 1 in order to partially fit the model and enable online learning:

```
>>> sgd_lr_online = SGDClassifier(loss='log_loss',
                                  penalty=None,
                                  fit_intercept=True,
                                  max_iter=1,
                                  learning_rate='constant',
                                  eta0=0.01,
                                  random_state=42)
```

Loop over every `100000` samples and partially fit the model:

```
>>> start_time = timeit.default_timer()
>>> for i in range(10):
...     x_train = X_train[i*100000:(i+1)*100000]
...     y_train = Y_train[i*100000:(i+1)*100000]
...     x_train_enc = enc.transform(x_train)
...     sgd_lr_online.partial_fit(x_train_enc.toarray(), y_train,
                                  classes=[0, 1])
```

Again, we use the `partial_fit` method for online learning. Also, we specify the `classes` parameter, which is required in online learning:

```
>>> print(f"--- {(timeit.default_timer() - start_time):.3f} seconds ---")
--- 87.399s seconds ---
```

Apply the trained model on the testing set, the next 100,000 samples, as follows:

```
>>> x_test_enc = enc.transform(X_test)
>>> pred = sgd_lr_online.predict_proba(x_test_enc.toarray())[:, 1]
>>> print(f'Training samples: {n_train * 10}, AUC on testing set: {roc_auc_
score(Y_test, pred):.3f}')
Training samples: 10000000, AUC on testing set: 0.762
```

With online learning, training based on a total of 1 million samples only takes 87 seconds and yields better accuracy.

We have been using logistic regression for binary classification so far. Can we use it for multiclass cases? Yes. However, we do need to make some small tweaks. Let's look at this in the next section.

Handling multiclass classification

One last thing worth noting is how logistic regression algorithms deal with multiclass classification. Although we interact with the scikit-learn classifiers in multiclass cases the same way as in binary cases, it is useful to understand how logistic regression works in multiclass classification.

Logistic regression for more than two classes is also called **multinomial logistic regression**, better known latterly as **softmax regression**. As you have seen in the binary case, the model is represented by one weight vector w, and the probability of the target being *1* or the positive class is written as follows:

$$\hat{y} = P(y = 1 \mid x) = \frac{1}{1 + \exp{(-w^T x)}}$$

In the K class case, the model is represented by K weight vectors, $w_1, w_2, ..., w_K$, and the probability of the target being class k is written as follows:

$$\widehat{y_k} = P(y = k \mid x) = \frac{exp(w_k{}^T x)}{\sum_{j=1}^{K} \exp{(w_j{}^T x)}}$$

See the following term:

$$\sum_{j=1}^{K} exp(w_j{}^T x)$$

The preceding term normalizes the following probabilities (k from *1* to K) so that they total *1*:

$$\widehat{y_k}$$

The cost function in the binary case is expressed as follows:

$$J(w) = \frac{1}{m} \sum_{i=1}^{m} -\left[y^{(i)} log\left(\hat{y}(x^{(i)})\right) + (1 - y^{(i)})log\left(1 - \hat{y}(x^{(i)})\right)\right] + \alpha \parallel w \parallel^q$$

Similarly, the cost function in the multiclass case becomes the following:

$$J(w) = \frac{1}{m} \sum_{i=1}^{m} -\left[\sum_{j=1}^{K} 1\{y^{(i)} = j\}log\left(\widehat{y_k}(x^{(i)})\right)\right]$$

Here, the $1\{y^{(i)} = j\}$ function is *1* only if $y^{(i)} = j$ is true, otherwise it's 0.

With the cost function defined, we obtain the Δw_j step for the j weight vector in the same way as we derived the step Δw in the binary case:

$$\Delta w_j = \frac{1}{m} \sum_{i=1}^{m} \left(-1\{y^{(i)} = j\} + \widehat{y_k}(x^{(i)})\right) x^{(i)}$$

In a similar manner, all K weight vectors are updated in each iteration. After sufficient iterations, the learned weight vectors, w_1, w_2, ..., w_K, are then used to classify a new sample x' by means of the following equation:

$$y' = argmax_k \ \widehat{y_k} = argmax_k \ P(y = k \ |x')$$

To have a better sense, let's experiment with it with a classic dataset, the handwritten digits for classification:

```
>>> from sklearn import datasets
>>> digits = datasets.load_digits()
>>> n_samples = len(digits.images)
```

As the image data is stored in 8*8 matrices, we need to flatten them, as follows:

```
>>> X = digits.images.reshape((n_samples, -1))
>>> Y = digits.target
```

We then split the data as follows:

```
>>> from sklearn.model_selection import train_test_split
>>> X_train, X_test, Y_train, Y_test = train_test_split(X, Y,
                                    test_size=0.2, random_state=42)
```

We then combine grid search and cross-validation to find the optimal multiclass logistic regression model, as follows:

```
>>> from sklearn.model_selection import GridSearchCV
>>> parameters = {'penalty': ['l2', None],
...               'alpha': [1e-07, 1e-06, 1e-05, 1e-04],
...               'eta0': [0.01, 0.1, 1, 10]}
>>> sgd_lr = SGDClassifier(loss='log_loss',
                    learning_rate='constant',
                    fit_intercept=True,
                    max_iter=50,
                    random_state=42)
>>> grid_search = GridSearchCV(sgd_lr, parameters,
                            n_jobs=-1, cv=5)
>>> grid_search.fit(X_train, Y_train)
>>> print(grid_search.best_params_)
{'alpha': 1e-05, 'eta0': 0.01, 'penalty': 'l2' }
```

We first define the hyperparameter grid we want to tune for the model. After initializing the classifier with some fixed parameters, we set up grid search cross-validation. We train on the training set and find the best set of hyperparameters.

To predict using the optimal model, we apply the following:

```
>>> sgd_lr_best = grid_search.best_estimator_
>>> accuracy = sgd_lr_best.score(X_test, Y_test)
>>> print(f'The accuracy on testing set is: {accuracy*100:.1f}%')
The accuracy on testing set is: 94.7%
```

It doesn't look much different from the previous example, since SGDClassifier handles multiclass internally. Feel free to compute the confusion matrix as an exercise. It will be interesting to see how the model performs on individual classes.

The next section will be a bonus section where we will implement logistic regression with TensorFlow and use click prediction as an example.

Implementing logistic regression using TensorFlow

We'll employ TensorFlow to implement logistic regression, utilizing click prediction as our illustrative example again. We use 90% of the first 100,000 samples for training and the remaining 10% for testing, and assume that X_train_enc, Y_train, X_test_enc, and Y_test contain the correct data:

1. First, we import TensorFlow, transform X_train_enc and X_test_enc into a NumPy array, and cast X_train_enc, Y_train, X_test_enc, and Y_test to float32:

    ```
    >>> import tensorflow as tf
    >>> X_train_enc = enc.fit_transform(X_train).toarray().astype('float32')
    >>> X_test_enc = enc.transform(X_test).toarray().astype('float32')
    >>> Y_train = Y_train.astype('float32')
    >>> Y_test = Y_test.astype('float32')
    ```

    In TensorFlow, it's common to work with data in the form of NumPy arrays. Additionally, TensorFlow operates with float32 by default for computational efficiency.

2. We use the tf.data module to shuffle and batch data:

    ```
    >>> batch_size = 1000
    >>> train_data = tf.data.Dataset.from_tensor_slices((X_train_enc, Y_train))
    >>> train_data = train_data.repeat().shuffle(5000).batch(batch_size).prefetch(1)
    ```

    For each weight update, only **one batch** of samples is consumed, instead of the one sample or the complete training set. The model moves a step based on the error calculated by a batch of samples. The batch size is 1,000 in this example.

    tf.data provides a set of tools and utilities for efficiently loading and preprocessing data for ML modeling. It is designed to handle large datasets and enables efficient data pipeline construction for training and evaluation.

3. Then, we define the weights and bias of the logistic regression model:

```
>>> n_features = X_train_enc.shape[1]
>>> W = tf.Variable(tf.zeros([n_features, 1]))
>>> b = tf.Variable(tf.zeros([1]))
```

4. We then create a gradient descent optimizer that searches for the best coefficients by minimizing the loss. We use Adam (Adam: *A method for stochastic optimization*, Kingma, D. P., & Ba, J. (2014)) as our optimizer, which is an advanced gradient descent with a learning rate (starting with 0.001) that is adaptive to gradients:

```
>>> learning_rate = 0.001
>>> optimizer = tf.optimizers.Adam(learning_rate)
```

5. We define the optimization process where we compute the current prediction and cost and update the model coefficients following the computed gradients:

```
>>> def run_optimization(x, y):
...     with tf.GradientTape() as tape:
...         logits = tf.add(tf.matmul(x, W), b)[:, 0]
...         loss = tf.reduce_mean(
                        tf.nn.sigmoid_cross_entropy_with_logits(
                                        labels=y, logits=logits))
        # Update the parameters with respect to the gradient calculations
...     gradients = tape.gradient(loss, [W, b])
...     optimizer.apply_gradients(zip(gradients, [W, b]))
```

Here, tf.GradientTape allows us to track TensorFlow computations and calculate gradients with respect to the given variables.

6. We run the training for 5,000 steps (one step is with one batch of random samples):

```
>>> training_steps = 5000
>>> for step, (batch_x, batch_y) in
                enumerate(train_data.take(training_steps), 1):
...     run_optimization(batch_x, batch_y)
...     if step % 500 == 0:
...         logits = tf.add(tf.matmul(batch_x, W), b)[:, 0]
...         loss = tf.reduce_mean(
                        tf.nn.sigmoid_cross_entropy_with_logits(
                                labels=batch_y, logits=logits))
...         print("step: %i, loss: %f" % (step, loss))
step: 500, loss: 0.448672
step: 1000, loss: 0.389186
step: 1500, loss: 0.413012
step: 2000, loss: 0.445663
```

```
step: 2500, loss: 0.361000
step: 3000, loss: 0.417154
step: 3500, loss: 0.359435
step: 4000, loss: 0.393363
step: 4500, loss: 0.402097
step: 5000, loss: 0.376734
```

For every 500 steps, we compute and print out the current cost to check the training performance. As you can see, the training loss is decreasing overall.

7. After the model is trained, we use it to make predictions on the testing set and report the AUC metric:

```
>>> logits = tf.add(tf.matmul(X_test_enc, W), b)[:, 0]
>>> pred = tf.nn.sigmoid(logits)
>>> auc_metric = tf.keras.metrics.AUC()
>>> auc_metric.update_state(Y_test, pred)
>>> print(f'AUC on testing set: {auc_metric.result().numpy():.3f}')
AUC on testing set: 0.736
```

We are able to achieve an AUC of 0.736 with the TensorFlow-based logistic regression model. You can also tweak the learning rate, the number of training steps, and other hyperparameters to obtain a better performance. This will be a fun exercise at the end of the chapter.

Best practice

The choice of the batch size in SGD can significantly impact the training process and the performance of the model. Here are some best practices for choosing the batch size:

- **Consider computational resources:** Larger batch sizes require more memory and computational resources, while smaller batch sizes may lead to slower convergence. Choose a batch size that fits within the memory constraints of your hardware while maximizing computational efficiency.

- **Empirical testing:** Experiment with different batch sizes and evaluate model performance on a validation dataset. Choose the batch size that yields the best trade-off between convergence speed and model performance.

- **Batch size versus learning rate:** The choice of batch size can interact with the learning rate. Larger batch sizes may require higher learning rates to prevent slow convergence, while smaller batch sizes may benefit from smaller learning rates to avoid instability.

- **Consider the nature of the data:** The nature of the data can also influence the choice of batch size. For example, in tasks where the samples are highly correlated or exhibit temporal dependencies (e.g., time series data), smaller batch sizes may be more effective.

You might be curious about how we can efficiently train the model on the entire dataset of 40 million samples. You will utilize tools such as **Spark** (https://spark.apache.org/) and the PySpark module to scale up our solution.

Summary

In this chapter, we continued working on the online advertising click-through prediction project. This time, we overcame the categorical feature challenge by means of the one-hot encoding technique. We then resorted to a new classification algorithm, logistic regression, for its high scalability to large datasets. The in-depth discussion of the logistic regression algorithm started with the introduction of the logistic function, which led to the mechanics of the algorithm itself. This was followed by how to train a logistic regression model using gradient descent.

After implementing a logistic regression classifier by hand and testing it on our click-through dataset, you learned how to train the logistic regression model in a more advanced manner, using SGD, and we adjusted our algorithm accordingly. We also practiced how to use the SGD-based logistic regression classifier from scikit-learn and applied it to our project.

We then continued to tackle the problems we might face in using logistic regression, including L1 and L2 regularization for eliminating overfitting, online learning techniques for training on large-scale datasets, and handling multi-class scenarios. You also learned how to implement logistic regression with TensorFlow. Finally, the chapter ended with applying the random forest model to feature selection, as an alternative to L1-regularized logistic regression.

Looking back on our learning journey, we have been working on classification problems since *Chapter 2, Building a Movie Recommendation Engine with Naïve Bayes*. We have now covered all the powerful and popular classification models in ML. We will move on to solving regression problems in the next chapter; regression is the sibling of classification in supervised learning. You will learn about regression models, including linear regression, and decision trees for regression.

Exercises

1. In the logistic regression-based click-through prediction project, can you also tweak hyperparameters such as penalty, eta0, and alpha in the SGDClassifier model? What is the highest testing AUC you are able to achieve?

2. Can you try to use more training samples, for instance, 10 million samples, in the online learning solution?

3. In the TensorFlow-based solution, can you tweak the learning rate, the number of training steps, and other hyperparameters to obtain better performance?

Unlock this book's exclusive benefits now

This book comes with additional benefits designed to elevate your learning experience.

Note: Have your purchase invoice ready before you begin.   https://www.packtpub.com/unlock/9781835085622

5

Predicting Stock Prices with Regression Algorithms

In the previous chapter, we predicted ad clicks using logistic regression. In this chapter, we will solve a problem that interests everyone—predicting stock prices. Getting wealthy by means of smart investment—who isn't interested?! Stock market movements and stock price predictions have been actively researched by a large number of financial, trading, and even technology corporations. A variety of methods have been developed to predict stock prices using machine learning techniques. Herein, we will focus on learning several popular regression algorithms, including linear regression, regression trees and regression forests, and support vector regression, utilizing them to tackle this billion (or trillion)-dollar problem.

We will cover the following topics in this chapter:

- What is regression?
- Mining stock price data
- Getting started with feature engineering
- Estimating with linear regression
- Estimating with decision tree regression
- Implementing a regression forest
- Evaluating regression performance
- Predicting stock prices with the three regression algorithms

What is regression?

Regression is one of the main types of supervised learning in machine learning. In regression, the training set contains observations (also called features) and their associated **continuous** target values. The process of regression has two phases:

- The first phase is exploring the relationships between the observations and the targets. This is the training phase.

- The second phase is using the patterns from the first phase to generate the target for a future observation. This is the prediction phase.

The overall process is depicted in the following diagram:

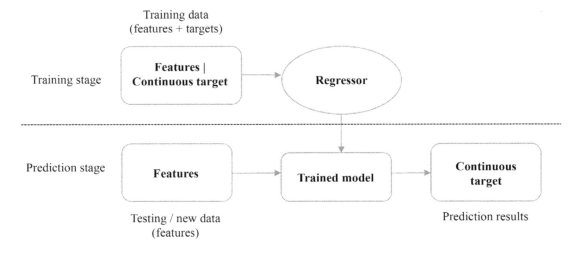

Figure 5.1: Training and prediction phase in regression

The major difference between regression and classification is that the output values in regression are continuous, while in classification they are discrete. This leads to different application areas for these two supervised learning methods. Classification is basically used to determine desired memberships or characteristics, as you've seen in previous chapters, such as email being spam or not, newsgroup topics, or ad click-through. Conversely, regression mainly involves estimating an outcome or forecasting a response.

An example of estimating continuous targets with linear regression is depicted as follows, where we try to fit a line against a set of two-dimensional data points:

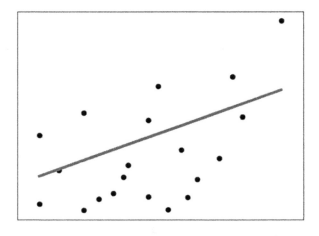

Figure 5.2: Linear regression example

Typical machine learning regression problems include the following:

- Predicting house prices based on location, square footage, and the number of bedrooms and bathrooms
- Estimating power consumption based on information about a system's processes and memory
- Forecasting demand in retail
- Predicting stock prices

I've talked about regression in this section and will briefly introduce its use in the stock market and trading in the next one.

Mining stock price data

In this chapter, we'll work as a stock quantitative analyst/researcher, exploring how to predict stock prices with several typical machine learning regression algorithms. Let's start with a brief overview of the stock market and stock prices.

A brief overview of the stock market and stock prices

The stock of a corporation signifies ownership in the corporation. A single share of the stock represents a claim on the fractional assets and the earnings of the corporation in proportion to the total number of shares. Stocks can be traded between shareholders and other parties via stock exchanges and organizations. Major stock exchanges include the New York Stock Exchange, the NASDAQ, London Stock Exchange Group, and the Hong Kong Stock Exchange. The prices that a stock is traded at fluctuate essentially due to the law of supply and demand.

In general, investors want to buy low and sell high. This sounds simple enough, but it's very challenging to implement, as it's monumentally difficult to say whether a stock price will go up or down. There are two main streams of studies that attempt to understand factors and conditions that lead to price changes, or even forecast future stock prices, **fundamental analysis** and **technical analysis**:

- **Fundamental analysis:** This stream focuses on underlying factors that influence a company's value and business, including overall economy and industry conditions from macro perspectives, the company's financial conditions, management, and competitors from micro perspectives.
- **Technical analysis:** Conversely, this stream predicts future price movements through the statistical study of past trading activity, including price movement, volume, and market data. Predicting prices via machine learning techniques is an important topic in technical analysis nowadays.

Many quantitative, or quant, trading firms use machine learning to empower automated and algorithmic trading.

In theory, we can apply regression techniques to predict the prices of a particular stock. However, it's difficult to ensure the stock we pick is suitable for learning purposes—its price should follow some learnable patterns, and it can't have been affected by unprecedented instances or irregular events. Hence, herein we'll focus on one of the most popular **stock indexes** to better illustrate and generalize our price regression approach.

Let's first cover what an index is. A stock index is a statistical measure of the value of a portion of the overall stock market. An index includes several stocks that are diverse enough to represent a section of the whole market. Also, the price of an index is typically computed as the weighted average of the prices of selected stocks.

The **NASDAQ Composite** is one of the longest-established and most commonly watched indexes in the world. It includes all the stocks listed on the NASDAQ exchange, covering a wide range of sectors. NASDAQ primarily lists stocks of technology companies, including established giants like Apple, Amazon, Microsoft, and Google (Alphabet), as well as emerging growth companies.

You can view its daily prices and performance on Yahoo Finance at `https://finance.yahoo.com/quote/%5EIXIC/history?p=%5EIXIC`. For example:

Time Period: Jul 08, 2022 - Jul 08, 2023 ⌄   Show: Historical Prices ⌄   Frequency: Daily ⌄   Apply

Currency in USD   ⬇ Download

Date	Open	High	Low	Close*	Adj Close**	Volume
Jul 07, 2023	13,668.07	13,804.51	13,656.73	13,660.72	13,660.72	5,098,120,000
Jul 06, 2023	13,653.17	13,689.52	13,567.25	13,679.04	13,679.04	6,113,080,000
Jul 05, 2023	13,772.10	13,844.50	13,764.25	13,791.65	13,791.65	5,339,340,000
Jul 03, 2023	13,798.70	13,839.09	13,773.41	13,816.77	13,816.77	2,902,300,000
Jun 30, 2023	13,719.98	13,816.68	13,716.16	13,787.92	13,787.92	4,661,120,000
Jun 29, 2023	13,592.36	13,618.53	13,540.26	13,591.33	13,591.33	4,388,140,000
Jun 28, 2023	13,506.02	13,654.14	13,495.73	13,591.75	13,591.75	4,533,270,000
Jun 27, 2023	13,389.25	13,578.80	13,366.97	13,555.67	13,555.67	5,053,660,000
Jun 26, 2023	13,468.74	13,573.57	13,334.42	13,335.78	13,335.78	4,430,600,000
Jun 23, 2023	13,484.10	13,572.19	13,442.65	13,492.52	13,492.52	7,734,500,000
Jun 22, 2023	13,443.41	13,631.85	13,441.56	13,630.61	13,630.61	4,499,550,000

Figure 5.3: Screenshot of daily prices and performance in Yahoo Finance

🔍 **Quick tip:** Need to see a high-resolution version of this image? Open this book in the next-gen Packt Reader or view it in the PDF/ePub copy.

🔒 **The next-gen Packt Reader** and a **free PDF/ePub copy** of this book are included with your purchase. Unlock them by scanning the QR code below or visiting `https://www.packtpub.com/unlock/9781835085622`.

On each trading day, the price of stock changes and is recorded in real time. Five values illustrating the movements in price over one unit of time (usually one day, but it can also be one week or one month) are key trading indicators. They are as follows:

- **Open:** The starting price for a given trading day
- **Close:** The final price on that day
- **High:** The highest prices at which the stock traded on that day
- **Low:** The lowest prices at which the stock traded on that day
- **Volume:** The total number of shares traded before the market closed on that day

We will focus on NASDAQ and use its historical prices and performance to predict future prices. In the following sections, we will explore how to develop price prediction models, specifically regression models, and what can be used as indicators or predictive features.

Getting started with feature engineering

When it comes to a machine learning algorithm, the first question to ask is usually what features are available or what the predictive variables are.

The driving factors that are used to predict future prices of NASDAQ, the **close** prices, include historical and current **open** prices as well as historical performance (**high, low,** and **volume**). Note that current or same-day performance (**high, low,** and **volume**) shouldn't be included because we simply can't foresee the highest and lowest prices at which the stock traded, or the total number of shares traded before the market closed on that day.

Predicting the close price with only those preceding four indicators doesn't seem promising and might lead to underfitting. So, we need to think of ways to generate more features in order to increase predictive power. In machine learning, **feature engineering** is the process of creating features in order to improve the performance of a machine learning algorithm. Feature engineering is essential in machine learning and is usually where we spend the most effort in solving a practical problem.

Feature engineering usually requires sufficient domain knowledge and can be very difficult and time-consuming. In reality, features used to solve a machine learning problem are not usually directly available and need to be specifically designed and constructed.

When making an investment decision, investors usually look at historical prices over a period of time, not just the price the day before. Therefore, in our stock price prediction case, we can compute the average close price over the past week (five trading days), over the past month, and over the past year as three new features. We can also customize the time window to the size we want, such as the past quarter or the past six months. On top of these three averaged price features, we can generate new features associated with the price trend by computing the ratios between each pair of average prices in the three different time frames, for instance, the ratio between the average price over the past week and the past year.

Besides prices, volume is another important factor that investors analyze. Similarly, we can generate new volume-based features by computing the average volumes in several different time frames and the ratios between each pair of averaged values.

Besides historical averaged values in a time window, investors also greatly consider stock volatility. Volatility describes the degree of variation of prices for a given stock or index over time. In statistical terms, it's basically the standard deviation of the close prices. We can easily generate new sets of features by computing the standard deviation of close prices in a particular time frame, as well as the standard deviation of volumes traded. Similarly, ratios between each pair of standard deviation values can be included in our engineered feature pool.

Last but not least, return is a significant financial metric that investors closely watch for. Return is the gain or loss percentage of a close price for a stock/index in a particular period. For example, daily return and annual return are financial terms we frequently hear.

They are calculated as follows:

$$return_{i:i-1} = \frac{price_i - price_{i-1}}{price_{i-1}}$$

$$return_{i:i-365} = \frac{price_i - price_{i-365}}{price_{i-365}}$$

Here, $price_i$ is the price on the ith day and $price_{i-1}$ is the price on the day before. Weekly and monthly returns can be computed similarly. Based on daily returns, we can produce a moving average over a particular number of days.

For instance, given the daily returns of the past week, $return_{i:i-1}$, $return_{i-1:i-2}$, $return_{i-2:i-3}$, $return_{i-3:i-4}$, and $return_{i-4:i-5}$, we can calculate the moving average over that week as follows:

$$MovingAvg_{i-5} = \frac{return_{i:i-1} + return_{i-1:i-2} + return_{i-2:i-3} + return_{i-3:i-4} + return_{i-4:i-5}}{5}$$

In summary, we can generate the following predictive variables by applying feature engineering techniques:

$AvgPrice_5$	The average close price over the past five days
$AvgPrice_{30}$	The average close price over the past month
$AvgPrice_{365}$	The average close price over the past year
$\dfrac{AvgPrice_5}{AvgPrice_{30}}$	The ratio between the average price over the past week and that over the past month
$\dfrac{AvgPrice_5}{AvgPrice_{365}}$	The ratio between the average price over the past week and that over the past year
$\dfrac{AvgPrice_{30}}{AvgPrice_{365}}$	The ratio between the average price over the past month and that over the past year
$AvgVolume_5$	The average volume over the past five days
$AvgVolume_{30}$	The average volume over the past month
$AvgVolume_{365}$	The average volume over the past year
$\dfrac{AvgVolume_5}{AvgVolume_{30}}$	The ratio between the average volume over the past week and that over the past month
$\dfrac{AvgVolume_5}{AvgVolume_{365}}$	The ratio between the average volume over the past week and that over the past year
$\dfrac{AvgVolume_{30}}{AvgVolume_{365}}$	The ratio between the average volume over the past month and that over the past year
$StdPrice_5$	The standard deviation of the close prices over the past five days
$StdPrice_{30}$	The standard deviation of the close prices over the past month
$StdPrice_{365}$	The standard deviation of the close prices over the past year

Figure 5.4: Generated features (1)

$\dfrac{\text{StdPrice}_5}{\text{StdPrice}_{30}}$	The ratio between the standard deviation of the prices over the past week and that over the past month
$\dfrac{\text{StdPrice}_5}{\text{StdPrice}_{365}}$	The ratio between the standard deviation of the prices over the past week and that over the past year
$\dfrac{\text{StdPrice}_{30}}{\text{StdPrice}_{365}}$	The ratio between the standard deviation of the prices over the past month and that over the past year
StdVolume_5	The standard deviation of the volumes over the past five days
StdVolume_{30}	The standard deviation of the volumes over the past month
StdVolume_{365}	The standard deviation of the volumes over the past year
$\dfrac{\text{StdVolume}_5}{\text{StdVolume}_{30}}$	The ratio between the standard deviation of the volumes over the past week and that over the past month
$\dfrac{\text{StdVolume}_5}{\text{StdVolume}_{365}}$	The ratio between the standard deviation of the volumes over the past week and that over the past year
$\dfrac{\text{StdVolume}_{30}}{\text{StdVolume}_{365}}$	The ratio between the standard deviation of the volumes over the past month and that over the past year
$\text{return}_{i:i-1}$	Daily return of the past day
$\text{return}_{i:i-5}$	Weekly return of the past week
$\text{return}_{i:i-30}$	Monthly return of the past month
$\text{return}_{i:i-365}$	Yearly return of the past year
MovingAvg_{i_5}	Moving average of the daily returns over the past week
MovingAvg_{i_30}	Moving average of the daily returns over the past month
MovingAvg_{i_365}	Moving average of the daily returns over the past year

Figure 5.5: Generated features (2)

Eventually, we are able to generate, in total, 31 sets of features, along with the following six original features:

- OpenPrice_i: This feature represents the open price
- OpenPrice_{i-1}: This feature represents the open price on the past day
- ClosePrice_{i-1}: This feature represents the close price on the past day
- HighPrice_{i-1}: This feature represents the highest price on the past day
- LowPrice_{i-1}: This feature represents the lowest price on the past day
- Volume_{i-1}: This feature represents the volume on the past day

Acquiring data and generating features

For easier reference, we will implement the code to generate features here rather than in later sections. We will start by obtaining the dataset we need for our project.

Throughout the project, we will acquire stock index price and performance data from Yahoo Finance. For example, on the Historical Data https://finance.yahoo.com/quote/%5EIXIC/history?p=%5EIXIC, we can change the Time Period to Dec 01, 2005 - Dec10, 2005, select Historical Prices in Show and Daily in Frequency (or open this link directly: https://finance.yahoo.com/quote/%5EIXIC/history?period1=1133395200&period2=1134172800&interval=1d&filter=history&frequency=1d&includeAdjustedClose=true), and then click on the **Apply** button. Click the **Download data** button to download the data and name the file 20051201_20051210.csv.

We can load the data we just downloaded as follows:

```
>>> mydata = pd.read_csv('20051201_20051210.csv', index_col='Date')
>>> mydata
                Open           High           Low            Close
Date
2005-12-01 2244.850098    2269.389893    2244.709961    2267.169922
2005-12-02 2266.169922    2273.610107    2261.129883    2273.370117
2005-12-05 2269.070068    2269.479980    2250.840088    2257.639893
2005-12-06 2267.760010    2278.159912    2259.370117    2260.760010
2005-12-07 2263.290039    2264.909912    2244.620117    2252.010010
2005-12-08 2254.800049    2261.610107    2233.739990    2246.459961
2005-12-09 2247.280029    2258.669922    2241.030029    2256.729980
                Adj Close   Volume
Date
2005-12-01 2267.169922    2010420000
2005-12-02 2273.370117    1758510000
2005-12-05 2257.639893    1659920000
2005-12-06 2260.760010    1788200000
2005-12-07 2252.010010    1733530000
2005-12-08 2246.459961    1908360000
2005-12-09 2256.729980    1658570000
```

Note that the output is a pandas DataFrame object. The Date column is the index column, and the rest of the columns are the corresponding financial variables. In the following lines of code, you will see how powerful pandas is at simplifying data analysis and transformation on **relational** (or table-like) data.

First, we implement feature generation by starting with a sub-function that directly creates features from the original six features, as follows:

```
>>> def add_original_feature(df, df_new):
...     df_new['open'] = df['Open']
...     df_new['open_1'] = df['Open'].shift(1)
...     df_new['close_1'] = df['Close'].shift(1)
...     df_new['high_1'] = df['High'].shift(1)
...     df_new['low_1'] = df['Low'].shift(1)
...     df_new['volume_1'] = df['Volume'].shift(1)
```

Then, we develop a sub-function that generates six features related to average close prices:

```
>>> def add_avg_price(df, df_new):
...     df_new['avg_price_5'] =
                    df['Close'].rolling(5).mean().shift(1)
...     df_new['avg_price_30'] =
                    df['Close'].rolling(21).mean().shift(1)
...     df_new['avg_price_365'] =
                    df['Close'].rolling(252).mean().shift(1)
...     df_new['ratio_avg_price_5_30'] =
                df_new['avg_price_5'] / df_new['avg_price_30']
...     df_new['ratio_avg_price_5_365'] =
                df_new['avg_price_5'] / df_new['avg_price_365']
...     df_new['ratio_avg_price_30_365'] =
                df_new['avg_price_30'] / df_new['avg_price_365']
```

Similarly, a sub-function that generates six features related to average volumes is as follows:

```
>>> def add_avg_volume(df, df_new):
...     df_new['avg_volume_5'] =
                    df['Volume'].rolling(5).mean().shift(1)
...     df_new['avg_volume_30'] =
                    df['Volume'].rolling(21).mean().shift(1)
...     df_new['avg_volume_365'] =
                    df['Volume'].rolling(252).mean().shift(1)
...     df_new['ratio_avg_volume_5_30'] =
                df_new['avg_volume_5'] / df_new['avg_volume_30']
...     df_new['ratio_avg_volume_5_365'] =
                df_new['avg_volume_5'] / df_new['avg_volume_365']
...     df_new['ratio_avg_volume_30_365'] =
                df_new['avg_volume_30'] / df_new['avg_volume_365']
```

As for the standard deviation, we develop the following sub-function for the price-related features:

```
>>> def add_std_price(df, df_new):
...     df_new['std_price_5'] =
                df['Close'].rolling(5).std().shift(1)
...     df_new['std_price_30'] =
                df['Close'].rolling(21).std().shift(1)
...     df_new['std_price_365'] =
                df['Close'].rolling(252).std().shift(1)
```

```
...       df_new['ratio_std_price_5_30'] =
                df_new['std_price_5'] / df_new['std_price_30']
...       df_new['ratio_std_price_5_365'] =
                df_new['std_price_5'] / df_new['std_price_365']
...       df_new['ratio_std_price_30_365'] =
                df_new['std_price_30'] / df_new['std_price_365']
```

Similarly, a sub-function that generates six volume-based standard deviation features is as follows:

```
>>> def add_std_volume(df, df_new):
...       df_new['std_volume_5'] =
                df['Volume'].rolling(5).std().shift(1)
...       df_new['std_volume_30'] =
                df['Volume'].rolling(21).std().shift(1)
...       df_new['std_volume_365'] =
                df['Volume'].rolling(252).std().shift(1)
...       df_new['ratio_std_volume_5_30'] =
                df_new['std_volume_5'] / df_new['std_volume_30']
...       df_new['ratio_std_volume_5_365'] =
                df_new['std_volume_5'] / df_new['std_volume_365']
...       df_new['ratio_std_volume_30_365'] =
                df_new['std_volume_30'] / df_new['std_volume_365']
```

Seven return-based features are generated using the following sub-function:

```
>>> def add_return_feature(df, df_new):
...       df_new['return_1'] = ((df['Close'] - df['Close'].shift(1))
                                / df['Close'].shift(1)).shift(1)
...       df_new['return_5'] = ((df['Close'] - df['Close'].shift(5))
                                / df['Close'].shift(5)).shift(1)
...       df_new['return_30'] = ((df['Close'] -
            df['Close'].shift(21)) / df['Close'].shift(21)).shift(1)
...       df_new['return_365'] = ((df['Close'] -
            df['Close'].shift(252)) / df['Close'].shift(252)).shift(1)
...       df_new['moving_avg_5'] =
                df_new['return_1'].rolling(5).mean().shift(1)
...       df_new['moving_avg_30'] =
                df_new['return_1'].rolling(21).mean().shift(1)
...       df_new['moving_avg_365'] =
                df_new['return_1'].rolling(252).mean().shift(1)
```

Finally, we put together the main feature generation function that calls all the preceding sub-functions:

```
>>> def generate_features(df):
...     """
...     Generate features for a stock/index based on historical price and
performance
...     @param df: dataframe with columns "Open", "Close", "High", "Low",
"Volume", "Adj Close"
...     @return: dataframe, data set with new features
...     """
...     df_new = pd.DataFrame()
...     # 6 original features
...     add_original_feature(df, df_new)
...     # 31 generated features
...     add_avg_price(df, df_new)
...     add_avg_volume(df, df_new)
...     add_std_price(df, df_new)
...     add_std_volume(df, df_new)
...     add_return_feature(df, df_new)
...     # the target
...     df_new['close'] = df['Close']
...     df_new = df_new.dropna(axis=0)
...     return df_new
```

 Note that the window sizes here are 5, 21, and 252, instead of 7, 30, and 365, representing the weekly, monthly, and yearly windows respectively. This is because there are 252 (rounded) trading days in a year, 21 trading days in a month, and 5 in a week.

We can apply this feature engineering strategy on the NASDAQ Composite data queried from 1990 to the first half of 2023, as follows (or directly download it from this page: https://finance.yahoo.com/quote/%5EIXIC/history?period1=631152000&period2=1688083200&interval=1d&filter=history&frequency=1d&includeAdjustedClose=true):

```
>>> data_raw = pd.read_csv('19900101_20230630.csv', index_col='Date')
>>> data = generate_features(data_raw)
```

Take a look at what the data with the new features looks like:

```
>>> print(data.round(decimals=3).head(5))
```

The preceding command line generates the following output:

	open	open_1	close_1	high_1	low_1	volume_1	avg_price_5 \
Date							
1991-01-03	371.2	373.0	372.2	373.5	371.8	92020000.0	372.14
1991-01-04	366.5	371.2	367.5	371.8	367.4	108390000.0	371.16
1991-01-07	363.5	366.5	367.2	367.9	365.9	103830000.0	370.38
1991-01-08	359.1	363.5	360.2	365.8	360.1	109460000.0	368.18
1991-01-09	362.4	359.1	359.0	360.5	358.2	111730000.0	365.22

	avg_price_30	avg_price_365	ratio_avg_price_5_30	... \
Date				...
1991-01-03	370.305	408.631	1.005	...
1991-01-04	370.600	408.266	1.002	...
1991-01-07	370.748	407.905	0.999	...
1991-01-08	370.238	407.514	0.994	...
1991-01-09	369.605	407.126	0.988	...

	ratio_std_volume_5_365	ratio_std_volume_30_365	return_1 \
Date			
1991-01-03	1.026	1.246	-0.004
1991-01-04	0.701	1.252	-0.013
1991-01-07	0.693	1.260	-0.001
1991-01-08	0.539	1.238	-0.019
1991-01-09	0.317	1.057	-0.003

	return_5	return_30	return_365	moving_avg_5	moving_avg_30 \
Date					
1991-01-03	-0.001	0.036	-0.192	0.000	0.002
1991-01-04	-0.013	0.017	-0.200	-0.000	0.002
1991-01-07	-0.011	0.009	-0.199	-0.003	0.001
1991-01-08	-0.030	-0.029	-0.215	-0.002	0.000
1991-01-09	-0.040	-0.036	-0.214	-0.006	-0.001

	moving_avg_365	close
Date		
1991-01-03	-0.001	367.5
1991-01-04	-0.001	367.2
1991-01-07	-0.001	360.2
1991-01-08	-0.001	359.0
1991-01-09	-0.001	357.5

[5 rows x 38 columns]

Figure 5.6: Printout of the first five rows of the DataFrame

Since all the features and driving factors are ready, we will now focus on regression algorithms that estimate the continuous target variables based on these predictive features.

Estimating with linear regression

The first regression model that comes to mind is **linear regression**. Does this mean fitting data points using a linear function, as its name implies? Let's explore it.

How does linear regression work?

In simple terms, linear regression tries to fit as many of the data points as possible, with a straight line in two-dimensional space or a plane in three-dimensional space. It explores the linear relationship between observations and targets, and the relationship is represented in a linear equation or weighted sum function. Given a data sample x with n features, $x_1, x_2, ..., x_n$ (x represents a feature vector and $x = (x_1, x_2, ..., x_n)$), and weights (also called **coefficients**) of the linear regression model w (w represents a vector $(w_1, w_2, ..., w_n)$), the target y is expressed as follows:

$$y = w_1 x_1 + w_2 x_2 + \cdots + w_n x_n = w^T x$$

Also, sometimes the linear regression model comes with an intercept (also called bias), w_0, so the preceding linear relationship becomes as follows:

$$y = w_0 + w_1 x_1 + w_2 x_2 + \cdots + w_n x_n = w^T x$$

Does it look familiar? The **logistic regression** algorithm you learned in *Chapter 4, Predicting Online Ad Click-Through with Logistic Regression*, is just an addition of logistic transformation on top of the linear regression, which maps the continuous weighted sum to the *0* (negative) or *1* (positive) class. Similarly, a linear regression model, or specifically its weight vector, w, is learned from the training data, with the goal of minimizing the estimation error defined as the **mean squared error** (MSE), which measures the average of squares of difference between the truth and prediction. Given m training samples, $(x^{(1)}, y^{(1)}), (x^{(2)}, y^{(2)}), ... (x^{(i)}, y^{(i)})..., (x^{(m)}, y^{(m)})$, the loss function $J(w)$ regarding the weights to be optimized is expressed as follows:

$$J(w) = \frac{1}{m} \sum_{i=1}^{m} \frac{1}{2} (\hat{y}(x^{(i)}) - y^{(i)})^2$$

Here, $(\hat{y}(x^{(i)})) = w^T x^{(i)}$ is the prediction.

Again, we can obtain the optimal w so that $J(w)$ is minimized using gradient descent. The first-order derivative, the gradient Δw, is derived as follows:

$$\Delta w = \frac{1}{m} \sum_{i=1}^{m} -(y^{(i)} - \hat{y}(x^{(i)})) x^{(i)}$$

Combined with the gradient and learning rate η, the weight vector w can be updated in each step as follows:

$$w := w + \eta \frac{1}{m} \sum_{i=1}^{m} \left(y^{(i)} - \hat{y}(x^{(i)}) \right) x^{(i)}$$

After a substantial number of iterations, the learned w is then used to predict a new sample x', as follows:

$$y' = w^T x'$$

After learning about the mathematical theory behind linear regression, let's implement it from scratch in the next section.

Implementing linear regression from scratch

Now that you have a thorough understanding of gradient-descent-based linear regression, we'll implement it from scratch.

We start by defining the function computing the prediction, $\hat{y}(x)$, with the current weights:

```
>>> def compute_prediction(X, weights):
...     """
...     Compute the prediction y_hat based on current weights
...     """
...     return np.dot(X, weights)
```

Then, we continue with the function updating the weight, w, with one step in a gradient descent manner, as follows:

```
>>> def update_weights_gd(X_train, y_train, weights,
learning_rate):
...     predictions = compute_prediction(X_train, weights)
...     weights_delta = np.dot(X_train.T, y_train - predictions)
...     m = y_train.shape[0]
...     weights += learning_rate / float(m) * weights_delta
...     return weights
```

Next, we add the function that calculates the loss $J(w)$ as well:

```
>>> def compute_loss(X, y, weights):
...     """
...     Compute the loss J(w)
...     """
...     predictions = compute_prediction(X, weights)
...     return np.mean((predictions - y) ** 2 / 2.0)
```

Now, put all functions together with a model training function by performing the following tasks:

1. Update the weight vector in each iteration
2. Print out the current cost for every 500 (or it can be any number) iterations to ensure cost is decreasing and things are on the right track

Let's see how it's done by executing the following commands:

```
>>> def train_linear_regression(X_train, y_train, max_iter, learning_rate, fit_
intercept=False, display_loss=500):
...     """
```

```
...       Train a linear regression model with gradient descent, and return
trained model
...       """
...       if fit_intercept:
...           intercept = np.ones((X_train.shape[0], 1))
...           X_train = np.hstack((intercept, X_train))
...       weights = np.zeros(X_train.shape[1])
...       for iteration in range(max_iter):
...           weights = update_weights_gd(X_train, y_train,
                                          weights, learning_rate)
...           # Check the cost for every 500 (by default) iterations
...           if iteration % 500 == 0:
...               print(compute_loss(X_train, y_train, weights))
...       return weights
```

Finally, predict the results of new input values using the trained model as follows:

```
>>> def predict(X, weights):
...       if X.shape[1] == weights.shape[0] - 1:
...           intercept = np.ones((X.shape[0], 1))
...           X = np.hstack((intercept, X))
...       return compute_prediction(X, weights)
```

Implementing linear regression is very similar to logistic regression, as you just saw. Let's examine it with a small example:

```
>>> X_train = np.array([[6], [2], [3], [4], [1],
                        [5], [2], [6], [4], [7]])
>>> y_train = np.array([5.5, 1.6, 2.2, 3.7, 0.8,
                        5.2, 1.5, 5.3, 4.4, 6.8])
```

Train a linear regression model with 100 iterations, at a learning rate of 0.01, based on intercept-included weights:

```
>>> weights = train_linear_regression(X_train, y_train,
            max_iter=100, learning_rate=0.01, fit_intercept=True)
```

Check the model's performance on new samples as follows:

```
>>> X_test = np.array([[1.3], [3.5], [5.2], [2.8]])
>>> predictions = predict(X_test, weights)
>>> import matplotlib.pyplot as plt
>>> plt.scatter(X_train[:, 0], y_train, marker='o', c='b')
>>> plt.scatter(X_test[:, 0], predictions, marker='*', c='k')
>>> plt.xlabel('x')
```

```
>>> plt.ylabel('y')
>>> plt.show()
```

Refer to the following screenshot for the result:

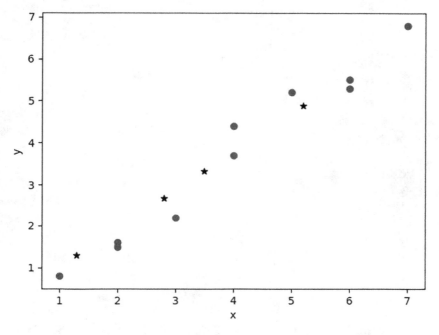

Figure 5.7: Linear regression on a toy dataset

The model we trained correctly predicts new samples (depicted by the stars).

Let's try it on another dataset, the diabetes dataset from scikit-learn:

```
>>> from sklearn import datasets
>>> diabetes = datasets.load_diabetes()
>>> print(diabetes.data.shape)
(442, 10)
>>> num_test = 30
>>> X_train = diabetes.data[:-num_test, :]
>>> y_train = diabetes.target[:-num_test]
```

Train a linear regression model with 5000 iterations, at a learning rate of 1, based on intercept-included weights (the loss is displayed every 500 iterations):

```
>>> weights = train_linear_regression(X_train, y_train,
            max_iter=5000, learning_rate=1, fit_intercept=True)
2960.1229915
1539.55080927
1487.02495658
```

```
 1480.27644342
 1479.01567047
 1478.57496091
 1478.29639883
 1478.06282572
 1477.84756968
 1477.64304737
>>> X_test = diabetes.data[-num_test:, :]
>>> y_test = diabetes.target[-num_test:]
>>> predictions = predict(X_test, weights)
>>> print(predictions)
[ 232.22305668 123.87481969 166.12805033 170.23901231
  228.12868839 154.95746522 101.09058779 87.33631249
  143.68332296 190.29353122 198.00676871 149.63039042
   169.56066651 109.01983998 161.98477191 133.00870377
   260.1831988 101.52551082 115.76677836 120.7338523
   219.62602446 62.21227353 136.29989073 122.27908721
   55.14492975 191.50339388 105.685612 126.25915035
   208.99755875 47.66517424]
>>> print(y_test)
[ 261. 113. 131. 174. 257. 55. 84. 42. 146. 212. 233.
   91. 111. 152. 120. 67. 310. 94. 183. 66. 173. 72.
   49. 64. 48. 178. 104. 132. 220. 57.]
```

The estimate is pretty close to the ground truth.

Next, let's utilize scikit-learn to implement linear regression.

Implementing linear regression with scikit-learn

So far, we have used gradient descent in weight optimization, but like with logistic regression, linear regression is also open to **Stochastic Gradient Descent (SGD)**. To use it, we can simply replace the update_weights_gd function with the update_weights_sgd function we created in *Chapter 4, Predicting Online Ad Click-Through with Logistic Regression*.

We can also directly use the SGD-based regression algorithm, SGDRegressor, from scikit-learn:

```
>>> from sklearn.linear_model import SGDRegressor
>>> regressor = SGDRegressor(loss='squared_error',
                             penalty='l2',
                             alpha=0.0001,
                             learning_rate='constant',
                             eta0=0.2,
                             max_iter=100,
                             random_state=42)
```

Here, `'squared_error'` for the `loss` parameter indicates that the cost function is MSE; penalty is the regularization term, and it can be None, l1, or l2, which is similar to SGDClassifier in *Chapter 4, Predicting Online Ad Click-Through with Logistic Regression*, in order to reduce overfitting; `max_iter` is the number of iterations; and the remaining two parameters mean the learning rate is 0.2 and unchanged during the course of training over, at most, 100 iterations. Train the model and output predictions on the testing set of the diabetes dataset, as follows:

```
>>> regressor.fit(X_train, y_train)
>>> predictions = regressor.predict(X_test)
>>> print(predictions)
[213.10213626 108.68382244 152.18820636 153.81308148 208.42650616 137.24771808
  88.91487772  73.83269079 131.35148348 173.65164632 178.16029669 135.26642772
 152.92346973  89.39394334 149.98088897 117.62875063 241.90665387  86.59992328
 101.90393228 105.13958969 202.13586812  50.60429115 121.43542595 106.34058448
  41.11664041 172.53683431  95.43229463 112.59395222 187.40792      36.1586737 ]
```

You can also implement linear regression with TensorFlow. Let's see this in the next section.

Implementing linear regression with TensorFlow

First, we import TensorFlow and construct the model:

```
>>> import tensorflow as tf
>>> layer0 = tf.keras.layers.Dense(units=1,
                        input_shape=[X_train.shape[1]])
>>> model = tf.keras.Sequential(layer0)
```

It uses a linear layer (or you can think of it as a linear function) to connect the input in the X_train.shape[1] dimension and the output in the 1 dimension.

Next, we specify the loss function, the MSE, and a gradient descent optimizer, Adam, with a learning rate of 1:

```
>>> model.compile(loss='mean_squared_error',
          optimizer=tf.keras.optimizers.Adam(1))
```

Now, we train the model on the diabetes dataset for 100 iterations, as follows:

```
>>> model.fit(X_train, y_train, epochs=100, verbose=True)
Epoch 1/100
412/412 [==============================] - 1s 2ms/sample - loss: 27612.9129
Epoch 2/100
412/412 [==============================] - 0s 44us/sample - loss: 23802.3043
Epoch 3/100
412/412 [==============================] - 0s 47us/sample - loss: 20383.9426
Epoch 4/100
412/412 [==============================] - 0s 51us/sample - loss: 17426.2599
```

```
Epoch 5/100
412/412 [==============================] - 0s 44us/sample - loss: 14857.0057
......
Epoch 96/100
412/412 [==============================] - 0s 55us/sample - loss: 2971.6798
Epoch 97/100
412/412 [==============================] - 0s 44us/sample - loss: 2970.8919
Epoch 98/100
412/412 [==============================] - 0s 52us/sample - loss: 2970.7903
Epoch 99/100
412/412 [==============================] - 0s 47us/sample - loss: 2969.7266
Epoch 100/100
412/412 [==============================] - 0s 46us/sample - loss: 2970.4180
```

This also prints out the loss for every iteration. Finally, we make predictions using the trained model:

```
>>> predictions = model.predict(X_test)[:, 0]
>>> print(predictions)
[231.52155  124.17711  166.71492  171.3975   227.70126  152.02522
 103.01532   91.79277  151.07457  190.01042  190.60373  152.52274
 168.92166  106.18033  167.02473  133.37477  259.24756  101.51256
 119.43106  120.893005 219.37921   64.873634 138.43217  123.665634
  56.33039  189.27441  108.67446  129.12535  205.06857   47.99469 ]
```

The next regression algorithm you will learn about is decision tree regression.

Estimating with decision tree regression

Decision tree regression is also called a **regression tree**. It is easy to understand a regression tree by comparing it with its sibling, the classification tree, which you are already familiar with. In this section, we will delve into employing decision tree algorithms for regression tasks.

Transitioning from classification trees to regression trees

In classification, a decision tree is constructed by recursive binary splitting and growing each node into left and right children. In each partition, it greedily searches for the most significant combination of features and its value as the optimal splitting point. The quality of separation is measured by the weighted purity of the labels of the two resulting children, specifically via Gini Impurity or Information Gain.

In regression, the tree construction process is almost identical to the classification one, with only two differences because the target becomes continuous:

- The quality of the splitting point is now measured by the weighted MSE of two children; the MSE of a child is equivalent to the variance of all target values, and the smaller the weighted MSE, the better the split

- The **average** value of targets in a terminal node becomes the leaf value, instead of the majority of labels in the classification tree

To make sure you understand regression trees, let's work on a small house price estimation example using the **house type** and **number of bedrooms**:

Type	Number of bedrooms	Price (thousand)
Semi	3	600
Detached	2	700
Detached	3	800
Semi	2	400
Semi	4	700

Figure 5.8: Toy dataset of house prices

We first define the MSE and weighted MSE computation functions that will be used in our calculation:

```
>>> def mse(targets):
...     # When the set is empty
...     if targets.size == 0:
...         return 0
...     return np.var(targets)
```

Then, we define the weighted MSE after a split in a node:

```
>>> def weighted_mse(groups):
...     total = sum(len(group) for group in groups)
...     weighted_sum = 0.0
...     for group in groups:
...         weighted_sum += len(group) / float(total) * mse(group)
...     return weighted_sum
```

Test things out by executing the following commands:

```
>>> print(f'{mse(np.array([1, 2, 3])):.4f}')
0.6667
>>> print(f'{weighted_mse([np.array([1, 2, 3]), np.array([1, 2])]):.4f}')
0.5000
```

To build the house price regression tree, we first exhaust all possible feature and value pairs, and we compute the corresponding MSE:

```
MSE(type, semi) = weighted_mse([[600, 400, 700], [700, 800]]) = 10333
MSE(bedroom, 2) = weighted_mse([[700, 400], [600, 800, 700]]) = 13000
MSE(bedroom, 3) = weighted_mse([[600, 800], [700, 400, 700]]) = 16000
MSE(bedroom, 4) = weighted_mse([[700], [600, 700, 800, 400]]) = 17500
```

The lowest MSE is achieved with the type, semi pair, and the root node is then formed by this splitting point. The result of this partition is as follows:

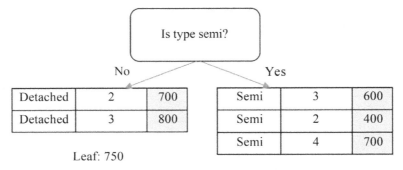

Figure 5.9: Splitting using (type=semi)

If we are satisfied with a one-level regression tree, we can stop here by assigning both branches as leaf nodes, with the value as the average of the targets of the samples included. Alternatively, we can go further down the road by constructing the second level from the right branch (the left branch can't be split further):

```
MSE(bedroom, 2) = weighted_mse([[], [600, 400, 700]]) = 15556
MSE(bedroom, 3) = weighted_mse([[400], [600, 700]]) = 1667
MSE(bedroom, 4) = weighted_mse([[400, 600], [700]]) = 6667
```

With the second splitting point specified by the bedroom, 3 pair (whether it has at least three bedrooms or not) with the lowest MSE, our tree becomes as shown in the following diagram:

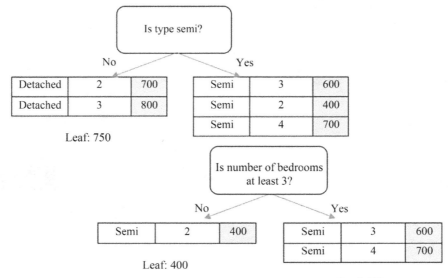

Figure 5.10: Splitting using (bedroom>=3)

We can finish up the tree by assigning average values to both leaf nodes.

Implementing decision tree regression

Now that you're clear about the regression tree construction process, it's time for coding.

The node splitting utility function we will define in this section is identical to what we used in *Chapter 3, Predicting Online Ad Click-Through with Tree-Based Algorithms*, which separates samples in a node into left and right branches, based on a feature and value pair:

```
>>> def split_node(X, y, index, value):
...     x_index = X[:, index]
...     # if this feature is numerical
...     if type(X[0, index]) in [int, float]:
...         mask = x_index >= value
...     # if this feature is categorical
...     else:
...         mask = x_index == value
...     # split into left and right child
...     left = [X[~mask, :], y[~mask]]
...     right = [X[mask, :], y[mask]]
...     return left, right
```

Next, we define the greedy search function, trying out all the possible splits and returning the one with the least weighted MSE:

```
>>> def get_best_split(X, y):
...     """
...     Obtain the best splitting point and resulting children for the data set X, y
...     @return: {index: index of the feature, value: feature value, children: left and right children}
...     """
...     best_index, best_value, best_score, children =
...                                     None, None, 1e10, None
...     for index in range(len(X[0])):
...         for value in np.sort(np.unique(X[:, index])):
...             groups = split_node(X, y, index, value)
...             impurity = weighted_mse(
...                             [groups[0][1], groups[1][1]])
...             if impurity < best_score:
...                 best_index, best_value, best_score, children
...                             = index, value, impurity, groups
...     return {'index': best_index, 'value': best_value,
...             'children': children}
```

The preceding selection and splitting process occurs recursively in each of the subsequent children. When a stopping criterion is met, the process at a node stops, and the mean value of the sample, `targets`, will be assigned to this terminal node:

```
>>> def get_leaf(targets):
...      # Obtain the leaf as the mean of the targets
...      return np.mean(targets)
```

And finally, here is the recursive function, `split`, that links it all together. It checks whether any stopping criteria are met and assigns the leaf node if so, proceeding with further separation otherwise:

```
>>> def split(node, max_depth, min_size, depth):
...      """
...      Split children of a node to construct new nodes or assign them
terminals
...      @param node: dict, with children info
...      @param max_depth: maximal depth of the tree
...      @param min_size: minimal samples required to further split a child
...      @param depth: current depth of the node
...      """
...      left, right = node['children']
...      del (node['children'])
...      if left[1].size == 0:
...          node['right'] = get_leaf(right[1])
...          return
...      if right[1].size == 0:
...          node['left'] = get_leaf(left[1])
...          return
...      # Check if the current depth exceeds the maximal depth
...      if depth >= max_depth:
...          node['left'], node['right'] = get_leaf(
...                          left[1]), get_leaf(right[1])
...          return
...      # Check if the left child has enough samples
...      if left[1].size <= min_size:
...          node['left'] = get_leaf(left[1])
...      else:
...          # It has enough samples, we further split it
...          result = get_best_split(left[0], left[1])
...          result_left, result_right = result['children']
...          if result_left[1].size == 0:
...              node['left'] = get_leaf(result_right[1])
...          elif result_right[1].size == 0:
```

```
...                     node['left'] = get_leaf(result_left[1])
...             else:
...                     node['left'] = result
...                     split(node['left'], max_depth, min_size, depth + 1)
...             # Check if the right child has enough samples
...             if right[1].size <= min_size:
...                     node['right'] = get_leaf(right[1])
...             else:
...                     # It has enough samples, we further split it
...                     result = get_best_split(right[0], right[1])
...                     result_left, result_right = result['children']
...                     if result_left[1].size == 0:
...                         node['right'] = get_leaf(result_right[1])
...                     elif result_right[1].size == 0:
...                         node['right'] = get_leaf(result_left[1])
...                     else:
...                         node['right'] = result
...                         split(node['right'], max_depth, min_size,
...                             depth + 1)
```

The entry point of the regression tree construction is as follows:

```
>>> def train_tree(X_train, y_train, max_depth, min_size):
...     root = get_best_split(X_train, y_train)
...     split(root, max_depth, min_size, 1)
...     return root
```

Now, let's test it with a hand-calculated example:

```
>>> X_train = np.array([['semi', 3],
...                     ['detached', 2],
...                     ['detached', 3],
...                     ['semi', 2],
...                     ['semi', 4]], dtype=object)
>>> y_train = np.array([600, 700, 800, 400, 700])
>>> tree = train_tree(X_train, y_train, 2, 2)
```

To verify that the trained tree is identical to what we constructed by hand, we write a function displaying the tree:

```
>>> CONDITION = {'numerical': {'yes': '>=', 'no': '<'},
...              'categorical': {'yes': 'is', 'no': 'is not'}}
>>> def visualize_tree(node, depth=0):
...     if isinstance(node, dict):
```

```
...             if type(node['value']) in [int, float]:
...                 condition = CONDITION['numerical']
...             else:
...                 condition = CONDITION['categorical']
...             print('{}|- X{} {} {}'.format(depth * ' ',
...                 node['index'] + 1, condition['no'],
...                 node['value']))
...             if 'left' in node:
...                 visualize_tree(node['left'], depth + 1)
...             print('{}|- X{} {} {}'.format(depth * ' ',
...                 node['index'] + 1, condition['yes'],
...                 node['value']))
...             if 'right' in node:
...                 visualize_tree(node['right'], depth + 1)
...     else:
...         print('{}[{}]'.format(depth * ' ', node))
>>> visualize_tree(tree)
|- X1 is not detached
  |- X2 < 3
    [400.0]
  |- X2 >= 3
    [650.0]
|- X1 is detached
  [750.0]
```

Now that you have a better understanding of the regression tree after implementing it from scratch, we can directly use the `DecisionTreeRegressor` package (https://scikit-learn.org/stable/modules/generated/sklearn.tree.DecisionTreeRegressor.html) from scikit-learn. Let's apply it to an example of predicting California house prices. The dataset contains a median house value as the target variable, median income, housing median age, total rooms, total bedrooms, population, households, latitude, and longitude as features. It was obtained from the StatLib repository (https://www.dcc.fc.up.pt/~ltorgo/Regression/cal_housing.html) and can be directly loaded using the `sklearn.datasets.fetch_california_housing` function, as follows:

```
>>> housing = datasets.fetch_california_housing()
```

We take the last 10 samples for testing and the rest to train a `DecisionTreeRegressor` decision tree, as follows:

```
>>> num_test = 10 # the last 10 samples as testing set
>>> X_train = housing.data[:-num_test, :]
>>> y_train = housing.target[:-num_test]
>>> X_test = housing.data[-num_test:, :]
>>> y_test = housing.target[-num_test:]
```

```
>>> from sklearn.tree import DecisionTreeRegressor
>>> regressor = DecisionTreeRegressor(max_depth=10,
                                      min_samples_split=3,
                                      random_state=42)
>>> regressor.fit(X_train, y_train)
```

We then apply the trained decision tree to the test set:

```
>>> predictions = regressor.predict(X_test)
>>> print(predictions)
[1.29568298 1.29568298 1.29568298 1.11946842 1.29568298 0.66193704 0.82554167
 0.8546936  0.8546936  0.8546936 ]
```

Compare predictions with the ground truth, as follows:

```
>>> print(y_test)
[1.12  1.072 1.156 0.983 1.168 0.781 0.771 0.923 0.847 0.894]
```

We see the predictions are quite accurate.

We have implemented a regression tree in this section. Is there an ensemble version of the regression tree? Let's see next.

Implementing a regression forest

In *Chapter 3, Predicting Online Ad Click-Through with Tree-Based Algorithms,* we explored **random forests** as an ensemble learning method, by combining multiple decision trees that are separately trained and randomly subsampling training features in each node of a tree. In classification, a random forest makes a final decision by a majority vote of all tree decisions. Applied to regression, a random forest regression model (also called a **regression forest**) assigns the average of regression results from all decision trees to the final decision.

Here, we will use the regression forest package, RandomForestRegressor, from scikit-learn and deploy it in our California house price prediction example:

```
>>> from sklearn.ensemble import RandomForestRegressor
>>> regressor = RandomForestRegressor(n_estimators=100,
                                      max_depth=10,
                                      min_samples_split=3,
                                      random_state=42)
>>> regressor.fit(X_train, y_train)
>>> predictions = regressor.predict(X_test)
>>> print(predictions)
[1.31785493 1.29359614 1.24146512 1.06039979 1.24015576 0.7915538  0.90307069
 0.83535894 0.8956997  0.91264529]
```

You've learned about three regression algorithms. So, how should we evaluate regression performance? Let's find out in the next section.

Evaluating regression performance

So far, we've covered three popular regression algorithms in depth and implemented them from scratch by using several prominent libraries. Instead of judging how well a model works on testing sets by printing out the prediction, we need to evaluate its performance with the following metrics, which give us better insights:

- The MSE, as I mentioned, measures the squared loss corresponding to the expected value. Sometimes, the square root is taken on top of the MSE in order to convert the value back into the original scale of the target variable being estimated. This yields the **Root Mean Squared Error (RMSE)**. Also, the RMSE has the benefit of penalizing large errors more, since we first calculate the square of an error.

- Conversely, the **Mean Absolute Error (MAE)** measures the absolute loss. It uses the same scale as the target variable and gives us an idea of how close the predictions are to the actual values.

  For both the MSE and MAE, the smaller the value, the better the regression model.

- R^2 (pronounced **r squared**) indicates the goodness of the fit of a regression model. It is the fraction of the dependent variable variation that a regression model is able to explain. It ranges from 0 to 1, representing from no fit to a perfect prediction. There is a variant of R^2 called **adjusted** R^2. It adjusts for the number of features in a model relative to the number of data points.

Let's compute these three measurements on a linear regression model, using corresponding functions from scikit-learn:

1. We will work on the diabetes dataset again and fine-tune the parameters of the linear regression model, using the grid search technique:

```
>>> diabetes = datasets.load_diabetes()
>>> num_test = 30 # the last 30 samples as testing set
>>> X_train = diabetes.data[:-num_test, :]
>>> y_train = diabetes.target[:-num_test]
>>> X_test = diabetes.data[-num_test:, :]
>>> y_test = diabetes.target[-num_test:]
>>> param_grid = {
...     "alpha": [1e-07, 1e-06, 1e-05],
...     "penalty": [None, "12"],
...     "eta0": [0.03, 0.05, 0.1],
...     "max_iter": [500, 1000]
... }
>>> from sklearn.model_selection import GridSearchCV
>>> regressor = SGDRegressor(loss='squared_error',
                             learning_rate='constant',
                             random_state=42)
>>> grid_search = GridSearchCV(regressor, param_grid, cv=3)
```

2. We obtain the optimal set of parameters:

```
>>> grid_search.fit(X_train, y_train)
>>> print(grid_search.best_params_)
{'alpha': 1e-07, 'eta0': 0.05, 'max_iter': 500, 'penalty': None}
>>> regressor_best = grid_search.best_estimator_
```

3. We predict the testing set with the optimal model:

```
>>> predictions = regressor_best.predict(X_test)
```

4. We evaluate the performance on testing sets based on the MSE, MAE, and R^2 metrics:

```
>>> from sklearn.metrics import mean_squared_error,
    mean_absolute_error, r2_score
>>> print(mean_squared_error(y_test, predictions))
1933.3953304460413
>>> print(mean_absolute_error(y_test, predictions))
35.48299900764652
>>> print(r2_score(y_test, predictions))
0.6247444629690868
```

Now that you've learned about three (or four, you could say) commonly used and powerful regression algorithms and performance evaluation metrics, let's utilize each of them to solve our stock price prediction problem.

Predicting stock prices with the three regression algorithms

Here are the steps to predict the stock price:

1. Earlier, we generated features based on data from 1990 to the first half of 2023, and we will now continue to construct the training set with data from 1990 to 2022 and the testing set with data from the first half of 2023:

```
>>> data_raw = pd.read_csv('19900101_20230630.csv', index_col='Date')
>>> data = generate_features(data_raw)
>>> start_train = '1990-01-01'
>>> end_train = '2022-12-31'
>>> start_test = '2023-01-01'
>>> end_test = '2023-06-30'
>>> data_train = data.loc[start_train:end_train]
>>> X_train = data_train.drop('close', axis=1).values
>>> y_train = data_train['close'].values
>>> print(X_train.shape)
(8061, 37)
```

```
>>> print(y_train.shape)
(8061,)
```

All fields in the `dataframe` data except `'close'` are feature columns, and `'close'` is the target column. We have `8,061` training samples and each sample is 37-dimensional. We also have `124` testing samples:

```
>>> data_train = data.loc[start_train:end_train]
>>> X_train = data_train.drop('close', axis=1).values
>>> y_train = data_train['close'].values
>>> print(X_test.shape)
(124, 37)
```

Best practice

Time series data often exhibits temporal dependencies, where values at one time point are influenced by previous values. Ignoring these dependencies can lead to poor model performance. We need to use a train-test split to evaluate models, ensuring that the test set contains data from a later time period than the training set to simulate real-world forecasting scenarios.

2. We will first experiment with SGD-based linear regression. Before we train the model, you should realize that SGD-based algorithms are sensitive to data with features at very different scales; for example, in our case, the average value of the open feature is around 3,777, while that of the `moving_avg_365` feature is 0.00052 or so. Hence, we need to normalize features into the same or a comparable scale. We do so by removing the mean and rescaling to unit variance with `StandardScaler`:

```
>>> from sklearn.preprocessing import StandardScaler
>>> scaler = StandardScaler()
```

3. We rescale both sets with `scaler`, taught by the training set:

```
>>> X_scaled_train = scaler.fit_transform(X_train)
>>> X_scaled_test = scaler.transform(X_test)
```

4. Now, we can search for the SGD-based linear regression with the optimal set of parameters. We specify `l2` regularization and `5000` maximal iterations and we tune the regularization term multiplier, `alpha`, and initial learning rate, `eta0`:

```
>>> param_grid = {
...     "alpha": [1e-4, 3e-4, 1e-3],
...     "eta0": [0.01, 0.03, 0.1],
... }
>>> lr = SGDRegressor(penalty='l2', max_iter=5000, random_state=42)
```

5. For cross-validation, we need to ensure that the training data in each split comes before the corresponding test data, preserving the temporal order of the time series. Here, we use the `TimeSeriesSplit` method from scikit-learn:

```
>>> from sklearn.model_selection import TimeSeriesSplit
>>> tscv = TimeSeriesSplit(n_splits=3)
>>> grid_search = GridSearchCV(lr, param_grid, cv=tscv, scoring='r2')
>>> grid_search.fit(X_scaled_train, y_train)
```

Here, we create a 3-fold time series-specific cross-validator and employ it in grid search.

6. Select the best linear regression model and make predictions of the testing samples:

```
>>> print(grid_search.best_params_)
{'alpha': 0.0001, 'eta0': 0.1}
>>> lr_best = grid_search.best_estimator_
>>> predictions_lr = lr_best.predict(X_scaled_test)
```

7. Measure the prediction performance via R^2:

```
>>> print(f'R^2: {r2_score(y_test, predictions_lr):.3f}')
R^2: 0.959
```

We achieve an R^2 of 0.959 with a fine-tuned linear regression model.

Best practice

With time series data, there is a risk of overfitting due to the potential complexity of temporal patterns. Models may capture noise instead of genuine patterns if not regularized properly. We need to apply regularization techniques like L1 or L2 regularization to prevent overfitting. Also, when you perform cross-validation for hyperparameter tuning, consider using time series-specific cross-validation methods to assess model performance while preserving temporal order.

8. Similarly, let's experiment with a decision tree. We tune the maximum depth of the tree, `max_depth`; the minimum number of samples required to further split a node, `min_samples_split`; and the minimum number of samples required to form a leaf node, `min_samples_leaf`, as follows:

```
>>> param_grid = {
...      'max_depth': [20, 30, 50],
...      'min_samples_split': [2, 5, 10],
...      'min_samples_leaf': [1, 3, 5]
... }
>>> dt = DecisionTreeRegressor(random_state=42)
>>> grid_search = GridSearchCV(dt, param_grid, cv=tscv,
```

```
                                              scoring='r2', n_jobs=-1)
>>> grid_search.fit(X_train, y_train)
```

Note this may take a while; hence, we use all available CPU cores for training.

9. Select the best regression forest model and make predictions of the testing samples:

```
>>> print(grid_search.best_params_)
{'max_depth': 30, 'min_samples_leaf': 3, 'min_samples_split': 2}
>>> dt_best = grid_search.best_estimator_
>>> predictions_dt = dt_best.predict(X_test)
```

10. Measure the prediction performance as follows:

```
>>> print(f'R^2: {r2_score(y_test, predictions_rf):.3f}')
R^2: 0.912
```

An R^2 of 0.912 is obtained with a tweaked decision tree.

11. Finally, we experiment with a random forest. We specify 30 decision trees to ensemble and tune the same set of hyperparameters used in each tree, as follows:

```
>>> param_grid = {
...      'max_depth': [20, 30, 50],
...      'min_samples_split': [2, 5, 10],
...      'min_samples_leaf': [1, 3, 5]
... }
>>> rf = RandomForestRegressor(n_estimators=30, n_jobs=-1, random_
state=42)
>>> grid_search = GridSearchCV(rf, param_grid, cv=tscv,
                                   scoring='r2', n_jobs=-1)
>>> grid_search.fit(X_train, y_train)
```

Note this may take a while; hence, we use all available CPU cores for training (indicated by n_jobs=-1).

12. Select the best regression forest model and make predictions of the testing samples:

```
>>> print(grid_search.best_params_)
{'max_depth': 30, 'min_samples_leaf': 1, 'min_samples_split': 5}
>>> rf_best = grid_search.best_estimator_
>>> predictions_rf = rf_best.predict(X_test)
```

13. Measure the prediction performance as follows:

```
>>> print(f'R^2: {r2_score(y_test, predictions_rf):.3f}')
R^2: 0.937
```

An R^2 of 0.937 is obtained with a tweaked forest regressor.

14. We also plot the prediction generated by each of the three algorithms, along with the ground truth:

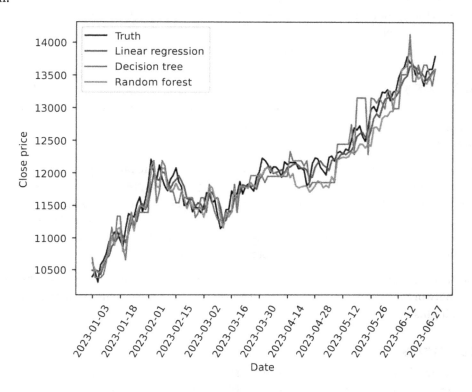

Figure 5.11: Predictions using the three algorithms versus the ground truth

The visualization is produced by the following code:

```
>>> plt.rc('xtick', labelsize=10)
>>> plt.rc('ytick', labelsize=10)
>>> plt.plot(data_test.index, y_test, c='k')
>>> plt.plot(data_test.index, predictions_lr, c='b')
>>> plt.plot(data_test.index, predictions_dt, c='g')
>>> plt.plot(data_test.index, predictions_rf, c='r')
>>> plt.xticks(range(0, 130, 10), rotation=60)
>>> plt.xlabel('Date', fontsize=10)
>>> plt.ylabel('Close price', fontsize=10)
>>> plt.legend(['Truth', 'Linear regression', 'Decision tree', 'Random
forest'], fontsize=10)
>>> plt.show()
```

We've built a stock predictor using three regression algorithms individually in this section. Overall, linear regression outperforms the other two algorithms.

Stock markets are known for their wild swings. Unlike more stable systems or a well-defined project in this chapter, stock prices are volatile and influenced by complex factors that are hard to quantify. Also, their behavior is not easily captured by even the most sophisticated models. Hence, it is notoriously difficult to accurately predict the stock market in the real world. This makes it a fascinating challenge to explore the capabilities of different machine learning models.

Summary

In this chapter, we worked on the project of predicting stock (specifically stock index) prices using machine learning regression techniques. Regression estimates a continuous target variable, as opposed to discrete output in classification

We started with a short introduction to the stock market and the factors that influence trading prices. We followed this with an in-depth discussion of three popular regression algorithms, linear regression, regression trees, and regression forests. We covered their definitions, mechanics, and implementations from scratch with several popular frameworks, including scikit-learn and TensorFlow, along with applications on toy datasets. You also learned the metrics used to evaluate a regression model. Finally, we applied what was covered in this chapter to solve our stock price prediction problem.

In the next chapter, we will continue working on the stock price prediction project, but with powerful **neural networks**. We will see whether they can beat what we have achieved with the three regression models in this chapter.

Exercises

1. As mentioned, can you add more signals to our stock prediction system, such as the performance of other major indexes? Does this improve prediction?
2. Try to ensemble those three regression models, for example, by averaging the predictions, and see whether you can perform better.

Join our book's Discord space

Join our community's Discord space for discussions with the authors and other readers:

```
https://packt.link/yuxi
```

6

Predicting Stock Prices with Artificial Neural Networks

Continuing the same project of stock price prediction from the last chapter, in this chapter, we will introduce and explain neural network models in depth. We will start by building the simplest neural network and go deeper by adding more computational units to it. We will cover neural network building blocks and other important concepts, including activation functions, feedforward, and backpropagation. We will also implement neural networks from scratch with scikit-learn, TensorFlow, and PyTorch. We will pay attention to how to learn with neural networks efficiently without overfitting, utilizing dropout and early stopping techniques. Finally, we will train a neural network to predict stock prices and see whether it can beat what we achieved with the three regression algorithms in the previous chapter.

We will cover the following topics in this chapter:

- Demystifying neural networks
- Building neural networks
- Picking the right activation functions
- Preventing overfitting in neural networks
- Predicting stock prices with neural networks

Demystifying neural networks

Here comes probably the most frequently mentioned model in the media, **Artificial Neural Networks** (**ANNs**); more often, we just call them **neural networks**. Interestingly, the neural network has been (falsely) considered equivalent to machine learning or artificial intelligence by the general public.

 An ANN is just one type of algorithm among many in machine learning, and machine learning is a branch of artificial intelligence. It is one of the ways we achieve **Artificial General Intelligence** (**AGI**), which is a hypothetical type of AI that can think, learn, and solve problems like a human.

Regardless, it is one of the most important machine learning models and has been rapidly evolving along with the revolution of **Deep Learning (DL)**.

Let's first understand how neural networks work.

Starting with a single-layer neural network

We start by explaining different layers in a network, then move on to the activation function, and finally, training a network with backpropagation.

Layers in neural networks

A simple neural network is composed of three layers—the **input layer, hidden layer**, and **output layer**—as shown in the following diagram:

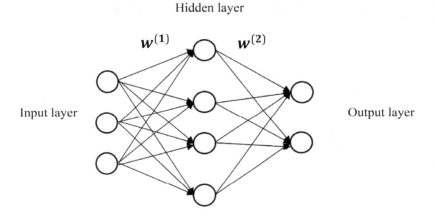

Figure 6.1: A simple shallow neural network

A **layer** is a conceptual collection of **nodes** (also called **units**), which simulate neurons in a biological brain. The input layer represents the input features, **x**, and each node is a predictive feature, x. The output layer represents the target variable(s).

In binary classification, the output layer contains only one node, whose value is the probability of the positive class. In multiclass classification, the output layer consists of n nodes, where n is the number of possible classes and the value of each node is the probability of predicting that class. In regression, the output layer contains only one node, the value of which is the prediction result.

The hidden layer can be considered a composition of latent information extracted from the previous layer. There can be more than one hidden layer. Learning with a neural network with two or more hidden layers is called **deep learning**. In this chapter, we will focus on one hidden layer to begin with.

Two adjacent layers are connected by conceptual edges (sort of like the synapses in a biological brain), which transmit signals from one neuron in a layer to another neuron in the next layer. The **edges** are parameterized by the weights, W, of the model. For example, $W^{(1)}$ in the preceding diagram connects the input and hidden layers and $W^{(2)}$ connects the hidden and output layers.

In a standard neural network, data is conveyed only from the input layer to the output layer, through a hidden layer(s). Hence, this kind of network is called a **feedforward** neural network. Basically, logistic regression is a feedforward neural network with no hidden layer where the output layer connects directly with the input layer. Adding hidden layers between the input and output layers introduces non-linearity. This allows the neural networks to learn more about the underlying relationship between the input data and the target.

Activation functions

An **activation function** is a mathematical operation applied to the output of each neuron in a neural network. It determines whether the neuron should be activated (i.e., its output value should be propagated forward to the next layer) based on the input it receives.

Suppose the input, x, is of n dimensions, and the hidden layer is composed of H hidden units. The weight matrix, $W^{(1)}$, connecting the input and hidden layers is of size n by H, where each column, $w_h^{(1)}$, represents the coefficients associating the input with the h-th hidden unit. The output (also called **activation**) of the hidden layer can be expressed mathematically as follows:

$$a^{(2)} = f(z^{(2)}) = f(W^{(1)}x)$$

Here, $f(z)$ is an activation function. As its name implies, the activation function checks how activated each neuron is, simulating the way our brains work. Their primary purpose is to introduce non-linearity into the output of a neuron, allowing the network to learn and perform complex mappings between inputs and outputs. Typical activation functions include the logistic function (more often called the **sigmoid** function in neural networks) and the **tanh** function, which is considered a rescaled version of the logistic function, as well as **ReLU** (short for **Rectified Linear Unit**), which is often used in DL:

$$sigmoid(z) = \frac{1}{1 + e^{-z}}$$

$$\tanh(z) = \frac{e^z - e^{-z}}{e^z + e^{-z}} = \frac{2}{1 + e^{-2z}} - 1$$

$$relu(z) = z^+ = \max(0, z)$$

We plot these three activation functions as follows:

- The **logistic** (**sigmoid**) function where the output value is in the range of (0, 1):

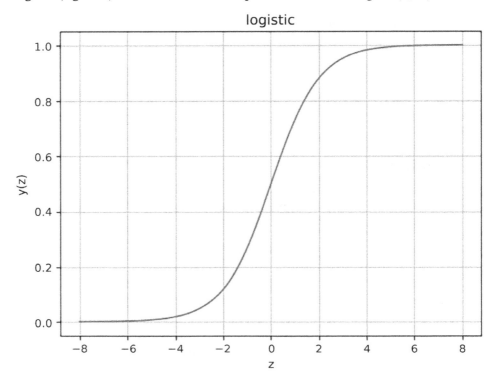

Figure 6.2: The logistic function

The visualization is produced by the following code:

```
>>> import numpy as np
>>> import matplotlib.pyplot as plt
>>> def sigmoid(z):
        return 1.0 / (1 + np.exp(-z))
>>> z = np.linspace(-8, 8, 1000)
>>> y = sigmoid(z)
>>> plt.plot(z, y)
>>> plt.xlabel('z')
>>> plt.ylabel('y(z)')
>>> plt.title('logistic')
>>> plt.grid()
>>> plt.show()
```

- The **tanh** function plot where the output value is in the range of (-1, 1):

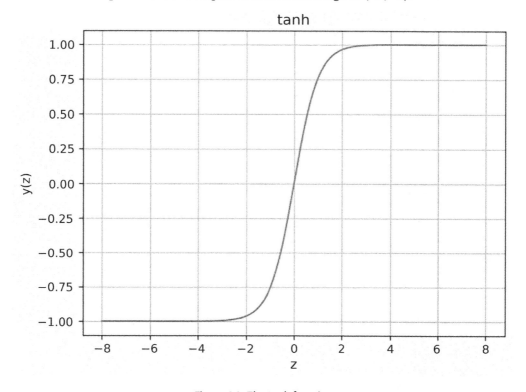

Figure 6.3: The tanh function

The visualization is produced by the following code:

```
>>> def tanh(z):
        return (np.exp(z) - np.exp(-z)) / (np.exp(z) + np.exp(-z))
>>> z = np.linspace(-8, 8, 1000)
>>> y = tanh(z)
>>> plt.plot(z, y)
>>> plt.xlabel('z')
>>> plt.ylabel('y(z)')
>>> plt.title('tanh')
>>> plt.grid()
>>> plt.show()
```

- The **ReLU** function plot where the output value is in the range of `(0, +inf)`:

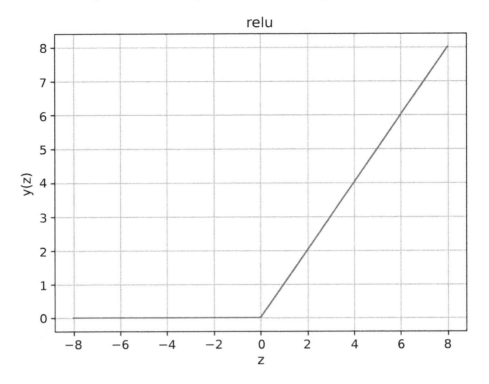

Figure 6.4: The ReLU function

The visualization is produced by the following code:

```
>>> relu(z):
        return np.maximum(np.zeros_like(z), z)
>>> z = np.linspace(-8, 8, 1000)
>>> y = relu(z)
>>> plt.plot(z, y)
>>> plt.xlabel('z')
>>> plt.ylabel('y(z)')
>>> plt.title('relu')
>>> plt.grid()
>>> plt.show()
```

As for the output layer, let's assume that there is one output unit (regression or binary classification) and that the weight matrix, $W^{(2)}$, connecting the hidden layer to the output layer is of size H by 1. In regression, the output can be expressed mathematically as follows (for consistency, I here denote it as $a^{(3)}$ instead of y):

$$a^{(3)} = f(z^{(3)}) = W^{(2)}a^{(2)}$$

The **Universal Approximation Theorem** is a key concept in understanding how neural networks enable learning. It states that a feedforward neural network with a single hidden layer containing a finite number of neurons can approximate any continuous function to arbitrary precision, given a sufficiently large number of neurons in the hidden layer. During the training process, the neural network learns to approximate the target function by adjusting its parameters (weights). This is typically done using optimization algorithms, such as gradient descent, which iteratively update the parameters to minimize the difference between the predicted outputs and the true targets. Let's see this process in detail in the next section.

Backpropagation

So, how can we obtain the optimal weights, $W = \{W(1), W(2)\}$, of the model? Similar to logistic regression, we can learn all weights using gradient descent with the goal of minimizing the **mean squared error (MSE)** cost or other loss function, $J(W)$. The difference is that the gradients, ΔW, are computed through **backpropagation**. After each forward pass through a network, a backward pass is performed to adjust the model's parameters.

As the word *back* in the name implies, the computation of the gradient proceeds backward: the gradient of the final layer is computed first and the gradient of the first layer is computed last. As for *propagation*, it means that partial computations of the gradient on one layer are reused in the computation of the gradient on the previous layer. Error information is propagated layer by layer, instead of being calculated separately.

In a single-layer network, the detailed steps of backpropagation are as follows:

1. We travel through the network from the input to the output and compute the output values, $a^{(2)}$, of the hidden layer as well as the output layer, $a^{(3)}$. This is the feedforward step.

2. For the last layer, we calculate the derivative of the cost function with regard to the input to the output layer:

$$\delta^{(3)} = \frac{\partial}{\partial z^{(3)}} J(W) = -(y - a^{(3)}) \cdot f'(z^{(3)}) = a^{(3)} - y$$

3. For the hidden layer, we compute the derivative of the cost function with regard to the input to the hidden layer:

$$\delta^{(2)} = \frac{\partial}{\partial z^{(2)}} J(W) = \frac{\partial z^{(3)}}{\partial z^{(2)}} \frac{\partial}{\partial z^{(3)}} J(W) = ((W^2)\delta^{(3)}) \cdot f'(z^{(2)})$$

4. We compute the gradients by applying the **chain rule**:

$$\Delta W^{(2)} = \frac{\partial J(W)}{\partial z^{(3)}} \frac{\partial z^{(3)}}{\partial W^{(2)}} = \delta^{(3)} a^{(2)}$$

$$\Delta W^{(1)} = \frac{\partial J(W)}{\partial z^{(2)}} \frac{\partial z^{(2)}}{\partial W^{(1)}} = \delta^{(2)} x$$

5.   We update the weights with the computed gradients and learning rate α:

$$W^{(1)} := W^{(1)} - \frac{1}{m}\alpha \Delta W^{(1)}$$

$$W^{(2)} := W^{(2)} - \frac{1}{m}\alpha \Delta W^{(2)}$$

Here, m is the number of samples.

We repeatedly update all the weights by taking these steps with the latest weights until the cost function converges or the model goes through enough iterations.

 The chain rule is a fundamental concept in calculus. It allows you to find the derivative of a composite function. You can read more in the mathematics course from Stanford University (`https://mathematics.stanford.edu/events/chain-rule-calculus`), or the differential calculus course, *Module 6, Applications of Differentiation*, from MIT (`https://ocw.mit.edu/courses/18-03sc-differential-equations-fall-2011/`).

This might not be easy to digest at first glance, so right after the next section, we will implement it from scratch, which will help you understand neural networks better.

Adding more layers to a neural network: DL

In real applications, a neural network usually comes with multiple hidden layers. That is how DL got its name—learning using neural networks with "stacked" hidden layers. An example of a DL model is as follows:

A Deep Learning Model

Figure 6.5: A deep neural network

In a stack of multiple hidden layers, the input of one hidden layer is the output of its previous layer, as you can see from *Figure 6.5*. Features (signals) are extracted from each hidden layer. Features from different layers represent patterns from different levels. Going beyond shallow neural networks (usually with only one hidden layer), a DL model (usually with two or more hidden layers) with the right network architectures and parameters can better learn complex non-linear relationships from data.

Let's see some typical applications of DL so that you will be more motivated to get started with upcoming DL projects.

Computer vision is widely considered the area with massive breakthroughs in DL. You will learn more about this in *Chapter 11, Categorizing Images of Clothing with Convolutional Neural Networks*, and *Chapter 14, Building an Image Search Engine Using CLIP: A Multimodal Approach*. For now, here is a list of common applications in computer vision:

- Image recognition, such as face recognition and handwritten digit recognition. Handwritten digit recognition, along with the common evaluation dataset MNIST, has become a "Hello, World!" project in DL.

- Image-based search engines heavily utilize DL techniques in their image classification and image similarity encoding components.

- Machine vision, which is a critical part of autonomous vehicles, perceives camera views to make real-time decisions.

- Color restoration from black and white photos and art transfer that ingeniously blends two images of different styles. The artificial masterpieces in Google Arts & Culture (https://artsandculture.google.com/) are impressive.

- Realistic image generation based on textual descriptions. This has applications in creating visual storytelling content and assisting in content creation for marketing and advertising.

Natural Language Processing (NLP) is another field where you can see the dominant use of DL in its modern solutions. You will learn more about this in *Chapter 12, Making Predictions with Sequences Using Recurrent Neural Networks,* and *Chapter 13, Advancing Language Understanding and Generation with the Transformer Models*. But let's quickly look at some examples now:

- Machine translation, where DL has dramatically improved accuracy and fluency, for example, the sentence-based **Google Neural Machine Translation (GNMT)** system.

- Text generation reproduces text by learning the intricate relationships between words in sentences and paragraphs with deep neural networks. You can become a virtual J. K. Rowling or Shakespeare if you train a model well on their works.

- Image captioning, also known as image-to-text, leverages deep neural networks to detect and recognize objects in images and "describe" those objects in a comprehensible sentence. It couples recent breakthroughs in computer vision and NLP. Examples can be found at https://cs.stanford.edu/people/karpathy/deepimagesent/generationdemo/ (developed by Andrej Karpathy from Stanford University).

- In other common NLP tasks such as sentiment analysis and information retrieval and extraction, DL models have achieved state-of-the-art performance.

- **Artificial Intelligence-Generated Content (AIGC)** is one of the recent breakthroughs. It uses DL technologies to create or assist in creating various types of content, such as articles, product descriptions, music, images, and videos.

Similar to shallow networks, we learn all the weights in a deep neural network using gradient descent with the goal of minimizing the MSE cost, $J(W)$. And gradients, ΔW, are computed through backpropagation. The difference is that we backpropagate more than one hidden layer. In the next section, we will implement neural networks by starting with shallow networks and then moving on to deep ones.

Building neural networks

This practical section will start with implementing a shallow network from scratch, followed by a deep network with two layers using scikit-learn. We will then implement a deep network with TensorFlow and PyTorch.

Implementing neural networks from scratch

To demonstrate how activation functions work, we will use sigmoid as the activation function in this example.

We first define the `sigmoid` function and its derivative function:

```
>>> def sigmoid_derivative(z):
...         return sigmoid(z) * (1.0 - sigmoid(z))
```

You can derive the derivative yourself if you are interested in verifying it.

We then define the training function, which takes in the training dataset, the number of units in the hidden layer (we will only use one hidden layer as an example), and the number of iterations:

```
>>> def train(X, y, n_hidden, learning_rate, n_iter):
...         m, n_input = X.shape
...         W1 = np.random.randn(n_input, n_hidden)
...         b1 = np.zeros((1, n_hidden))
...         W2 = np.random.randn(n_hidden, 1)
...         b2 = np.zeros((1, 1))
...         for i in range(1, n_iter+1):
...             Z2 = np.matmul(X, W1) + b1
...             A2 = sigmoid(Z2)
...             Z3 = np.matmul(A2, W2) + b2
...             A3 = Z3
...
...             dZ3 = A3 - y
...             dW2 = np.matmul(A2.T, dZ3)
...             db2 = np.sum(dZ3, axis=0, keepdims=True)
...
```

```
...          dZ2 = np.matmul(dZ3, W2.T) * sigmoid_derivative(Z2)
...          dW1 = np.matmul(X.T, dZ2)
...          db1 = np.sum(dZ2, axis=0)
...
...          W2 = W2 - learning_rate * dW2 / m
...          b2 = b2 - learning_rate * db2 / m
...          W1 = W1 - learning_rate * dW1 / m
...          b1 = b1 - learning_rate * db1 / m
...
...          if i % 100 == 0:
...              cost = np.mean((y - A3) ** 2)
...              print('Iteration %i, training loss: %f' %
                                                (i, cost))
...      model = {'W1': W1, 'b1': b1, 'W2': W2, 'b2': b2}
...      return model
```

Note that besides weights, *W*, we also employ bias, *b*. Before training, we first randomly initialize weights and biases. In each iteration, we feed all layers of the network with the latest weights and biases, then calculate the gradients using the backpropagation algorithm, and finally, update the weights and biases with the resulting gradients. For training performance inspection, we print out the loss and the MSE for every 100 iterations.

To test the model, we will use California house prices as the example dataset again. As a reminder, data normalization is usually recommended whenever gradient descent is used. Hence, we will standardize the input data by removing the mean and scaling to unit variance:

```
>>> from sklearn import datasets
>>> housing = datasets.fetch_california_housing()
>>> num_test = 10 # the last 10 samples as testing set
>>> from sklearn import preprocessing
>>> scaler = preprocessing.StandardScaler()
>>> X_train = housing.data[:-num_test, :]
>>> X_train = scaler.fit_transform(X_train)
>>> y_train = housing.target[:-num_test].reshape(-1, 1)
>>> X_test = housing.data[-num_test:, :]
>>> X_test = scaler.transform(X_test)
>>> y_test = housing.target[-num_test:]
```

With the scaled dataset, we can now train a one-layer neural network with 20 hidden units, a 0.1 learning rate, and 2000 iterations:

```
>>> n_hidden = 20
>>> learning_rate = 0.1
>>> n_iter = 2000
```

```
>>> model = train(X_train, y_train, n_hidden, learning_rate, n_iter)
Iteration 100, training loss: 0.557636
Iteration 200, training loss: 0.519375
Iteration 300, training loss: 0.501025
Iteration 400, training loss: 0.487536
Iteration 500, training loss: 0.476553
Iteration 600, training loss: 0.467207
Iteration 700, training loss: 0.459076
Iteration 800, training loss: 0.451934
Iteration 900, training loss: 0.445621
Iteration 1000, training loss: 0.440013
Iteration 1100, training loss: 0.435024
Iteration 1200, training loss: 0.430558
Iteration 1300, training loss: 0.426541
Iteration 1400, training loss: 0.422920
Iteration 1500, training loss: 0.419653
Iteration 1600, training loss: 0.416706
Iteration 1700, training loss: 0.414049
Iteration 1800, training loss: 0.411657
Iteration 1900, training loss: 0.409502
Iteration 2000, training loss: 0.407555
```

Then, we define a prediction function, which will take in a model and produce the regression results:

```
>>> def predict(x, model):
...     W1 = model['W1']
...     b1 = model['b1']
...     W2 = model['W2']
...     b2 = model['b2']
...     A2 = sigmoid(np.matmul(x, W1) + b1)
...     A3 = np.matmul(A2, W2) + b2
...     return A3
```

Finally, we apply the trained model on the testing set:

```
>>> predictions = predict(X_test, model)
```

Print out the predictions and their ground truths to compare them:

```
>>> print(predictions[:, 0])
[1.11805681 1.1387508  1.06071523 0.81930286 1.21311999 0.6199933 0.92885765
0.81967297 0.90882797 0.87857088]
>>> print(y_test)
[1.12  1.072 1.156 0.983 1.168 0.781 0.771 0.923 0.847 0.894]
```

After successfully building a neural network model from scratch, we will move on to the implementation with scikit-learn.

Implementing neural networks with scikit-learn

We will utilize the MLPRegressor class (**MLP** stands for **multi-layer perceptron**, a nickname for neural networks) to implement neural networks:

```
>>> from sklearn.neural_network import MLPRegressor
>>> nn_scikit = MLPRegressor(hidden_layer_sizes=(16, 8),
                             activation='relu',
                             solver='adam',
                             learning_rate_init=0.001,
                             random_state=42,
                             max_iter=2000)
```

The hidden_layer_sizes hyperparameter represents the number of hidden neurons. In this example, the network contains two hidden layers with 16 and 8 nodes, respectively. ReLU activation is used.

> The Adam optimizer is a replacement for the stochastic gradient descent algorithm. It updates the gradients adaptively based on training data. For more information about Adam, check out the paper at https://arxiv.org/abs/1412.6980.

We fit the neural network model on the training set and predict on the testing data:

```
>>> nn_scikit.fit(X_train, y_train.ravel())
>>> predictions = nn_scikit.predict(X_test)
>>> print(predictions)
[1.19968791 1.2725324  1.30448323 0.88688675 1.18623612 0.72605956 0.87409406
0.85671201 0.93423154 0.94196305]
```

And we calculate the MSE on the prediction:

```
>>> from sklearn.metrics import mean_squared_error
>>> print(mean_squared_error(y_test, predictions))
0.010613171947751738
```

We've implemented a neural network with scikit-learn. Let's do so with TensorFlow in the next section.

Implementing neural networks with TensorFlow

In TensorFlow 2.x, it is simple to initiate a deep neural network model using the Keras (https://keras.io/) module. Let's implement neural networks with TensorFlow by following these steps:

1. First, we import the necessary modules and set a random seed, which is recommended for reproducible modeling:

```
>>> import tensorflow as tf
>>> from tensorflow import keras
>>> tf.random.set_seed(42)
```

2. Next, we create a Keras Sequential model by passing a list of layer instances to the constructor, including two fully connected hidden layers with 16 nodes and 8 nodes, respectively. And again, ReLU activation is used:

```
>>> model = keras.Sequential([
...     keras.layers.Dense(units=16, activation='relu'),
...     keras.layers.Dense(units=8, activation='relu'),
...     keras.layers.Dense(units=1)
... ])
```

3. We compile the model by using Adam as the optimizer with a learning rate of `0.01` and MSE as the learning goal:

```
>>> model.compile(loss='mean_squared_error',
...               optimizer=tf.keras.optimizers.Adam(0.01))
```

4. After defining the model, we now train it against the training set:

```
>>> model.fit(X_train, y_train, epochs=300)
Train on 496 samples
Epoch 1/300
645/645 [==============================] - 1s 1ms/step - loss: 0.6494
Epoch 2/300
645/645 [==============================] - 1s 1ms/step - loss: 0.3827
Epoch 3/300
645/645 [==============================] - 1s 1ms/step - loss: 0.3700
......

......
Epoch 298/300
645/645 [==============================] - 1s 1ms/step - loss: 0.2724
Epoch 299/300
645/645 [==============================] - 1s 1ms/step - loss: 0.2735
Epoch 300/300
645/645 [==============================] - 1s 1ms/step - loss: 0.2730
1/1 [==============================] - 0s 82ms/step
```

We fit the model with `300` iterations. In each iteration, the training loss (MSE) is displayed.

5. Finally, we use the trained model to predict the testing cases and print out the predictions and their MSE:

```
>>> predictions = model.predict(X_test)[:, 0]
```

```
>>> print(predictions)
[1.2387774  1.2480505  1.229521   0.8988129  1.1932802  0.75052583
0.75052583 0.88086814 0.9921638  0.9107932 ]
>>> print(mean_squared_error(y_test, predictions))
0.008271122735361234
```

As you can see, we add layer by layer to the neural network model in the TensorFlow Keras API. We start from the first hidden layer (with 16 nodes), then the second hidden layer (with 8 nodes), and finally, the output layer (with 1 unit, the target variable). It is quite similar to building with LEGO.

In the industry, neural networks are often implemented with PyTorch. Let's see how to do it in the next section.

Implementing neural networks with PyTorch

We will now implement neural networks with PyTorch by following these steps:

1. First, we import the necessary modules and set a random seed, which is recommended for reproducible modeling:

```
>>> import torch
>>> import torch.nn as nn
>>> torch.manual_seed(42)
```

2. Next, we create a torch.nn Sequential model by passing a list of layer instances to the constructor, including two fully connected hidden layers with 16 nodes and 8 nodes, respectively. Again, ReLU activation is used in each fully connected layer:

```
>>> model = nn.Sequential(nn.Linear(X_train.shape[1], 16),
                          nn.ReLU(),
                          nn.Linear(16, 8),
                          nn.ReLU(),
                          nn.Linear(8, 1))
```

3. We initialize an Adam optimizer with a learning rate of 0.01 and MSE as the learning goal:

```
>>> loss_function = nn.MSELoss()
>>> optimizer = torch.optim.Adam(model.parameters(), lr=0.01)
```

4. After defining the model, we need to create tensor objects from the input NumPy arrays before using them to train the PyTorch model:

```
>>> X_train_torch = torch.from_numpy(X_train.astype(np.float32))
>>> y_train_torch = torch.from_numpy(y_train.astype(np.float32))
```

5. Now we can train the model against the PyTorch-compatible training set. We first define a training function that will be called in each epoch as follows:

```
>>> def train_step(model, X_train, y_train, loss_function, optimizer):
```

```
        pred_train = model(X_train)
        loss = loss_function(pred_train, y_train)

        model.zero_grad()
        loss.backward()
        optimizer.step()
        return loss.item()
```

6. We fit the model with 500 iterations. In every 100 iterations, the training loss (MSE) is displayed as follows:

```
>>> for epoch in range(500):
        loss = train_step(model, X_train_torch, y_train_torch,
                          loss_function, optimizer)
        if epoch % 100 == 0:
            print(f"Epoch {epoch} - loss: {loss}")
Epoch 0 - loss: 4.908532619476318
Epoch 100 - loss: 0.5002815127372742
Epoch 200 - loss: 0.40820521116256714
Epoch 300 - loss: 0.3870624303817749
Epoch 400 - loss: 0.3720889091491699
```

7. Finally, we use the trained model to predict the testing cases and print out the predictions and their MSE:

```
>>> X_test_torch = torch.from_numpy(X_test.astype(np.float32))
>>> predictions = model(X_test_torch).detach().numpy()[:, 0]
>>> print(predictions)
[1.171479  1.130001  1.1055213 0.8627995 1.0910968 0.6725116 0.8869568
 0.8009699 0.8529027 0.8760005]
>>> print(mean_squared_error(y_test, predictions))
0.006939044434639928
```

It turns out that developing a neural network model with PyTorch is as simple as building with LEGO too.

 Both PyTorch and TensorFlow are popular deep learning frameworks, and their popularity can vary depending on different factors such as application domains, research communities, industry adoption, and personal preferences. However, as of 2023, PyTorch has been more widely adopted and has a larger user base overall, according to Papers With Code (https://paperswithcode.com/trends) and Google Trends (https://trends.google.com/trends/explore?geo=US&q=tensorflow,pytorch&hl=en). Hence, we will focus on PyTorch implementations for DL throughout the rest of the book.

Next, we will look at how to choose the right activation functions.

Picking the right activation functions

So far, we have used the ReLU and sigmoid activation functions in our implementations. You may wonder how to pick the right activation function for your neural networks. Detailed advice on when to choose a particular activation function is given next:

- **Linear:** $f(z) = z$. You can interpret this as no activation function. We usually use it in the output layer in regression networks as we don't need any transformation to the outputs.
- **Sigmoid** (logistic) transforms the output of a layer to a range between 0 and 1. You can interpret it as the probability of an output prediction. Therefore, we usually use it in the output layer in **binary classification** networks. Besides that, we sometimes use it in hidden layers. However, it should be noted that the sigmoid function is monotonic but its derivative is not. Hence, the neural network may get stuck at a suboptimal solution.
- **Softmax:** As was mentioned in *Chapter 4, Predicting Online Ad Click-Through with Logistic Regression*, softmax is a generalized logistic function used for multiclass classification. Hence, we use it in the output layer in **multiclass classification** networks.
- **Tanh** is a better version of the sigmoid function with stronger gradients. As you can see in the plots earlier in the chapter, the derivatives in the tanh function are steeper than those for the sigmoid function. It has a range of -1 to 1. It is common to use the tanh function in hidden layers.
- **ReLU** is probably the most frequently used activation function nowadays. It is the "default" one in hidden layers in feedforward networks. Its range is from 0 to infinity, and both the function itself and its derivative are monotonic. It has several benefits over tanh. One is sparsity, meaning that only a subset of neurons are activated at any given time. This can help reduce the computational cost of training and inference, as fewer neurons need to be computed. ReLU also mitigates the vanishing gradient problem, which occurs when gradients become very small during backpropagation, leading to slow or stalled learning. ReLU does not saturate for positive inputs, allowing gradients to flow more freely during training. One drawback of the ReLU function is the inability to appropriately map the negative part of the input where all negative inputs are transformed to 0. To fix the "dying negative" problem in ReLU, **Leaky ReLU** was invented to introduce a small slope in the negative part. When $z < 0$, $f(z) = az$, where a is usually a small value, such as 0.01.

To recap, ReLU is usually in hidden layer activation. You can try Leaky ReLU if ReLU doesn't work well. Sigmoid and tanh can be used in hidden layers but are not recommended in deep networks with many layers. For the output layer, linear activation (or no activation) is used in the regression network; sigmoid is for the binary classification network and softmax is for the multiple classification case.

Picking the right activation is important, and so is avoiding overfitting in neural networks. Let's see how to do this in the next section.

Preventing overfitting in neural networks

A neural network is powerful as it can derive hierarchical features from data with the right architecture (the right number of hidden layers and hidden nodes). It offers a great deal of flexibility and can fit a complex dataset. However, this advantage will become a weakness if the network is not given enough control over the learning process. Specifically, it may lead to overfitting if a network is only good at fitting to the training set but is not able to generalize to unseen data. Hence, preventing overfitting is essential to the success of a neural network model.

There are mainly three ways to impose restrictions on our neural networks: L1/L2 regularization, dropout, and early stopping. We practiced the first method in *Chapter 4, Predicting Online Ad Click-Through with Logistic Regression*, and will discuss the other two in this section.

Dropout

Dropout means ignoring a certain set of hidden nodes during the learning phase of a neural network. And those hidden nodes are chosen randomly given a specified probability. In the forward pass during a training iteration, the randomly selected nodes are temporarily not used in calculating the loss; in the backward pass, the randomly selected nodes are not updated temporarily.

In the following diagram, we choose three nodes in the network to ignore during training:

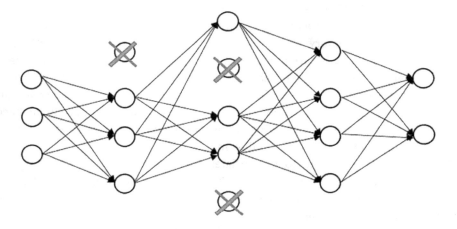

A neural network with dropout

Figure 6.6: Three nodes to ignore in a neural network

Recall that a regular layer has nodes fully connected to nodes from the previous layer and the following layer. It will lead to overfitting if a large network develops and memorizes the co-dependency between individual pairs of nodes. Dropout breaks this co-dependency by temporarily deactivating certain nodes in each iteration. Therefore, it effectively reduces overfitting and won't disrupt learning at the same time.

The fraction of nodes being randomly chosen in each iteration is also called the dropout rate. In practice, we usually set a dropout rate no greater than 50%. If the dropout rate is too high, it can excessively hinder the model's learning capacity, slowing down training and reducing the model's ability to extract useful patterns from the data.

Best practice

Determining the dropout rate empirically involves experimenting with different dropout rates and evaluating their effects on the model's performance. Here's a typical approach:

1. Start with a low rate (e.g., `0.1` or `0.2`) and train the model on your dataset. Monitor the model's performance metrics on a validation set.
2. Gradually increase the dropout rate in small increments (e.g., by `0.1`) and retrain the model each time. Monitor the performance metrics after each training run.
3. Evaluate performance obtained with different dropout rates. Be mindful of overfitting, as too high of a dropout rate can hinder model performance; if the dropout rate is too low, the model may not effectively prevent overfitting.

In PyTorch, we use the `torch.nn.Dropout` object to add dropout to a layer. An example is as follows:

```
>>> model_with_dropout = nn.Sequential(nn.Linear(X_train.shape[1], 16),
                              nn.ReLU(),
                              nn.Dropout(0.1),
                              nn.Linear(16, 8),
                              nn.ReLU(),
                              nn.Linear(8, 1))
```

In the preceding example, 10% of nodes randomly picked from the first hidden layer are ignored in an iteration during training.

Keep in mind that dropout should only occur in the training phase. In the prediction phase, all nodes should be fully connected again. Hence, we have to switch the model to evaluation mode with the `.eval()` method to disable dropout before we evaluate the model or make predictions with the trained model. Let's see it in the following California housing example:

1. First, we compile the model (with dropout) by using Adam as the optimizer with a learning rate of `0.01` and MSE as the learning goal:

```
>>> optimizer = torch.optim.Adam(model_with_dropout.parameters(),
    lr=0.01)
```

2. Next, we can train the model (with dropout) for `1,000` iterations:

```
>>> for epoch in range(1000):
        loss = train_step(model_with_dropout, X_train_torch, y_train_
    torch,
                          loss_function, optimizer)
```

```
        if epoch % 100 == 0:
            print(f"Epoch {epoch} - loss: {loss}")
Epoch 0 - loss: 4.921249866485596
Epoch 100 - loss: 0.5313398838043213
Epoch 200 - loss: 0.4458008408546448
Epoch 300 - loss: 0.4264270067214966
Epoch 400 - loss: 0.4085545539855957
Epoch 500 - loss: 0.3640516400337219
Epoch 600 - loss: 0.35677382349967957
Epoch 700 - loss: 0.35208994150161743
Epoch 800 - loss: 0.34980857372283936
Epoch 900 - loss: 0.3431631028652191
```

In every 100 iterations, the training loss (MSE) is displayed.

3. Finally, we use the trained model (with dropout) to predict the testing cases and print out the MSE:

```
>>> model_with_dropout.eval()
>>> predictions = model_with_dropout (X_test_torch).detach().numpy()[:,
0]
>>> print(mean_squared_error(y_test, predictions))
 0.005699420832357341
```

As mentioned earlier, don't forget to run model_with_dropout.eval() before evaluating the model with dropout. Otherwise, the dropout layers will continue to randomly deactivate neurons, leading to inconsistent results between different model evaluations on the same data.

Early stopping

As the name implies, training a network with **early stopping** will end if the model performance doesn't improve for a certain number of iterations. The model performance is measured on a validation set that is different from the training set, in order to assess how well it generalizes. During training, if the performance degrades after several (let's say 50) iterations, it means the model is overfitting and not able to generalize well anymore. Hence, stopping the learning early in this case helps prevent overfitting. Usually, we evaluate the model against a validation set. If the metric on the validation set is not improving for more than n epochs, we stop the training process.

We will demonstrate how to apply early stopping in PyTorch using the California housing example as well:

1. First, we recreate the model and optimizer as we did previously:

```
>>> model = nn.Sequential(nn.Linear(X_train.shape[1], 16),
                nn.ReLU(),
                nn.Linear(16, 8),
```

```
                        nn.ReLU(),
                        nn.Linear(8, 1))
>>> optimizer = torch.optim.Adam(model.parameters(), lr=0.01)
```

2. Next, we define the early stopping criterion as the test loss doesn't improve for more than `100` epochs:

```
>>> patience = 100
>>> epochs_no_improve = 0
>>> best_test_loss = float('inf')
```

3. Now we adopt early stopping and train the model for, at most, `500` iterations:

```
>>> import copy
>>> best_model = model
>>> for epoch in range(500):
        loss = train_step(model, X_train_torch, y_train_torch,
                           loss_function, optimizer)
        predictions = model(X_test_torch).detach().numpy()[:, 0]
        test_loss = mean_squared_error(y_test, predictions)
        if test_loss > best_test_loss:
            epochs_no_improve += 1
            if epochs_no_improve > patience:
                print(f"Early stopped at epoch {epoch}")
                break
        else:
            epochs_no_improve = 0
            best_test_loss = test_loss
            best_model = copy.deepcopy(model)
Early stopped at epoch 224
```

Following every training step, we compute the test loss and compare it to the previously recorded best one. If it shows improvement, we save the current model using the copy module and reset the epochs_no_improve counter. However, if there is no improvement in the test loss for up to 100 consecutive iterations, we stop the training process as we have reached the tolerance threshold (patience). In our example, training stopped after epoch 224.

4. Finally, we use the previously recorded best model to predict the testing cases and print out the predictions and their MSE:

```
>>> predictions = best_model(X_test_torch).detach().numpy()[:, 0]
>>> print(mean_squared_error(y_test, predictions))
0.005459465255681108
```

This is better than `0.0069`, which we obtained in the vanilla approach, and `0.0057`, which we achieved using dropout for overfitting prevention.

While the Universal Approximation Theorem guarantees that neural networks can represent any function, it doesn't guarantee good generalization performance. Overfitting can occur if the model has too much capacity relative to the complexity of the underlying data distribution. Therefore, controlling the capacity of the model through techniques like regularization and early stopping is essential to ensure that the learned function generalizes well to unseen data.

Now that you've learned about neural networks and their implementation, let's utilize them to solve our stock price prediction problem.

Predicting stock prices with neural networks

We will build the stock predictor with PyTorch in this section. We will start with feature generation and data preparation, followed by network building and training. After that, we will fine-tune the network to boost the stock predictor.

Training a simple neural network

We prepare data and train a simple neural work with the following steps:

1. We load the stock data, generate features, and label the `generate_features` function we developed in *Chapter 5, Predicting Stock Prices with Regression Algorithms*:

    ```
    >>> data_raw = pd.read_csv('19900101_20230630.csv', index_col='Date')
    >>> data = generate_features(data_raw)
    ```

2. We construct the training set using data from 1990 to 2022 and the testing set using data from the first half of 2023:

    ```
    >>> start_train = '1990-01-01'
    >>> end_train = '2022-12-31'
    >>> start_test = '2023-01-01'
    >>> end_test = '2023-06-30'
    >>> data_train = data.loc[start_train:end_train]
    >>> X_train = data_train.drop('close', axis=1).values
    >>> y_train = data_train['close'].values
    >>> data_test = data.loc[start_test:end_test]
    >>> X_test = data_test.drop('close', axis=1).values
    >>> y_test = data_test['close'].values
    ```

3. We need to normalize features into the same or a comparable scale. We do so by removing the mean and rescaling to unit variance:

    ```
    >>> from sklearn.preprocessing import StandardScaler
    >>> scaler = StandardScaler()
    ```

We rescale both sets with the scaler taught by the training set:

```
>>> X_scaled_train = scaler.fit_transform(X_train)
>>> X_scaled_test = scaler.transform(X_test)
```

4. Next, we need to create tensor objects from the input NumPy arrays before using them to train the PyTorch model:

```
>>> X_train_torch = torch.from_numpy(X_scaled_train.astype(np.float32))
>>> X_test_torch = torch.from_numpy(X_scaled_test.astype(np.float32))
>>> y_train = y_train.reshape(y_train.shape[0], 1)
>>> y_train_torch = torch.from_numpy(y_train.astype(np.float32))
```

5. We now build a neural network using the `torch.nn` module:

```
>>> torch.manual_seed(42)
>>> model = nn.Sequential(nn.Linear(X_train.shape[1], 32),
                          nn.ReLU(),
                          nn.Linear(32, 1))
```

The network we begin with has one hidden layer with 32 nodes followed by a ReLU function.

6. We compile the model by using Adam as the optimizer with a learning rate of `0.3` and MSE as the learning goal:

```
>>> loss_function = nn.MSELoss()
>>> optimizer = torch.optim.Adam(model.parameters(), lr=0.3)
```

7. After defining the model, we perform training for `1,000` iterations:

```
>>> for epoch in range(1000):
        loss = train_step(model, X_train_torch, y_train_torch,
                          loss_function, optimizer)
        if epoch % 100 == 0:
            print(f"Epoch {epoch} - loss: {loss}")
Epoch 0 - loss: 24823446.0
Epoch 100 - loss: 189974.171875
Epoch 200 - loss: 52102.01171875
Epoch 300 - loss: 17849.333984375
Epoch 400 - loss: 8928.6689453125
Epoch 500 - loss: 6497.75927734375
Epoch 600 - loss: 5670.634765625
Epoch 700 - loss: 5265.48828125
Epoch 800 - loss: 5017.7021484375
Epoch 900 - loss: 4834.28466796875
```

8. Finally, we use the trained model to predict the testing data and display metrics:

```
>>> predictions = model(X_test_torch).detach().numpy()[:, 0]
>>> from sklearn.metrics import mean_squared_error, mean_absolute_error,
r2_score
>>> print(f'MSE: {mean_squared_error(y_test, predictions):.3f}')
MSE: 30051.643
>>> print(f'MAE: {mean_absolute_error(y_test, predictions):.3f}')
MAE: 137.096
>>> print(f'R^2: {r2_score(y_test, predictions):.3f}')
R^2: 0.954
```

We achieve an R^2 of 0.954 with a simple neural network model.

Fine-tuning the neural network

Can we do better? Of course, we haven't fine-tuned the hyperparameters yet. We perform model fine-tuning in PyTorch with the following steps:

1. TensorBoard provides functionality for logging various metrics and visualizations during model training and evaluation. You can use TensorBoard with PyTorch to track and visualize metrics such as loss, accuracy, gradients, and model architectures, among others. We rely on the tensorboard module in PyTorch utils, so we import it first:

```
>>> from torch.utils.tensorboard import SummaryWriter
```

2. We want to tweak the number of hidden nodes in the hidden layer (again, we are using one hidden layer for this example), the number of training iterations, and the learning rate. We pick the following values of hyperparameters to experiment on:

```
>>> hparams_config = {
        "hidden_size": [16, 32],
        "epochs": [1000, 3000],
        "lr": [0.1, 0.3],
    }
```

Here, we experiment with two options for the number of hidden nodes, 16 and 32; we use two options for the number of iterations, 300 and 1000; and we use two options, 0.1 and 0.3, for the learning rate.

3. After initializing the hyperparameters to optimize, we will iterate each hyperparameter combination, and train and validate the model using a given set of hyperparameters by calling the helper function train_validate_model as follows:

```
>>> def train_validate_model(hidden_size, epochs, lr):
        model = nn.Sequential(nn.Linear(X_train.shape[1], hidden_size),
                              nn.ReLU(),
                              nn.Linear(hidden_size, 1))
```

```
            optimizer = torch.optim.Adam(model.parameters(), lr=lr)

        # Create the TensorBoard writer
        writer_path = f"runs/{experiment_num}/{hidden_size}/{epochs}/
{lr}"
        writer = SummaryWriter(log_dir=writer_path)

        for epoch in range(epochs):
            loss = train_step(model, X_train_torch, y_train_torch, loss_
function,
                            optimizer)

            predictions = model(X_test_torch).detach().numpy()[:, 0]
            test_mse = mean_squared_error(y_test, predictions)

            writer.add_scalar(
                tag="train loss",
                scalar_value=loss,
                global_step=epoch,
            )
            writer.add_scalar(
                tag="test loss",
                scalar_value=test_mse,
                global_step=epoch,
            )

        test_r2 = r2_score(y_test, predictions)
        print(f'R^2: {test_r2:.3f}\n')

        # Add the hyperparameters and metrics to TensorBoard
        writer.add_hparams(
            {
                "hidden_size": hidden_size,
                "epochs": epochs,
                "lr": lr,
            },
            {
                "test MSE": test_mse,
                "test R^2": test_r2,
            },
        )
```

Here, in each hyperparameter combination, we build and fit a neural network model based on the given hyperparameters, including the number of hidden nodes, the learning rate, and the number of training iterations. There's nothing much different here from what we did before. But when we train the model, we also update TensorBoard by logging the hyperparameters and metrics including the train loss and test loss with the `add_scalar` method.

The TensorBoard writer object is straightforward. It provides visualization for the model graph and metrics during training and validation.

At the end, we compute and display the R^2 of the prediction on the test set. We also log the test MSE and R^2 using the `add_hparams` method along with the given hyperparameter combination.

4. Now we fine-tune the neural network by iterating eight hyperparameter combinations:

```
>>> torch.manual_seed(42)
>>> experiment_num = 0
>>> for hidden_size in hparams_config["hidden_size"]:
        for epochs in hparams_config["epochs"]:
            for lr in hparams_config["lr"]:
                experiment_num += 1
                print(f"Experiment {experiment_num}: hidden_size =
{hidden_size},
                    epochs = {epochs}, lr = {lr}")
                train_validate_model(hidden_size, epochs, lr)
```

You will see the following output:

```
Experiment 1: hidden_size = 16, epochs = 1000, lr = 0.1
R^2: 0.771


Experiment 2: hidden_size = 16, epochs = 1000, lr = 0.3
R^2: 0.952


Experiment 3: hidden_size = 16, epochs = 3000, lr = 0.1
R^2: 0.969


Experiment 4: hidden_size = 16, epochs = 3000, lr = 0.3
R^2: 0.977


Experiment 5: hidden_size = 32, epochs = 1000, lr = 0.1
R^2: 0.877


Experiment 6: hidden_size = 32, epochs = 1000, lr = 0.3
R^2: 0.957
```

```
Experiment 7: hidden_size = 32, epochs = 3000, lr = 0.1
R^2: 0.970


Experiment 8: hidden_size = 32, epochs = 3000, lr = 0.3
R^2: 0.959
```

Experiment 4 with the combination of (`hidden_size=16, epochs=3000, learning_rate=0.3`) is the best performing one, where we achieve an R^2 of 0.977.

Best practice

Hyperparameter tuning for neural networks can significantly impact model performance. Here are some best practices for hyperparameter tuning in neural networks:

- **Define a search space:** Determine which hyperparameters to tune and their ranges. Common hyperparameters include learning rate, batch size, number of hidden layers, number of neurons per layer, activation functions, dropout rates, and so on.

- **Use cross-validation:** This helps prevent overfitting and provides a more robust estimate of model performance.

- **Monitor performance metrics:** Track relevant metrics such as loss, accuracy, precision, recall, MSE, R^2, and so on during training and validation.

- **Early stopping:** Monitor the validation loss during training, and stop training when it starts to increase consistently while the training loss decreases.

- **Regularization:** Use regularization techniques such as L1 and L2 regularization and dropout to prevent overfitting and improve generalization performance.

- **Experiment with different architectures:** Try different network architectures, including the number of layers, the number of neurons per layer, and the activation functions. Experiment with deep vs. shallow networks and wide vs. narrow networks.

- **Use parallelism:** If computational resources allow, parallelize the hyperparameter search process to speed up experimentation. Tools like TensorFlow's `tf.distribute.Strategy` or PyTorch's `torch.nn.DataParallel` can be used to distribute training across multiple GPUs or machines.

5. You will notice that a new folder, `runs`, is created after these experiments start. It contains the training and validation performance for each experiment. After 8 experiments finish, it's time to launch TensorBoard. We use the following command:

```
tensorboard --host 0.0.0.0 --logdir runs
```

Once it is launched, you will see the beautiful dashboard at `http://localhost:6006/`. You can see a screenshot of the expected result here:

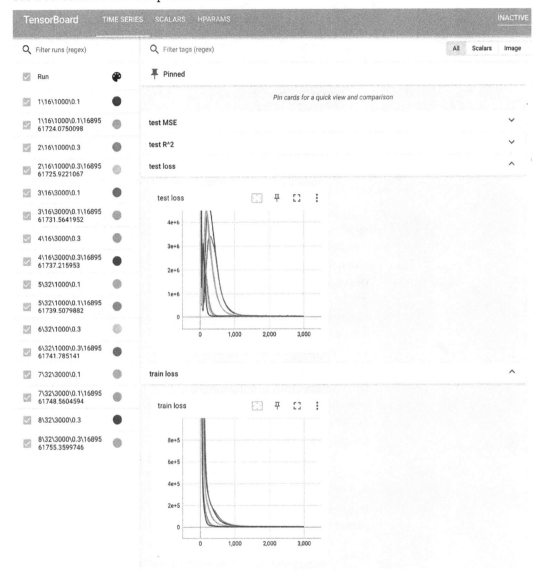

Figure 6.7: Screenshot of TensorBoard

The time series of train and test loss provide valuable insights. They allow us to assess the progress of training and identify signs of overfitting or underfitting. Overfitting can be identified when the train loss decreases over time while the test loss remains stagnant or increases. On the other hand, underfitting is indicated by relatively high train and test loss values, indicating that the model fails to adequately fit the training data.

6. Next, we click on the **HPARAMS** tab to see the hyperparameter logs. You can see all the hyperparameter combinations and the respective metrics (MSE and R^2) displayed in a table, as shown here:

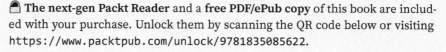

Trial ID	Show Metrics	hidden_size	epochs	lr	test MSE	test R^2
1\16\1000\0.1\1...	☐	16.000	1000.0	0.10000	1.4910e+5	0.77095
2\16\1000\0.3\1...	☐	16.000	1000.0	0.30000	30998	0.95238
3\16\3000\0.1\1...	☐	16.000	3000.0	0.10000	19885	0.96945
4\16\3000\0.3\1...	☐	16.000	3000.0	0.30000	15226	0.97661
5\32\1000\0.1\1...	☐	32.000	1000.0	0.10000	80296	0.87665
6\32\1000\0.3\1...	☐	32.000	1000.0	0.30000	28097	0.95684
7\32\3000\0.1\1...	☐	32.000	3000.0	0.10000	19518	0.97002
8\32\3000\0.3\...	☐	32.000	3000.0	0.30000	26506	0.95928

Figure 6.8: Screenshot of TensorBoard for hyperparameter tuning

 🔍 **Quick tip:** Need to see a high-resolution version of this image? Open this book in the next-gen Packt Reader or view it in the PDF/ePub copy.

🔒 **The next-gen Packt Reader** and a **free PDF/ePub copy** of this book are included with your purchase. Unlock them by scanning the QR code below or visiting `https://www.packtpub.com/unlock/9781835085622`.

Again, you can see experiment 4 yields the best performance.

7. Finally, we use the optimal model to make predictions:

```
>>> hidden_size = 16
>>> epochs = 3000
>>> lr = 0.3
```

```
>>> best_model = nn.Sequential(nn.Linear(X_train.shape[1], hidden_size),
                               nn.ReLU(),
                               nn.Linear(hidden_size, 1))
>>> optimizer = torch.optim.Adam(best_model.parameters(), lr=lr)
>>> for epoch in range(epochs):
    train_step(best_model, X_train_torch, y_train_torch, loss_function,
               optimizer
>>> predictions = best_model(X_test_torch).detach().numpy()[:, 0]
```

8. Plot the prediction along with the ground truth as follows:

```
>>> import matplotlib.pyplot as plt
>>> plt.rc('xtick', labelsize=10)
>>> plt.rc('ytick', labelsize=10)
>>> plt.plot(data_test.index, y_test, c='k')
>>> plt.plot(data_test.index, predictions, c='b')
>>> plt.xticks(range(0, 130, 10), rotation=60)
>>> plt.xlabel('Date' , fontsize=10)
>>> plt.ylabel('Close price' , fontsize=10)
>>> plt.legend(['Truth', 'Neural network'] , fontsize=10)
>>> plt.show()
```

Refer to the following screenshot for the result:

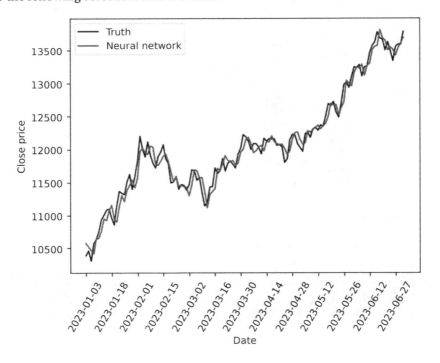

Figure 6.9: Prediction and ground truth of stock prices

The fine-tuned neural network does a good job of predicting stock prices.

In this section, we further improved the neural network stock predictor by fine-tuning the hyperparameters. Feel free to use more hidden layers, or apply dropout or early stopping to see whether you can get a better result.

Summary

In this chapter, we worked on the stock prediction project again, but with neural networks this time. We started with a detailed explanation of neural networks, including the essential components (layers, activations, feedforward, and backpropagation), and transitioned to DL. We moved on to implementations from scratch with scikit-learn, TensorFlow, and PyTorch. We also learned about ways to avoid overfitting, such as dropout and early stopping. Finally, we applied what we covered in this chapter to solve our stock price prediction problem.

In the next chapter, we will explore NLP techniques and unsupervised learning.

Exercises

1. As mentioned, can you use more hidden layers in the neural network stock predictor and rerun the model fine-tuning? Can you get a better result?
2. Following the first exercise, can you apply dropout and/or early stopping and see if you can beat the current best R^2 of 0.977?

7

Mining the 20 Newsgroups Dataset with Text Analysis Techniques

In previous chapters, we went through a bunch of fundamental machine learning concepts and supervised learning algorithms. Starting from this chapter, as the second step of our learning journey, we will be covering in detail several important unsupervised learning algorithms and techniques related to text analysis. To make our journey more interesting, we will start with a **Natural Language Processing (NLP)** problem—exploring the 20 newsgroups data. You will gain hands-on experience and learn how to work with text data, especially how to convert words and phrases into machine-readable values and how to clean up words with little meaning. We will also visualize text data by mapping it into a two-dimensional space in an unsupervised learning manner.

We will go into detail on each of the following topics:

- How computers understand language – NLP
- Touring popular NLP libraries and picking up NLP basics
- Getting the newsgroups data
- Exploring the newsgroups data
- Thinking about features for text data
- Visualizing the newsgroups data with t-SNE
- Representing words with dense vectors – word embedding

How computers understand language — NLP

In *Chapter 1, Getting Started with Machine Learning and Python*, I mentioned that machine learning-driven programs or computers are good at discovering event patterns by processing and working with data. When the data is well structured or well defined, such as in a Microsoft Excel spreadsheet table or a relational database table, it is intuitively obvious why machine learning is better at dealing with it than humans. Computers read such data the same way as humans—for example, revenue: 5,000,000 as the revenue being 5 million, and age: 30 as the age being 30; then computers crunch assorted data and generate insights in a faster way than humans. However, when the data is unstructured, such as words with which humans communicate, news articles, or someone's speech in another language, it seems that computers cannot understand words as well as humans do (yet). While computers have made significant progress in understanding words and natural language, they still fall short of human-level understanding in many aspects.

What is NLP?

There is a lot of information in the world about words, raw text, or, broadly speaking, **natural language**. This refers to any language that humans use to communicate with each other. Natural language can take various forms, including, but not limited to, the following:

- Text, such as a web page, SMS, emails, and menus
- Audio, such as speech and commands to Siri
- Signs and gestures
- Many other forms, such as songs, sheet music, and Morse code

The list is endless, and we are all surrounded by natural language all of the time (that's right, right now as you are reading this book). Given the importance of this type of unstructured data (natural language data), we must have methods to get computers to understand and reason with natural language and to extract data from it. Programs equipped with NLP techniques can already do a lot in certain areas, which already seems magical!

NLP is a significant subfield of machine learning that deals with the interactions between machines (computers) and human (natural) languages. The data for NLP tasks can be in different forms, for example, text from social media posts, web pages, or even medical prescriptions, or audio from voice mails, commands to control systems, or even a favorite song or movie. Nowadays, NLP is broadly involved in our daily lives: we cannot live without machine translation, weather forecast scripts are automatically generated, we find voice search convenient, we get the answer to a question (such as "What is the population of Canada?") quickly thanks to intelligent question-answering systems, speech-to-text technology helps people with special needs, and so on.

Generative AI and its applications like ChatGPT are pushing the boundaries of NLP even further. Imagine a world where you can have a conversation with a virtual assistant that can not only answer your questions in a comprehensive way but also generate different creative text formats, like poems, code, scripts, musical pieces, emails, letters, and so on. By analyzing massive amounts of text data, it can learn the underlying patterns and structures of language, allowing it to generate human-quality text content. For instance, you could ask ChatGPT to write a funny birthday poem for your friend, craft a compelling marketing email for your business, or even brainstorm ideas for a new blog post.

The history of NLP

If machines are able to understand language like humans do, we consider them intelligent. In 1950, the famous mathematician Alan Turing proposed in an article, *Computing Machinery and Intelligence*, a test as a criterion of machine intelligence. It's now called the **Turing test** (https://plato.stanford.edu/entries/turing-test/), and its goal is to examine whether a computer is able to adequately understand languages so as to fool humans into thinking that the machine is another human. It is probably no surprise to you that no computer has passed the Turing test yet, but the 1950s is considered to be when the history of NLP started.

Understanding language might be difficult, but would it be easier to automatically translate texts from one language to another? On my first ever programming course, the lab booklet had the algorithm for coarse-grained machine translation. This type of translation involved looking up words in dictionaries and generating text in a new language. A more practically feasible approach would be to gather texts that are already translated by humans and train a computer program on these texts. In 1954, in the Georgetown–IBM experiment (https://en.wikipedia.org/wiki/Georgetown%E2%80%93IBM_experiment), scientists claimed that machine translation would be solved in three to five years. Unfortunately, a machine translation system that can beat human expert translators does not exist yet. But machine translation has been greatly evolving since the introduction of deep learning and has seen incredible achievements in certain areas, for example, social media (Facebook open sourced a neural machine translation system, https://ai.facebook.com/tools/translate/), real-time conversation (Microsoft Translator, SwiftKey Keyboard, and Google Pixel Buds), and image-based translation, such as Google Translate.

Conversational agents, or chatbots, are another hot topic in NLP. The fact that computers are able to have a conversation with us has reshaped the way businesses are run. In 2016, Microsoft's AI chatbot, Tay (https://blogs.microsoft.com/blog/2016/03/25/learning-tays-introduction/), was unleashed to mimic a teenage girl and converse with users on Twitter (now X) in real time. She learned how to speak from all the things users posted and commented on Twitter. However, she was overwhelmed by tweets from trolls and automatically learned their bad behaviors and started to output inappropriate things on her feeds. She ended up being terminated within 24 hours. Generative AI models like ChatGPT are another area of active research, pushing the boundaries of what's possible. They can be helpful for creative text formats or specific tasks, but achieving true human-level understanding in conversation remains an ongoing pursuit.

NLP applications

There are also several text analysis tasks that attempt to organize knowledge and concepts in such a way that they become easier for computer programs to manipulate.

The way we organize and represent concepts is called **ontology**. An ontology defines concepts and relationships between concepts. For instance, we can have a so-called triple, such as ("python", "language", "is-a") representing the relationship between two concepts, such as *Python is a language*.

An important use case for NLP at a much lower level, compared to the previous cases, is **part-of-speech (PoS) tagging**. A PoS is a grammatical word category such as a noun or verb. PoS tagging tries to determine the appropriate tag for each word in a sentence or a larger document.

The following table gives examples of English PoSs:

Part of speech	Examples
Noun	David, machine
Pronoun	They, her
Adjective	Awesome, amazing
Verb	Read, write
Adverb	Very, quite
Preposition	Out, at
Conjunction	And, but
Interjection	Phew, oops
Article	A, the

Table 7.1: PoS examples

There are a variety of real-world NLP applications involving supervised learning, such as PoS tagging, mentioned earlier, and **sentiment analysis**. A typical example is identifying news sentiment, which could be positive or negative in the binary case, or positive, neutral, or negative in multiclass classification. News sentiment analysis provides a significant signal to trading in the stock market.

Another example we can easily think of is news topic classification, where classes may or may not be mutually exclusive. In the newsgroup example that we just discussed, classes are mutually exclusive (despite slight overlapping), such as technology, sports, and religion. It is, however, good to realize that a news article can be occasionally assigned multiple categories (multi-label classification). For example, an article about the Olympic Games may be labeled sports and politics if there is unexpected political involvement.

Finally, an interesting application that is perhaps unexpected is **Named Entity Recognition** (NER). Named entities are phrases of definitive categories, such as names of persons, companies, geographic locations, dates and times, quantities, and monetary values. NER is an important subtask of information extraction to seek and identify such entities. For example, we can conduct NER on the following sentence: `SpaceX[Organization]`, a `California[Location]`-based company founded by a famous tech entrepreneur `Elon Musk[Person]`, announced that it would manufacture the next-generation, `9[Quantity]`-meter-diameter launch vehicle and spaceship for the first orbital flight in `2020[Date]`.

Other key NLP applications include:

- **Language translation**: NLP powers machine translation systems, enabling automatic translation of text or speech from one language to another. Platforms like Google Translate and Microsoft Translator utilize NLP to provide real-time translation services.
- **Speech recognition**: NLP is essential in speech recognition systems, converting spoken language into written text. Virtual assistants like Siri, Alexa, and Google Assistant rely on NLP to understand user commands and respond appropriately.

- **Text summarization:** NLP can automatically generate concise summaries of lengthy texts, providing a quick overview of the content. Text summarization is useful for information retrieval and content curation.

- **Language generation:** NLP models, such as **Generative Pre-trained Transformers (GPTs)**, can generate human-like text, including creative writing, poetry, and dialogue generation.

- **Information retrieval:** NLP assists in information retrieval from large volumes of unstructured data, such as web pages, documents, and news articles. Search engines use NLP techniques to understand user queries and retrieve relevant results.

- **Chatbots, question answering, and virtual assistants:** NLP powers chatbots and virtual assistants to provide interactive and conversational experiences. These systems can answer queries, assist with tasks, and guide users through various processes.

In the next chapter, we will discuss how unsupervised learning, including clustering and topic modeling, is applied to text data. We will begin by covering NLP basics in the upcoming sections of this chapter.

Touring popular NLP libraries and picking up NLP basics

Now that we have covered a short list of real-world applications of NLP, we will be touring the essential stack of Python NLP libraries. These packages handle a wide range of NLP tasks, as mentioned previously, including sentiment analysis, text classification, and NER.

Installing famous NLP libraries

The most famous NLP libraries in Python include the **Natural Language Toolkit (NLTK), spaCy, Gensim,** and **TextBlob.** The scikit-learn library also has impressive NLP-related features. Let's take a look at them in more detail:

- **NLTK:** This library (http://www.nltk.org/) was originally developed for educational purposes and is now widely used in industry as well. It is said that you can't talk about NLP without mentioning NLTK. It is one of the most famous and leading platforms for building Python-based NLP applications. You can install it simply by running the following command line in the terminal:

```
sudo pip install -U nltk
```

If you're using conda, execute the following command line:

```
conda install nltk
```

- **spaCy:** This library (https://spacy.io/) is a more powerful toolkit in the industry than NLTK. This is mainly for two reasons: first, spaCy is written in Cython, which is much more memory-optimized (now you can see where the Cy in spaCy comes from) and excels in NLP tasks; second, spaCy uses state-of-the-art algorithms for core NLP problems, such as **convolutional neural network (CNN)** models for tagging and NER. However, it could seem advanced for beginners. In case you're interested, here are the installation instructions.

Run the following command line in the terminal:

```
pip install -U spacy
```

For conda, execute the following command line:

```
conda install -c conda-forge spacy
```

- **Gensim:** This library (https://radimrehurek.com/gensim/), developed by Radim Rehurek, has been gaining popularity over recent years. It was initially designed in 2008 to generate a list of similar articles given an article, hence the name of this library (generate similar—> Gensim). It was later drastically improved by Radim Rehurek in terms of its efficiency and scalability. Again, you can easily install it via pip by running the following command line:

```
pip install --upgrade gensim
```

In the case of conda, you can execute the following command line in the terminal:

```
conda install -c conda-forge gensim
```

 You should make sure that the dependencies, NumPy and SciPy, are already installed before Gensim.

- **TextBlob:** This library (https://textblob.readthedocs.io/en/dev/) is a relatively new one built on top of NLTK. It simplifies NLP and text analysis with easy-to-use built-in functions and methods, as well as wrappers around common tasks. We can install TextBlob by running the following command line in the terminal:

```
pip install -U textblob
```

Or, for conda:

```
conda install -c conda-forge textblob
```

TextBlob has some useful features that are not available in NLTK (currently), such as spell checking and correction, language detection, and translation.

Corpora

NLTK comes with over 100 collections of large and well-structured text datasets, which are called **corpora** in NLP. Here are some of the main corpora that NLTK provides:

- **Gutenberg Corpus:** A collection of literary works from Project Gutenberg, containing thousands of books in various languages.
- **Reuters Corpus:** A collection of news articles from the Reuters newswire service, widely used for text classification and topic modeling tasks.

- **Web and Chat Text:** A collection of web text and chat conversations, providing a glimpse into informal language and internet slang.
- **Movie Reviews Corpus:** A collection of movie reviews, often used for sentiment analysis and text classification tasks.
- **Treebank Corpus:** A collection of parsed and tagged sentences from the Penn Treebank, used for training and evaluating syntactic parsers.
- **WordNet:** A lexical database of English words, containing synsets (groups of synonymous words) and hypernyms (is-a relationships).

Corpora can be used as dictionaries for checking word occurrences and as training pools for model learning and validating. Some more useful and interesting corpora include the Web Text corpus, Twitter (X) samples, the Shakespeare corpus, Sentiment Polarity, the Names corpus (this contains lists of popular names, which we will be exploring very shortly), WordNet, and the Reuters benchmark corpus. The full list can be found at http://www.nltk.org/nltk_data.

Before using any of these corpus resources, we need to first download them by running the following code in the Python interpreter:

```
>>> import nltk
>>> nltk.download()
```

A new window will pop up and ask you which collections (the **Collections** tab in the following screenshot) or corpus (the **Corpora** tab in the following screenshot) to download, and where to keep the data:

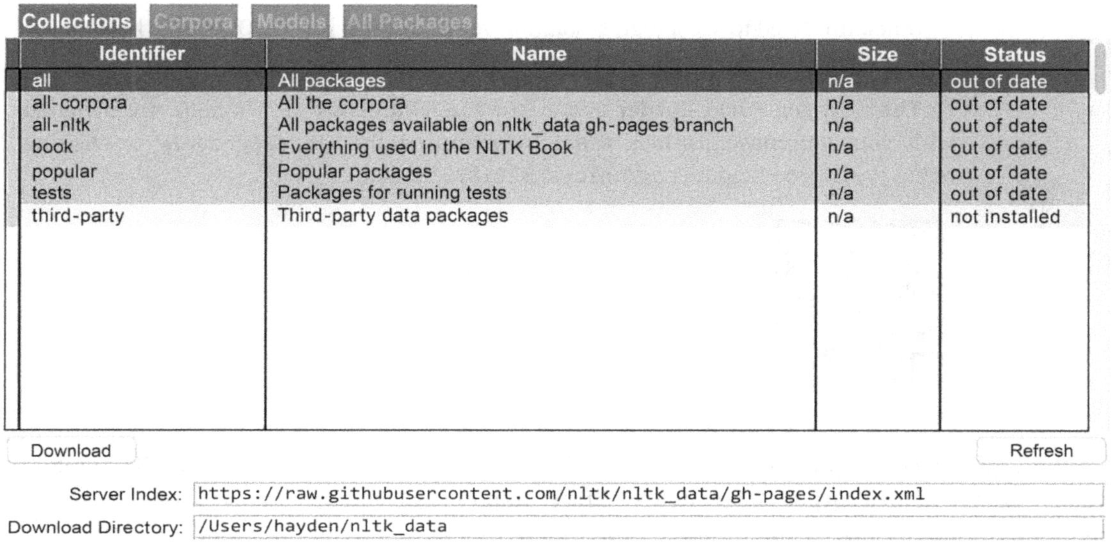

Collections	Corpora	Models	All Packages		
Identifier	Name	Size	Status
all	All packages	n/a	out of date
all-corpora	All the corpora	n/a	out of date
all-nltk	All packages available on nltk_data gh-pages branch	n/a	out of date
book	Everything used in the NLTK Book	n/a	out of date
popular	Popular packages	n/a	out of date
tests	Packages for running tests	n/a	out of date
third-party	Third-party data packages	n/a	not installed

| Download | | | Refresh |

Server Index: `https://raw.githubusercontent.com/nltk/nltk_data/gh-pages/index.xml`

Download Directory: `/Users/hayden/nltk_data`

Figure 7.1: Collections tab in the NLTK installation

Installing the whole popular package is the quickest solution since it contains all the important corpora needed for your current study and future research. Installing a particular corpus, as shown in the following screenshot, is also fine:

Identifier	Name	Size	Status
lin_thesaurus	Lin's Dependency Thesaurus	85.0 MB	installed
mac_morpho	MAC-MORPHO: Brazilian Portuguese news text with part-of-s	2.9 MB	installed
machado	Machado de Assis -- Obra Completa	5.9 MB	installed
masc_tagged	MASC Tagged Corpus	1.5 MB	installed
movie_reviews	Sentiment Polarity Dataset Version 2.0	3.8 MB	installed
mte_teip5	MULTEXT-East 1984 annotated corpus 4.0	14.1 MB	installed
names	Names Corpus, Version 1.3 (1994-03-29)	20.8 KB	installed
nombank.1.0	NomBank Corpus 1.0	6.4 MB	installed
nonbreaking_prefixes	Non-Breaking Prefixes (Moses Decoder)	24.8 KB	out of date
nps_chat	NPS Chat	294.3 KB	installed
omw	Open Multilingual Wordnet	11.5 MB	out of date
opinion_lexicon	Opinion Lexicon	24.4 KB	installed
panlex_swadesh	PanLex Swadesh Corpora	2.7 MB	out of date
paradigms	Paradigm Corpus	24.3 KB	installed
pe08	Cross-Framework and Cross-Domain Parser Evaluation Shared	78.8 KB	not installed
pil	The Patient Information Leaflet (PIL) Corpus	1.4 MB	installed

Collections | **Corpora** | Models | All Packages

Download   Refresh

Server Index: `https://raw.githubusercontent.com/nltk/nltk_data/gh-pages/index.xml`
Download Directory: `/Users/hayden/nltk_data`

Figure 7.2: Corpora tab in the NLTK installation

🔍 **Quick tip:** Need to see a high-resolution version of this image? Open this book in the next-gen Packt Reader or view it in the PDF/ePub copy.

 📖 **The next-gen Packt Reader** and a free **PDF/ePub copy** of this book are included with your purchase. Unlock them by scanning the QR code below or visiting `https://www.packtpub.com/unlock/9781835085622`.

Once the package or corpus you want to explore is installed, you can take a look at the **Names** corpus (make sure the names corpus is installed for this example).

First, import the names corpus:

```
>>> from nltk.corpus import names
```

We can check out the first 10 names in the list:

```
>>> print(names.words()[:10])
['Abagael', 'Abagail', 'Abbe', 'Abbey', 'Abbi', 'Abbie',
'Abby', 'Abigael', 'Abigail', 'Abigale']
```

There are, in total, 7944 names, as shown in the following output derived by executing the following command:

```
>>> print(len(names.words()))
7944
```

Other corpora are also fun to explore.

Besides the easy-to-use and abundant corpora pool, more importantly, NLTK is also good at many NLP and text analysis tasks, including tokenization, PoS tagging, NER, word stemming, and lemmatization. We'll look at these tasks next.

Tokenization

Given a text sequence, **tokenization** is the task of breaking it into fragments, which can be words, characters, or sentences. Certain characters are usually removed, such as punctuation marks, digits, and emoticons. The remaining fragments are the so-called **tokens** used for further processing.

Tokens composed of one word are also called **unigrams** in computational linguistics; **bigrams** are composed of two consecutive words; **trigrams** of three consecutive words; and **n-grams** of n consecutive words. Here is an example of tokenization:

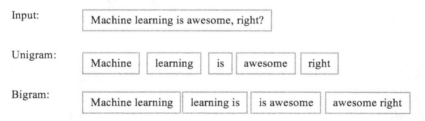

Figure 7.3: Tokenization example

We can implement word-based tokenization using the word_tokenize function in NLTK. We will use the input text `'''I am reading a book., and on the next line, It is Python Machine Learning By Example,, then 4th edition.'''`, as an example, as shown in the following commands:

```
>>> from nltk.tokenize import word_tokenize
>>> sent = '''I am reading a book.
...             It is Python Machine Learning By Example,
...             4th edition.'''
>>> print(word_tokenize(sent))
['I', 'am', 'reading', 'a', 'book', '.', 'It', 'is', 'Python', 'Machine',
'Learning', 'By', 'Example', ',', '3rd', 'edition', '.']
```

Word tokens are obtained.

The word_tokenize function keeps punctuation marks and digits, and only discards whitespaces and newlines.

You might think word tokenization is simply splitting a sentence by space and punctuation. Here's an interesting example showing that tokenization is more complex than you think:

```
>>> sent2 = 'I have been to U.K. and U.S.A.'
>>> print(word_tokenize(sent2))
['I', 'have', 'been', 'to', 'U.K.', 'and', 'U.S.A', '.']
```

The tokenizer accurately recognizes the words 'U.K.' and 'U.S.A' as tokens instead of 'U' and '.' followed by 'K', for example.

spaCy also has an outstanding tokenization feature. It uses an accurately trained model that is constantly updated. To install it, we can run the following command:

```
python -m spacy download en_core_web_sm
```

Then, we load the en_core_web_sm model (if you have not downloaded the model, you can run python -m spacy download en_core_web_sm to do so) and parse the sentence using this model:

```
>>> import spacy
>>> nlp = spacy.load('en_core_web_sm')
>>> tokens2 = nlp(sent2)
>>> print([token.text for token in tokens2])
['I', 'have', 'been', 'to', 'U.K.', 'and', 'U.S.A.']
```

We can also segment text based on sentences. For example, in the same input text, using the sent_tokenize function from NLTK, we have the following commands:

```
>>> from nltk.tokenize import sent_tokenize
>>> print(sent_tokenize(sent))
['I am reading a book.',
'It's Python Machine Learning By Example,\n          4th edition.']
```

Two sentence-based tokens are returned, as there are two sentences in the input text.

PoS tagging

We can apply an off-the-shelf tagger from NLTK or combine multiple taggers to customize the tagging process. It is easy to directly use the built-in tagging function, pos_tag, as in pos_tag(input_tokens), for instance, but behind the scenes, it is actually a prediction from a pre-built supervised learning model. The model is trained based on a large corpus composed of words that are correctly tagged.

Reusing an earlier example, we can perform PoS tagging as follows:

```
>>> import nltk
>>> tokens = word_tokenize(sent)
>>> print(nltk.pos_tag(tokens))
[('I', 'PRP'), ('am', 'VBP'), ('reading', 'VBG'), ('a', 'DT'), ('book', 'NN'),
('.', '.'), ('It', 'PRP'), ('is', 'VBZ'), ('Python', 'NNP'), ('Machine',
'NNP'), ('Learning', 'NNP'), ('By', 'IN'), ('Example', 'NNP'), (',', ','),
('4th', 'CD'), ('edition', 'NN'), ('.', '.')]
```

The PoS tag following each token is returned. We can check the meaning of a tag using the `help` function. Looking up `PRP` and `VBP`, for example, gives us the following output:

```
>>> nltk.help.upenn_tagset('PRP')
PRP: pronoun, personal
    hers herself him himself hisself it itself me myself one oneself ours
ourselves ownself self she thee theirs them themselves they thou thy us
>>> nltk.help.upenn_tagset('VBP')
VBP: verb, present tense, not 3rd person singular
    predominate wrap resort sue twist spill cure lengthen brush terminate appear
tend stray glisten obtain comprise detest tease attract emphasize mold postpone
sever return wag ...
```

In spaCy, getting a PoS tag is also easy. The `token` object parsed from an input sentence has an attribute called `pos_`, which is the tag we are looking for. Let's print `pos_` for each token, as follows:

```
>>> print([(token.text, token.pos_) for token in tokens2])
[('I', 'PRON'), ('have', 'VERB'), ('been', 'VERB'), ('to', 'ADP'), ('U.K.',
'PROPN'), ('and', 'CCONJ'), ('U.S.A.', 'PROPN')]
```

We have just played around with PoS tagging with NLP packages. What about NER? Let's see in the next section.

NER

Given a text sequence, the NER task is to locate and identify words or phrases that are of definitive categories, such as names of persons, companies, locations, and dates. Let's take a peep at an example of using spaCy for NER.

First, tokenize an input sentence, `The book written by Hayden Liu in 2024 was sold at $30 in America`, as usual, as shown in the following command:

```
>>> tokens3 = nlp('The book written by Hayden Liu in 2024 was sold at $30 in
America')
```

The resultant `token` object contains an attribute called `ents`, which are the named entities. We can extract the tagging for each recognized named entity as follows:

```
>>> print([(token_ent.text, token_ent.label_) for token_ent in tokens3.ents])
[('Hayden Liu', 'PERSON'), ('2024', 'DATE'), ('30', 'MONEY'), ('America',
'GPE')]
```

We can see from the results that `Hayden  Liu` is `PERSON`, `2024` is `DATE`, `30` is `MONEY`, and `America` is `GPE` (country). Please refer to `https://spacy.io/api/annotation#section-named-entities` for a full list of named entity tags.

Stemming and lemmatization

Word **stemming** is a process of reverting an inflected or derived word to its root form. For instance, *machine* is the stem of *machines*, and *learning* and *learned* are generated from *learn* as their stem.

The word **lemmatization** is a cautious version of stemming. It considers the PoS of a word when conducting stemming. Also, it traces back to the lemma (base or canonical form) of the word. We will discuss these two text preprocessing techniques, stemming and lemmatization, in further detail shortly. For now, let's take a quick look at how they're implemented respectively in NLTK by performing the following steps:

1. Import `porter` as one of the three built-in stemming algorithms (`LancasterStemmer` and `SnowballStemmer` are the other two) and initialize the stemmer as follows:

    ```
    >>> from nltk.stem.porter import PorterStemmer
    >>> porter_stemmer = PorterStemmer()
    ```

2. Stem `machines` and `learning`, as shown in the following code:

    ```
    >>> porter_stemmer.stem('machines')
    'machin'
    >>> porter_stemmer.stem('learning')
    'learn'
    ```

 Stemming sometimes involves the chopping of letters if necessary, as you can see in `machin` in the preceding command output.

3. Now, import a lemmatization algorithm based on the built-in WordNet corpus and initialize a lemmatizer:

    ```
    >>> from nltk.stem import WordNetLemmatizer
    >>> lemmatizer = WordNetLemmatizer()
    ```

Similar to stemming, we lemmatize `machines` and `learning`:

```
>>> lemmatizer.lemmatize('machines')
'machine'
>>> lemmatizer.lemmatize('learning')
'learning'
```

Why is `learning` unchanged? The algorithm defaults to finding the lemma for nouns unless you specify otherwise. If you want to treat `learning` as a verb, you can specify it in `lemmatizer.lemmatize('learning', nltk.corpus.wordnet.VERB)`, which will return `learn`.

Semantics and topic modeling

Gensim is famous for its powerful semantic and topic-modeling algorithms. Topic modeling is a typical text-mining task of discovering the hidden semantic structures in a document. A semantic structure in plain English is the distribution of word occurrences. It is obviously an unsupervised learning task. What we need to do is to feed in plain text and let the model figure out the abstract topics. For example, we can use topic modeling to group product reviews on an e-commerce site based on the common themes expressed in the reviews. We will study topic modeling in detail in *Chapter 8, Discovering Underlying Topics in the Newsgroups Dataset with Clustering and Topic Modeling*.

In addition to robust semantic modeling methods, gensim also provides the following functionalities:

- **Word embedding**: Also known as **word vectorization**, this is an innovative way to represent words while preserving words' co-occurrence features. Later in this chapter, we will delve into a comprehensive exploration of word embedding.

- **Similarity querying**: This functionality retrieves objects that are similar to the given query object. It's a feature built on top of word embedding.

- **Distributed computing**: This functionality makes it possible to efficiently learn from millions of documents.

Last but not least, as mentioned in the first chapter, *Getting Started with Machine Learning and Python*, scikit-learn is the main package we have used throughout this entire book. Luckily, it provides all the text processing features we need, such as tokenization, along with comprehensive machine learning functionalities. Plus, it comes with a built-in loader for the 20 newsgroups dataset.

Now that the tools are available and properly installed, what about the data?

Getting the newsgroups data

The project in this chapter is about the 20 newsgroups dataset. It's composed of text taken from newsgroup articles, as its name implies. It was originally collected by Ken Lang and now has been widely used for experiments in text applications of machine learning techniques, specifically NLP techniques.

The data contains approximately 20,000 documents across 20 online newsgroups. A newsgroup is a place on the internet where people can ask and answer questions about a certain topic. The data is already cleaned to a certain degree and already split into training and testing sets. The cutoff point is at a certain date.

The original data comes from `http://qwone.com/~jason/20Newsgroups/`, with 20 different topics listed, as follows:

- `comp.graphics`
- `comp.os.ms-windows.misc`
- `comp.sys.ibm.pc.hardware`
- `comp.sys.mac.hardware`
- `comp.windows.x`
- `rec.autos`
- `rec.motorcycles`
- `rec.sport.baseball`
- `rec.sport.hockey`
- `sci.crypt`
- `sci.electronics`
- `sci.med`
- `sci.space`
- `misc.forsale`
- `talk.politics.misc`
- `talk.politics.guns`
- `talk.politics.mideast`
- `talk.religion.misc`
- `alt.atheism`
- `soc.religion.christian`

All of the documents in the dataset are in English. And we can easily deduce the topics from the newsgroups' names.

The dataset is labeled and each document is composed of text data and a group label. This also makes it a perfect fit for supervised learning, such as text classification. At the end of the chapter, feel free to practice classification on this dataset using what you've learned so far in this book.

Some of the newsgroups are closely related or even overlapping – for instance, those five computer groups (`comp.graphics`, `comp.os.ms-windows.misc`, `comp.sys.ibm.pc.hardware`, `comp.sys.mac.hardware`, and `comp.windows.x`). Some are not closely related to each other, such as Christian (`soc.religion.christian`) and baseball (`rec.sport.baseball`).

Hence, it's a perfect use case for unsupervised learning such as clustering, with which we can see whether similar topics are grouped together and unrelated ones are far apart. Moreover, we can even discover abstract topics beyond the original 20 labels using topic modeling techniques.

For now, let's focus on exploring and analyzing the text data. We will get started with acquiring the data.

It is possible to download the dataset manually from the original website or many other online repositories. However, there are also many versions of the dataset—some are cleaned in a certain way and some are in raw form. To avoid confusion, it is best to use a consistent acquisition method. The scikit-learn library provides a utility function that loads the dataset. Once the dataset is downloaded, it's automatically cached. We don't need to download the same dataset twice.

In most cases, caching the dataset, especially for a relatively small one, is considered a good practice. Other Python libraries also provide data download utilities, but not all of them implement automatic caching. This is another reason why we love scikit-learn.

As always, we first import the loader function for the 20 newsgroups data, as follows:

```
>>> from sklearn.datasets import fetch_20newsgroups
```

Then, we download the dataset with all the default parameters, as follows:

```
>>> groups = fetch_20newsgroups()
Downloading 20news dataset. This may take a few minutes.
Downloading dataset from https://ndownloader.figshare.com/files/5975967 (14 MB)
```

We can also specify one or more certain topic groups and particular sections (training, testing, or both) and just load such a subset of data in the program. The full list of parameters and options for the loader function is summarized in the following table:

Parameter	Default value	Example values	Description
subset	'train'	'train', 'test', 'all'	The dataset to load: the training set, the testing set, or both.
data_home	~/scikit_learn_data	~/myfolder	Directory where the files are stored and cached.
categories	None	['sci.space", alt.atheism']	List of newsgroups to load. If None, all newsgroups will be loaded.
shuffle	True	True, False	Boolean indicating whether to shuffle the data.
random_state	42	7, 43	Random seed integer used to shuffle the data.
remove	0	('headers', 'footers', 'quotes')	Tuple indicating the part(s) among "header, footer, and quote" of each newsgroup post to omit. Nothing is removed by default.
download_if_missing	True	True, False	Boolean indicating whether to download the data if it is not found locally.

Table 7.2: List of parameters of the fetch_20newsgroups() function

Remember that `random_state` is useful for the purpose of reproducibility. You are able to get the same dataset every time you run the script. Otherwise, working on datasets shuffled under different orders might bring in unnecessary variations.

In this section, we loaded the newsgroups data. Let's explore it next.

Exploring the newsgroups data

After we download the 20 newsgroups dataset by whatever means we prefer, the `data` object of `groups` is cached in memory. The `data` object is in the form of a key-value dictionary. Its keys are as follows:

```
>>> groups.keys()
dict_keys(['data', 'filenames', 'target_names', 'target', 'DESCR'])
```

The `target_names` key gives the 20 newsgroups names:

```
>>> groups['target_names']
    ['alt.atheism', 'comp.graphics', 'comp.os.ms-windows.misc', 'comp.sys.ibm.
pc.hardware', 'comp.sys.mac.hardware', 'comp.windows.x', 'misc.forsale', 'rec.
autos', 'rec.motorcycles', 'rec.sport.baseball', 'rec.sport.hockey', 'sci.
crypt', 'sci.electronics', 'sci.med', 'sci.space', 'soc.religion.christian',
'talk.politics.guns', 'talk.politics.mideast', 'talk.politics.misc', 'talk.
religion.misc']
```

The `target` key corresponds to a newsgroup, but is encoded as an integer:

```
>>> groups.target
array([7, 4, 4, ..., 3, 1, 8])
```

Then, what are the distinct values for these integers? We can use the `unique` function from NumPy to figure it out:

```
>>> import numpy as np
>>> np.unique(groups.target)
array([ 0, 1, 2, 3, 4, 5, 6, 7, 8, 9, 10, 11, 12, 13, 14, 15, 16, 17, 18, 19])
```

They range from 0 to 19, representing the 1st, 2nd, 3rd, ..., and 20th newsgroup topics in `groups['target_names']`.

In the context of multiple topics or categories, it is important to know what the distribution of topics is. A balanced class distribution is the easiest to deal with because there are no under-represented or over-represented categories. However, frequently, we have a skewed distribution with one or more categories dominating.

We will use the seaborn package (https://seaborn.pydata.org/) to compute the histogram of categories and plot it utilizing the matplotlib package (https://matplotlib.org/). We can install both packages via pip as follows:

```
python -m pip install -U matplotlib
pip install seaborn
```

In the case of conda, you can execute the following command line:

```
conda install -c conda-forge matplotlib
conda install seaborn
```

Remember to install matplotlib before seaborn as matplotlib is one of the dependencies of the seaborn package.

Now, let's display the distribution of the classes, as follows:

```
>>> import seaborn as sns
>>> import matplotlib.pyplot as plt
>>> sns.histplot(groups.target, bins=20)
>>> plt.xticks(range(0, 20, 1))
>>> plt.show()
```

Refer to the following screenshot for the result:

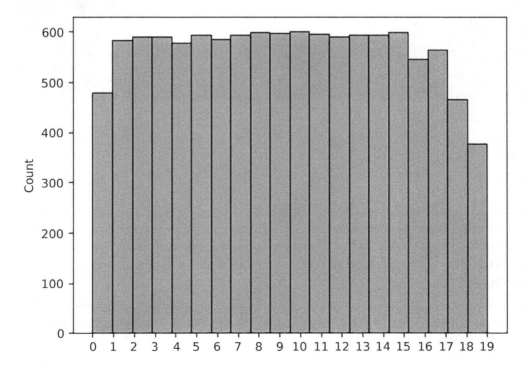

Figure 7.4: Distribution of newsgroup classes

As you can see, the distribution is approximately uniform so that's one less thing to worry about.

 It's good to visualize data to get a general idea of how the data is structured, what possible issues may arise, and whether there are any irregularities that we have to take care of.

Other keys are quite self-explanatory: data contains all newsgroup documents and filenames stores the path where each document is located in your filesystem.

Now, let's have a look at the first document and its topic number and name by executing the following command:

```
>>> groups.data[0]
"From: lerxst@wam.umd.edu (where's my thing)\nSubject: WHAT car is this!?\
nNntp-Posting-Host: rac3.wam.umd.edu\nOrganization: University of Maryland,
College Park\nLines: 15\n\n I was wondering if anyone out there could enlighten
me on this car I saw\nthe other day. It was a 2-door sports car, looked to be
from the late 60s/\nearly 70s. It was called a Bricklin. The doors were really
small. In addition,\nthe front bumper was separate from the rest of the body.
This is \nall I know. If anyone can tellme a model name, engine specs, years\
nof production, where this car is made, history, or whatever info you\nhave on
this funky looking car, please e-mail.\n\nThanks,\n- IL\n ---- brought to you
by your neighborhood Lerxst ----\n\n\n\n\n"
>>> groups.target[0]
7
>>> groups.target_names[groups.target[0]]
'rec.autos'
```

 If random_state isn't fixed (42 by default), you may get different results running the preceding scripts.

As you can see, the first document is from the rec.autos newsgroup, which was assigned the number 7. Reading this post, we can easily figure out that it's about cars. The word car actually occurs a number of times in the document. Words such as bumper also seem very car-oriented. However, words such as doors may not necessarily be car-related, as they may also be associated with home improvement or another topic.

As a side note, it makes sense to not distinguish between doors and door, or the same word with different capitalization, such as Doors. There are some rare cases where capitalization does matter – for instance, if we're trying to find out whether a document is about the band called The Doors or the more common concept, the doors (made of wood or another material).

Thinking about features for text data

From the preceding analysis, we can safely conclude that if we want to figure out whether a document was from the rec.autos newsgroup, the presence or absence of words such as car, doors, and bumper can be very useful features. The presence or not of a word is a Boolean variable, and we can also look at the count of certain words. For instance, car occurs multiple times in the document. Maybe the more times such a word is found in a text, the more likely it is that the document has something to do with cars.

Counting the occurrence of each word token

It seems that we are only interested in the occurrence of certain words, their count, or a related measure, and not in the order of the words. We can therefore view a text as a collection of words. This is called the **Bag of Words (BoW)** model. This is a very basic model but it works pretty well in practice. We can optionally define a more complex model that takes into account the order of words and PoS tags. However, such a model is going to be more computationally expensive and more difficult to program. In reality, the basic BoW model, in most cases, suffices. We can give it a shot and see whether the BoW model makes sense.

We begin by converting documents into a matrix where each row represents each newsgroup document and each column represents a word token, or specifically, a unigram to begin with. The value of each element in the matrix is the number of times the word (column) occurs in the document (row). We are utilizing the CountVectorizer class from scikit-learn to do the work:

```
>>> from sklearn.feature_extraction.text import CountVectorizer
```

The important parameters and options for the count conversion function are summarized in the following table:

Constructor parameter	Default value	Example values	Description
ngram_range	(1,1)	(1,2), (2,2)	Lower and upper bound of the n-grams to be extracted in the input text, for example (1,1) means unigram, (1,2) means unigram and bigram.
stop_words	None	'english' or list ['a','the', 'of'] or None	Which stop word list to use: can be 'english' referring to the built-in list, or a customized input list. If None, no words will be removed.
lowercase	True	True, False	Whether or not to convert all characters to lowercase.
max_features	None	None, 200, 500	The number of top (most frequent) tokens to consider, or all tokens if None.

Constructor parameter	Default value	Example values	Description
binary	False	True, False	If true, all non-zero counts become 1s.

Table 7.3: List of parameters of the CountVectorizer() function

We first initialize the count vectorizer with the 500 top features (500 most frequent tokens):

```
>>> count_vector = CountVectorizer(max_features=500)
```

Use it to fit on the raw text data as follows:

```
>>> data_count = count_vector.fit_transform(groups.data)
```

Now the count vectorizer captures the top 500 features and generates a token count matrix out of the original text input:

```
>>> data_count
<11314x500 sparse matrix of type '<class 'numpy.int64'>'
    with 798221 stored elements in Compressed Sparse Row format>
>>> data_count[0]
<1x500 sparse matrix of type '<class 'numpy.int64'>'
    with 53 stored elements in Compressed Sparse Row format>
```

The resulting count matrix is a sparse matrix where each row only stores non-zero elements (hence, only 798,221 elements instead of 11314 * 500 = 5,657,000). For example, the first document is converted into a sparse vector composed of 53 non-zero elements.

If you are interested in seeing the whole matrix, feel free to run the following:

```
>>> data_count.toarray()
```

If you just want the first row, run the following:

```
>>> data_count.toarray()[0]
```

Let's take a look at the following output derived from the preceding command:

```
array([0, 0, 0, 0, 0, 0, 0, 0, 0, 1, 0, 0, 0, 0, 0, 0, 0, 0, 0, 0, 0, 0,
       0, 0, 0, 0, 0, 0, 0, 0, 0, 0, 0, 0, 0, 0, 0, 0, 0, 0, 0, 1, 0, 0,
       0, 0, 0, 0, 0, 0, 0, 0, 0, 2, 0, 0, 0, 0, 0, 0, 0, 0, 0, 0, 0, 0,
       0, 0, 0, 0, 0, 0, 0, 1, 0, 0, 0, 0, 0, 0, 0, 0, 0, 0, 0, 0, 0, 0,
       0, 0, 1, 0, 0, 1, 0, 1, 0, 0, 5, 0, 0, 0, 0, 0, 0, 0, 0, 0, 0, 0,
       0, 0, 0, 0, 0, 1, 0, 0, 0, 0, 0, 1, 0, 0, 0, 0, 0, 0, 0, 0, 0, 0,
       0, 0, 0, 0, 0, 0, 0, 0, 0, 2, 0, 0, 0, 0, 0, 0, 0, 0, 0, 0, 0, 0,
       0, 0, 0, 0, 0, 0, 0, 0, 0, 0, 0, 0, 0, 3, 0, 0, 0, 0, 0, 0, 0, 0,
       0, 0, 0, 0, 0, 0, 0, 0, 0, 0, 0, 0, 0, 0, 0, 0, 0, 1, 0, 0, 0, 0,
       0, 0, 0, 0, 0, 0, 0, 1, 0, 0, 0, 0, 0, 0, 2, 0, 0, 1, 0, 1, 0, 0,
       0, 0, 0, 3, 0, 0, 0, 2, 0, 0, 0, 0, 0, 0, 0, 0, 0, 0, 0, 1, 0, 0, 0,
       0, 0, 0, 0, 0, 0, 0, 1, 0, 0, 0, 0, 0, 0, 0, 0, 1, 0, 0, 1, 1, 0, 0,
       0, 0, 0, 0, 0, 0, 1, 0, 0, 0, 0, 0, 0, 0, 0, 0, 0, 0, 0, 0, 0, 0, 1,
       1, 0, 0, 0, 0, 0, 0, 0, 0, 0, 1, 0, 0, 0, 0, 0, 0, 0, 3, 0, 0, 2,
       0, 0, 0, 0, 0, 1, 0, 0, 1, 1, 0, 0, 1, 0, 0, 0, 0, 0, 0, 0, 0, 0,
       0, 0, 0, 1, 0, 0, 0, 1, 0, 0, 0, 0, 0, 0, 0, 0, 0, 0, 0, 0, 0, 0,
       0, 0, 0, 1, 0, 0, 0, 0, 0, 0, 0, 0, 0, 0, 0, 0, 0, 0, 0, 0, 0, 0,
       0, 0, 0, 0, 0, 0, 0, 0, 0, 0, 0, 0, 0, 1, 0, 0, 0, 0, 0, 0, 0, 0,
       0, 0, 0, 0, 0, 0, 0, 1, 0, 0, 0, 0, 0, 0, 0, 0, 0, 0, 0, 1, 0, 6,
       0, 0, 0, 1, 0, 0, 1, 0, 0, 5, 0, 0, 0, 0, 0, 0, 0, 2, 0, 0, 0, 0,
       0, 0, 0, 0, 0, 0, 0, 0, 1, 0, 0, 0, 0, 0, 0, 0, 0, 0, 0, 0, 0, 0,
       0, 0, 4, 0, 0, 0, 0, 1, 1, 0, 2, 0, 0, 0, 0, 0, 0, 0, 0, 0, 0, 0,
       0, 0, 0, 0, 0, 0, 0, 0, 0, 0, 0, 1, 0, 0, 2, 1], dtype=int64)
```

Figure 7.5: Output of count vectorization

So, what are those 500 top features? They can be found in the following output:

```
>>> print(count_vector. get_feature_names_out())
['00' '000' '10' '100' '11' '12' '13' '14' '145' '15' '16' '17' '18' '19'
 '1993' '20' '21' '22' '23' '24' '25' '26' '27' '30' '32' '34' '40' '50' '93'
 'a86' 'able' 'about' 'above' 'ac' 'access' 'actually' 'address' 'after'
 ......
 ......
 ......
 'well' 'were' 'what' 'when' 'where' 'whether' 'which' 'while' 'who' 'whole'
 'why' 'will' 'win' 'window' 'windows' 'with' 'without' 'won' 'word' 'work'
 'works' 'world' 'would' 'writes' 'wrong' 'wrote' 'year' 'years' 'yes' 'yet'
 'you' 'your']
```

Our first trial doesn't look perfect. Obviously, the most popular tokens are numbers, or letters with numbers such as a86, which do not convey important information. Moreover, there are many words that have no actual meaning, such as you, the, them, and then. Also, some words contain identical information, for example, tell and told, use and used, and time and times. Let's tackle these issues.

Text preprocessing

We begin by retaining letter-only words so that numbers such as 00 and 000 and combinations of letters and numbers such as b8f will be removed. The filter function is defined as follows:

```
>>> data_cleaned = []
>>> for doc in groups.data:
...     doc_cleaned = ' '.join(word for word in doc.split()
                                      if word.isalpha())
...     data_cleaned.append(doc_cleaned)
```

This will generate a cleaned version of the newsgroups data.

Dropping stop words

We didn't talk about stop_words as an important parameter in CountVectorizer. **Stop words** are those common words that provide little value in helping to differentiate documents. In general, stop words add noise to the BoW model and can be removed.

There's no universal list of stop words. Hence, depending on the tools or packages you are using, you will remove different sets of stop words. Take scikit-learn as an example—you can check the list that follows:

```
>>> from sklearn.feature_extraction import _stop_words
>>> print(_stop_words.ENGLISH_STOP_WORDS)
frozenset({'latter', 'somewhere', 'further', 'full', 'de', 'under', 'beyond',
'than', 'must', 'has', 'him', 'hereafter', 'they', 'third', 'few', 'most',
'con', 'thereby', 'ltd', 'take', 'five', 'alone', 'yours', 'above', 'hereupon',
'seeming', 'least', 'over', 'amongst', 'everyone', 'anywhere', 'yourself',
'these', 'name', 'even', 'in', 'forty', 'part', 'perhaps', 'sometimes',
'seems', 'down', 'among', 'still', 'own', 'wherever', 'same', 'about',
'because', 'four', 'none', 'nothing', 'could'
......

......
'myself', 'except', 'whom', 'up', 'six', 'get', 'sixty', 'those', 'whither',
'once', 'something', 'elsewhere', 'my', 'both', 'another', 'one', 'a', 'hasnt',
'everywhere', 'thin', 'not', 'eg', 'someone', 'seem', 'detail', 'either',
'being'})
```

To drop stop words from the newsgroups data, we simply just need to specify the stop_words parameter:

```
>>> count_vector_sw = CountVectorizer(stop_words="english", max_features=500)
```

Besides stop words, you may notice that names are included in the top features, such as andrew. We can filter names with the Names corpus from NLTK we just worked with.

Reducing inflectional and derivational forms of words

As mentioned earlier, we have two basic strategies to deal with words from the same root—stemming and lemmatization. Stemming is a quicker approach that involves, if necessary, chopping off letters; for example, *words* becomes *word* after stemming. The result of stemming doesn't have to be a valid word. For instance, *trying* and *try* become *tri*. Lemmatizing, on the other hand, is slower but more accurate. It performs a dictionary lookup and guarantees to return a valid word. Recall that we implemented both stemming and lemmatization using NLTK previously.

Putting all of these (preprocessing, dropping stop words, lemmatizing, and count vectorizing) together, we obtain the following:

```python
>>> all_names = set(names.words())
>>> def get_cleaned_data(groups, lemmatizer, remove_words):
        data_cleaned = []
        for doc in groups.data:
...         doc = doc.lower()
...         doc_cleaned = ' '.join(lemmatizer.lemmatize(word)
                                   for word in doc.split()
                                   if word.isalpha() and
                                   word not in remove_words)
...         data_cleaned.append(doc_cleaned)
        return data_cleaned
>>> data_cleaned = get_cleaned_data(groups, lemmatizer, all_names)
>>> data_cleaned_count = count_vector_sw.fit_transform(data_cleaned)
```

Now the features are much more meaningful:

```python
>>> print(count_vector_sw.get_feature_names_out())
['able', 'accept', 'access', 'according', 'act', 'action', 'actually', 'add',
'address', 'ago', 'agree', 'algorithm', 'allow', 'american', 'anonymous',
'answer', 'anybody', 'apple', 'application', 'apr', 'april', 'arab', 'area',
'argument', 'armenian', 'article', 'ask', 'asked',
......
......
'video', 'view', 'wa', 'want', 'wanted', 'war', 'water', 'way', 'weapon',
'week', 'went', 'western', 'white', 'widget', 'win', 'window', 'woman', 'word',
'work', 'working', 'world', 'worth', 'write', 'written', 'wrong', 'year',
'york', 'young']
```

We have just converted text from each raw newsgroup document into a sparse vector of size 500. For a vector from a document, each element represents the number of times a word token occurs in this document. Also, these 500 word tokens are selected based on their overall occurrences after text pre-processing, the removal of stop words, and lemmatization. Now, you may ask questions such as, "Is such an occurrence vector representative enough, or does such an occurrence vector convey enough information that can be used to differentiate the document from documents on other topics?" You will see the answer in the next section.

Visualizing the newsgroups data with t-SNE

We can answer these questions easily by visualizing those representation vectors. If we can see that the document vectors from the same topic form a cluster, we did a good job mapping the documents into vectors. But how? They are of 500 dimensions, while we can visualize data of, **at most**, three dimensions. We can resort to t-SNE for dimensionality reduction.

What is dimensionality reduction?

Dimensionality reduction is an important machine learning technique that reduces the number of features and, at the same time, retains as much information as possible. It is usually performed by obtaining a set of new principal features.

As mentioned before, it is difficult to visualize data of high dimensions. Given a three-dimensional plot, we sometimes don't find it straightforward to observe any findings, not to mention 10, 100, or 1,000 dimensions. Moreover, some of the features in high-dimensional data may be correlated and, as a result, bring in redundancy. This is why we need dimensionality reduction.

Dimensionality reduction is not simply taking out a pair of two features from the original feature space. It is transforming the original feature space into a new space of fewer dimensions. The data transformation can be linear, such as the famous one, **Principal Component Analysis** (**PCA**), which maps the data in a higher dimensional space to a lower dimensional space where the variance of the data is maximized, which we will talk about in *Chapter 9, Recognizing Faces with Support Vector Machine*, or nonlinear, such as neural networks and t-SNE, which is coming up shortly. **Non-negative Matrix Factorization** (**NMF**) is another powerful algorithm, which we will study in detail in *Chapter 8, Discovering Underlying Topics in the Newsgroups Dataset with Clustering and Topic Modeling*.

At the end of the day, most dimensionality reduction algorithms are in the family of **unsupervised learning** as the target or label information (if available) is not used in data transformation.

t-SNE for dimensionality reduction

t-SNE stands for **t-distributed Stochastic Neighbor Embedding**. It is a popular nonlinear dimensionality reduction technique developed by Laurens van der Maaten and Geoffrey Hinton (`https://www.cs.toronto.edu/~hinton/absps/tsne.pdf`). t-SNE has been widely used for data visualization in various domains, including computer vision, NLP, bioinformatics, and computational genomics.

As its name implies, t-SNE embeds high-dimensional data into a low-dimensional (usually two-dimensional or three-dimensional) space while preserving the local structure and pairwise similarities of the data as much as possible. It first models a probability distribution over neighbors around data points by assigning a high probability to similar data points and an extremely small probability to dissimilar ones. Note that similarity and neighbor distances are measured by Euclidean distance or other metrics. Then, t-SNE constructs a projection onto a low-dimensional space where the divergence between the input distribution and output distribution is minimized. The original high-dimensional space is modeled as a Gaussian distribution, while the output low-dimensional space is modeled as a t-distribution.

We'll herein implement t-SNE using the TSNE class from scikit-learn:

```
>>> from sklearn.manifold import TSNE
```

Now, let's use t-SNE to verify our count vector representation.

We pick three distinct topics, talk.religion.misc, comp.graphics, and sci.space, and visualize document vectors from these three topics.

First, just load documents of these three labels, as follows:

```
>>> categories_3 = ['talk.religion.misc', 'comp.graphics', 'sci.space']
>>> groups_3 = fetch_20newsgroups(categories=categories_3)
>>> data_cleaned = get_cleaned_data(groups_3, lemmatizer, all_names)
>>> data_cleaned_count_3 = count_vector_sw.fit_transform(data_cleaned)
```

We go through the same process and generate a count matrix, data_cleaned_count_3, with 500 features from the input, groups_3. You can refer to the steps in previous sections as you just need to repeat the same code.

Next, we apply t-SNE to reduce the 500-dimensional matrix to a two-dimensional matrix:

```
>>> tsne_model = TSNE(n_components=2, perplexity=40,
                        random_state=42, learning_rate=500)
>>> data_tsne = tsne_model.fit_transform(data_cleaned_count_3.toarray())
```

The parameters we specify in the TSNE object are as follows:

- n_components: The output dimension
- perplexity: The number of nearest data points considered neighbors in the algorithm with a typical value of between 5 and 50
- random_state: The random seed for program reproducibility
- learning_rate: The factor affecting the process of finding the optimal mapping space with a typical value of between 10 and 1,000

Note that the TSNE object only takes in a dense matrix, hence we convert the sparse matrix, data_cleaned_count_3, into a dense one using toarray().

We just successfully reduced the input dimension from 500 to 2. Finally, we can easily visualize it in a two-dimensional scatter plot where the x axis is the first dimension, the y axis is the second dimension, and the color, c, is based on the topic label of each original document:

```
>>> plt.scatter(data_tsne[:, 0], data_tsne[:, 1], c=groups_3.target)
>>> plt.show()
```

Refer to the following screenshot for the end result:

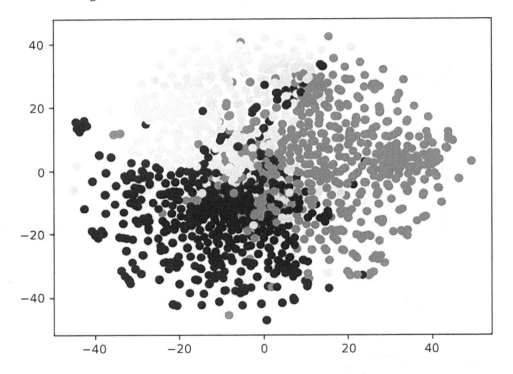

Figure 7.6: Applying t-SNE to data from three different topics

Data points from the three topics are in different colors – green, purple, and yellow. We can observe three clear clusters. Data points from the same topic are close to each other, while those from different topics are far away. Obviously, count vectors are great representations of original text data as they preserve the distinction between three different topics.

You can also play around with the parameters and see whether you can obtain a nicer plot where the three clusters are better separated.

Count vectorization does well in keeping document disparity. How about maintaining similarity? We can also check that using documents from overlapping topics, such as these five topics—comp.graphics, comp.os.ms-windows.misc, comp.sys.ibm.pc.hardware, comp.sys.mac.hardware, and comp.windows.x:

```
>>> categories_5 = ['comp.graphics', 'comp.os.ms-windows.misc', 'comp.sys.ibm.
pc.hardware', 'comp.sys.mac.hardware', 'comp.windows.x']
>>> groups_5 = fetch_20newsgroups(categories=categories_5)
```

Similar processes (including text clean-up, count vectorization, and t-SNE) are repeated and the resulting plot is displayed as follows:

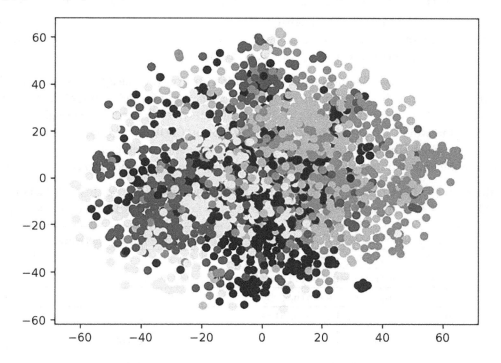

Figure 7.7: Applying t-SNE to data from five similar topics

Data points from those five computer-related topics are all over the place, which means they are contextually similar. To conclude, count vectors are simple yet great representations of original text data as they are also good at preserving similarity among related topics. The question now arises: can we improve upon word (term) count representations? Let's progress to the next section, where we will explore dense vector representations.

Representing words with dense vectors — word embedding

Word count representation results in a high-dimensional, sparse vector where each element represents the frequency of a specific word. Recall that we only looked at the 500 most frequent words previously to avoid this issue. Otherwise, we would have to represent each document with a vector of more than 1 million dimensions (depending on the size of the vocabulary). Also, word count representation lacks the ability to capture the semantics or context of words. It only considers the frequency of words in a document or corpus. On the contrary, **word embedding** represents words in a **dense** (**continuous**) vector space.

Building embedding models using shallow neural networks

Word embedding maps each word to a dense vector of fixed dimensions. Its dimensionality is a lot lower than the size of the vocabulary and is usually several hundred only. For example, the word *machine* can be represented as a vector [1.4, 2.1, 10.3, 0.2, 6.81].

So, how can we embed a word into a vector? One solution is **word2vec** (see *Efficient Estimation of Word Representations in Vector Space*, by Tomas Mikolov, Kai Chen, Greg Corrado, and Jeff Dean, https://arxiv.org/pdf/1301.3781); this trains a **shallow neural network** to predict a word given the other words around it, which is called **Continuous Bag of Words** (**CBOW**), or to predict the other words around a word, which is called the **skip-gram** approach. The **weights** (**coefficients**) of the trained neural network are the embedding vectors for the corresponding words. Let's look at a concrete example.

Given the sentence *I love reading python machine learning by example* in a corpus and 5 as the size of the **word window**, we can have the following training sets for the CBOW neural network:

Input of neural network	Output of neural network
(I, love, python, machine)	(reading)
(love, reading, machine, learning)	(python)
(reading, python, learning, by)	(machine)
(python, machine, by, example)	(learning)

Table 7.4: Input and output of the neural network for CBOW

During training, the inputs and outputs of the neural network are one-hot encoding vectors, where values are either 1 for present words or 0 for absent words. And we can have millions of training samples constructed from a corpus, sentence by sentence. After the network is trained, the weights that connect the input layer and hidden layer embed individual input words.

A skip-gram-based neural network embeds words in a similar way. But its input and output are an inverse version of CBOW. Given the same sentence, *I love reading python machine learning by example*, and 5 as the size of the word window, we can have the following training sets for the skip-gram neural network:

Input of neural network	Output of neural network
(reading)	(i)
(reading)	(love)
(reading)	(python)
(reading)	(machine)
(python)	(love)
(python)	(reading)
(python)	(machine)
(python)	(learning)
(machine)	(reading)
(machine)	(python)
(machine)	(learning)
(machine)	(by)
(learning)	(python)
(learning)	(machine)
(learning)	(by)
(learning)	(example)

Table 7.5: Input and output of the neural network for skip-gram

The embedding vectors are of real values, where each dimension encodes an aspect of meaning for the words in the vocabulary. This helps preserve the semantic information of the words, as opposed to discarding it, as in the dummy one-hot encoding approach using the word count approach. An interesting phenomenon is that vectors from semantically similar words are proximate to each other in geometric space. For example, both the words *clustering* and *grouping* refer to unsupervised clustering in the context of machine learning, hence their embedding vectors are close together. Word embedding is able to capture the meanings of words and their **context**.

Best practice

Visualizing word embeddings can be a helpful tool to explore patterns, identify relationships between words, and assess the effectiveness of your embedding model. Here are some best practices for visualizing word embeddings:

- **Dimensionality reduction:** Word embeddings typically have high-dimensional vectors. To visualize them, reduce their dimensionality. We can use techniques like PCA or t-SNE to project the high-dimensional data into 2D or 3D space while preserving distances between data points.

- **Clustering:** Cluster similar word embeddings together to identify groups of words with similar meanings or contexts.

Utilizing pre-trained embedding models

Training a word embedding neural network can be time-consuming and computationally expensive. Fortunately, many organizations and research institutions (such as Google, Meta AI Research, OpenAI, Stanford NLP Group, and Hugging Face) have developed pre-trained word embedding models based on different kinds of corpora and made them readily available for developers and researchers to use in various NLP tasks. We can simply use these **pre-trained** models to map words to vectors. Some popular pre-trained word embedding models are as follows:

Name	fasttext-wiki-news-subwords-300	
Corpus	Wikipedia 2017	
Vector size	300	
Vocabulary size	1 million	
File size	958 MB	
More information	https://fasttext.cc/docs/en/english-vectors.html	

Name	glove-twitter-100	glove-twitter-25
Corpus	Twitter (2 billion tweets)	
Vector size	100	25
Vocabulary size	1.2 million	
File size	387 MB	104 MB
More information	https://nlp.stanford.edu/projects/glove/	

Name	word2vec-google-news-300	
Corpus	Google News (about 100 billion words)	
Vector size	300	
Vocabulary size	3 million	
File size	1662 MB	
More information	https://code.google.com/archive/p/word2vec/	

Figure 7.8: Configurations of popular pre-trained word embedding models

Once we have embedding vectors for individual words, we can represent a document sample by averaging all of the vectors of words present in this document.

Best practice

Once you have the word embeddings for all words in the document, aggregate them into a single vector representation for the entire document. Common aggregation techniques include averaging and summation. More sophisticated methods include the following:

- Weighted average, where the weights are based on word importance, such as TF-IDF score
- Max/min pooling, where the maximum or minimum value for each dimension across all word embeddings is taken

The resulting vectors of document samples are then consumed by downstream predictive tasks, such as classification, similarity ranking in search engines, and clustering.

Now let's play around with `gensim`, a popular NLP package with powerful word embedding modules.

First, we import the package and load a pre-trained model, `glove-twitter-25`, as follows:

```
>>> import gensim.downloader as api
>>> model = api.load("glove-twitter-25")
[==================================================] 100.0%
104.8/104.8MB downloaded
```

You will see the process bar if you run this line of code. The `glove-twitter-25` model is one of the smallest ones so the download will not take very long.

We can obtain the embedding vector for a word (`computer`, for example), as follows:

```
>>> vector = model['computer']
>>> print('Word computer is embedded into:\n', vector)
Word computer is embedded into:
[ 0.64005 -0.019514 0.70148 -0.66123 1.1723 -0.58859 0.25917
 -0.81541 1.1708 1.1413 -0.15405 -0.11369 -3.8414 -0.87233
   0.47489 1.1541 0.97678 1.1107 -0.14572 -0.52013 -0.52234
  -0.92349 0.34651 0.061939 -0.57375 ]
```

The result is a 25-dimension float vector, as expected.

We can also get the top 10 words that are most contextually relevant to `computer` using the `most_similar` method, as follows:

```
>>> similar_words = model.most_similar("computer")
>>> print('Top ten words most contextually relevant to computer:\n',
           similar_words)
```

```
Top ten words most contextually relevant to computer:
 [('camera', 0.907833456993103), ('cell', 0.891890287399292),
('server', 0.8744666576385498), ('device', 0.869352400302887),
('wifi', 0.8631256818771362), ('screen', 0.8621907234191895), ('app',
0.8615544438362122), ('case', 0.8587921857833862), ('remote',
0.8583616018295288), ('file', 0.8575270771980286)]
```

The result looks promising.

Finally, we demonstrate how to generate embedding vectors for a document with a simple example, as follows:

```
>>> doc_sample = ['i', 'love', 'reading', 'python', 'machine',
                  'learning', 'by', 'example']
>>> doc_vector = np.mean([model[word] for word in doc_sample],
                                                    axis=0)
>>> print('The document sample is embedded into:\n', doc_vector)
The document sample is embedded into:
 [-0.17100249 0.1388764 0.10616798 0.200275 0.1159925 -0.1515975
  1.1621187 -0.4241785 0.2912 -0.28199488 -0.31453252 0.43692702
 -3.95395 -0.35544625 0.073975 0.1408525 0.20736426 0.17444688
  0.10602863 -0.04121475 -0.34942 -0.2736689 -0.47526264 -0.11842456
 -0.16284864]
```

The resulting vector is the **average** of embedding vectors of eight input words.

In traditional NLP applications, such as text classification and information retrieval tasks, where word frequency plays a significant role, word count representation is still an outstanding solution. In more complicated areas requiring understanding and semantic relationships between words, such as text summarization, machine translation, and question answering, word embedding is used extensively and extracts far better features than the traditional approach.

Summary

In this chapter, you learned the fundamental concepts of NLP as an important subfield in machine learning, including tokenization, stemming and lemmatization, and PoS tagging. We also explored three powerful NLP packages and worked on some common tasks using NLTK and spaCy. Then we continued with the main project, exploring the 20 newsgroups data. We began by extracting features with tokenization techniques and went through text preprocessing, stop word removal, and lemmatization. We then performed dimensionality reduction and visualization with t-SNE and proved that count vectorization is a good representation of text data. We proceeded with a more modern representation technique, word embedding, and illustrated how to utilize a pre-trained embedding model.

We had some fun mining the 20 newsgroups data using dimensionality reduction as an unsupervised approach. Moving forward, in the next chapter, we'll be continuing our unsupervised learning journey, specifically looking at topic modeling and clustering.

Exercises

1. Do you think all of the top 500 word tokens contain valuable information? If not, can you impose another list of stop words?

2. Can you use stemming instead of lemmatization to process the 20 newsgroups data?

3. Can you increase `max_features` in `CountVectorizer` from `500` to `5000` and see how the t-SNE visualization will be affected?

4. Experiment with representing the data for the three topics discussed in the chapter using the `word2vec-google-news-300` model in Gensim and visualize them with t-SNE. Assess whether the visualization appears more improved compared to the result shown in *Figure 7.6* using word count representation.

Join our book's Discord space

Join our community's Discord space for discussions with the authors and other readers:

`https://packt.link/yuxi`

8

Discovering Underlying Topics in the Newsgroups Dataset with Clustering and Topic Modeling

In the previous chapter, we went through a text visualization using t-SNE. t-SNE, or any dimensionality reduction algorithm, is a type of unsupervised learning. In this chapter, we will be continuing our unsupervised learning journey, specifically focusing on clustering and topic modeling. We will start with how unsupervised learning learns without guidance and how it is good at discovering hidden information underneath data.

Next, we will talk about clustering as an important branch of unsupervised learning, which identifies different groups of observations from data. For instance, clustering is useful for market segmentation, where consumers of similar behaviors are grouped into one segment for marketing purposes. We will perform clustering on the 20 newsgroups text dataset and see what clusters will be produced.

Another unsupervised learning route we will take is topic modeling, which is the process of extracting themes hidden in the dataset. You will be amused by how many interesting themes we are able to mine from the 20 newsgroups dataset.

We will cover the following topics:

- Leaning without guidance – unsupervised learning
- Getting started with k-means clustering
- Clustering newsgroups data
- Discovering underlying topics in newsgroups

Learning without guidance — unsupervised learning

In the previous chapter, we applied t-SNE to visualize the newsgroup text data, reduced to two dimensions. t-SNE, or dimensionality reduction in general, is a type of **unsupervised learning**. Instead of being guided by predefined labels or categories, such as a class or membership (classification), and a continuous value (regression), unsupervised learning identifies inherent structures or commonalities in the input data. Since there is no guidance in unsupervised learning, there is no clear answer on what is a right or wrong result. Unsupervised learning has the freedom to discover hidden information underneath input data.

An easy way to understand unsupervised learning is to think of going through many practice questions for an exam. In supervised learning, you are given answers to those practice questions. You basically figure out the relationship between the questions and answers and learn how to map the questions to the answers. Hopefully, you will do well in the actual exam in the end by giving the correct answers. However, in unsupervised learning, you are not provided with the answers to those practice questions. What you might do in this instance could include the following:

- Grouping similar practice questions so that you can later study related questions together at one time
- Finding questions that are highly repetitive so that you don't have to waste time working out the answer for each one individually
- Spotting rare questions so that you can be better prepared for them
- Extracting the key chunk of each question by removing boilerplate text so you can cut to the point

You will notice that the outcomes of all these tasks are pretty open-ended. They are correct as long as they are able to describe the commonality and the structure underneath the data.

Practice questions are the **features** in machine learning, which are also often called **attributes**, **observations**, or **predictive variables**. Answers to questions are the labels in machine learning, which are also called **targets** or **target variables**. Practice questions with answers provided are called **labeled data**, while practice questions without answers are called **unlabeled data**. Unsupervised learning works with unlabeled data and acts on that information without guidance.

Unsupervised learning can include the following types:

- **Clustering:** This means grouping data based on commonality, which is often used for exploratory data analysis. Grouping similar practice questions, as mentioned earlier, is an example of clustering. Clustering techniques are widely used in customer segmentation or for grouping similar online behaviors for a marketing campaign. We will learn more about the popular algorithm k-means clustering in this chapter.

- **Association:** This explores the co-occurrence of particular values of two or more features. Outlier detection (also called anomaly detection) is a typical case, where rare observations are identified. Spotting rare questions in the preceding example can be achieved using outlier detection techniques.

- **Projection:** This maps the original feature space to a reduced dimensional space retaining or extracting a set of principal variables. Extracting the key chunk of practice questions is an example projection or, specifically, a dimensionality reduction. The t-SNE we learned about previously is a good example.

Unsupervised learning is extensively employed in the area of NLP mainly because of the difficulty of obtaining labeled text data. Unlike numerical data (such as house prices, stock data, and online click streams), labeling text can sometimes be subjective, manual, and tedious. Unsupervised learning algorithms that do not require labels become effective when it comes to mining text data.

In *Chapter 7, Mining the 20 Newsgroups Dataset with Text Analysis Techniques*, you experienced using t-SNE to reduce the dimensionality of text data. Now, let's explore text mining with clustering algorithms and topic modeling techniques. We will start with clustering the newsgroups data.

Getting started with k-means clustering

The newsgroups data comes with labels, which are the categories of the newsgroups, and a number of categories that are closely related or even overlapping, for instance, the five computer groups: comp. graphics, comp.os.ms-windows.misc, comp.sys.ibm.pc.hardware, comp.sys.mac.hardware, and comp.windows.x, and the two religion-related ones: alt.atheism and talk.religion.misc.

Let's now pretend we don't know those labels or they don't exist. Will samples from related topics be clustered together? We will now resort to the k-means clustering algorithm.

How does k-means clustering work?

The goal of the k-means algorithm is to partition the data into k groups based on feature similarities. k is a predefined property of a k-means clustering model. Each of the k clusters is specified by a centroid (center of a cluster) and each data sample belongs to the cluster with the nearest centroid. During training, the algorithm iteratively updates the k centroids based on the data provided. Specifically, it involves the following steps:

1. **Specifying k:** The algorithm needs to know how many clusters to generate as an end result.
2. **Initializing centroids:** The algorithm starts with randomly selecting k samples from the dataset as centroids.

3.   **Assigning clusters:** Now that we have k centroids, samples that share the same closest centroid constitute one cluster. k clusters are created as a result. Note that closeness is usually measured by the **Euclidean distance**. Other metrics can also be used, such as the **Manhattan distance** and **Chebyshev distance**, which are listed in the following table:

Given two 2-dimension data points (x_1, y_1) and (x_2, y_2)

Distance metric	Calculation
Euclidean distance	$\sqrt{(x_1 - x_2)^2 + (y_1 - y_2)^2}$
Manhattan distance	$|x_1 - x_2| + |y_1 - y_2|$
Chebyshev distance	$\max(|x_1 - x_2|, |y_1 - y_2|)$

Figure 8.1: Distance metrics

4.   **Updating centroids:** For each cluster, we need to recalculate its center point, which is the mean of all the samples in the cluster. k centroids are updated to be the means of corresponding clusters. This is why the algorithm is called **k-means**.

5.   **Repeating steps 3 and 4:** We keep repeating assigning clusters and updating centroids until the model converges when no or a small enough update of centroids can be done, or enough iterations have been completed.

The outputs of a trained k-means clustering model include the following:

- The cluster ID of each training sample, ranging from 1 to k
- k centroids, which can be used to cluster new samples—a new sample will belong to the cluster of the closest centroid

It is easy to understand the k-means clustering algorithm and its implementation is also straightforward, as you will discover next.

Implementing k-means from scratch

We will use the `iris` dataset from scikit-learn as an example. Let's first load the data and visualize it. We herein only use two features out of the original four for simplicity:

```
>>> from sklearn import datasets
>>> iris = datasets.load_iris()
>>> X = iris.data[:, 2:4]
>>> y = iris.target
```

Since the dataset contains three iris classes, we plot it in three different colors, as follows:

```
>>> import numpy as np
>>> from matplotlib import pyplot as plt
>>> plt.scatter(X[:,0], X[:,1], c=y)
>>> plt.show()
```

This will give us the following output for the original data plot:

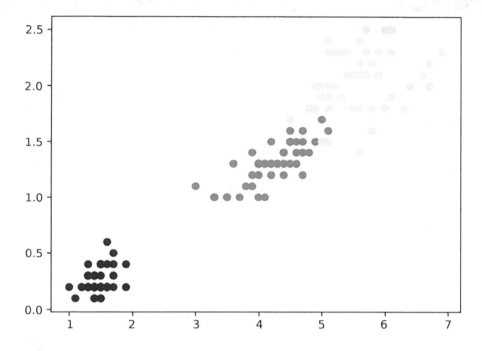

Figure 8.2: Plot of the original iris dataset

Assuming we know nothing about the label *y*, we try to cluster the data into three groups, as there seem to be three clusters in the preceding plot (or you might say two, which we will come back to later). Let's perform *step 1*, *specifying k*, and *step 2*, *initializing centroids*, by randomly selecting three samples as the initial centroids:

```
>>> k = 3
>>> np.random.seed(0)
>>> random_index = np.random.choice(range(len(X)), k)
>>> centroids = X[random_index]
```

We visualize the data (without labels anymore) along with the initial random centroids:

```
>>> def visualize_centroids(X, centroids):
...     plt.scatter(X[:, 0], X[:, 1])
...     plt.scatter(centroids[:, 0], centroids[:, 1], marker='*',
                                       s=200, c='#050505')
...     plt.show()
>>> visualize_centroids(X, centroids)
```

Refer to the following screenshot for the data, along with the initial random centroids:

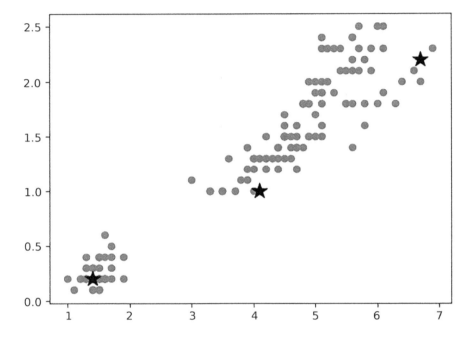

Figure 8.3: Data points with random centroids

Now we perform *step 3*, which entails assigning clusters based on the nearest centroids. First, we need to define a function calculating distance, which is measured by the Euclidean distance, as demonstrated here:

```
>>> def dist(a, b):
...     return np.linalg.norm(a - b, axis=1)
```

💡 **Quick tip:** Enhance your coding experience with the **AI Code Explainer** and **Quick Copy** features. Open this book in the next-gen Packt Reader. Click the **Copy** button (**1**) to quickly copy code into your coding environment, or click the **Explain** button (**2**) to get the AI assistant to explain a block of code to you.

```
                                                          Copy      Explain
function calculate(a, b) {                                 1          2
  return {sum: a + b};
};
```

📖 **The next-gen Packt Reader** is included for free with the purchase of this book. Unlock it by scanning the QR code below or visiting
https://www.packtpub.com/unlock/9781835085622.

Then, we develop a function that assigns a sample to the cluster of the nearest centroid:

```
>>> def assign_cluster(x, centroids):
...     distances = dist(x, centroids)
...     cluster = np.argmin(distances)
...     return cluster
```

With the clusters assigned, we perform *step 4*, which involves updating the centroids to the mean of all samples in the individual clusters:

```
>>> def update_centroids(X, centroids, clusters):
...     for i in range(k):
...         cluster_i = np.where(clusters == i)
...         centroids[i] = np.mean(X[cluster_i], axis=0)
```

Finally, we have *step 5*, which involves repeating *step 3* and *step 4* until the model converges and whichever of the following occurs:

- Centroids move less than the pre-specified threshold
- Sufficient iterations have been taken

We set the tolerance of the first condition and the maximum number of iterations as follows:

```
>>> tol = 0.0001
>>> max_iter = 100
```

Initialize the clusters' starting values, along with the starting clusters for all samples, as follows:

```
>>> iter = 0
>>> centroids_diff = 100000
>>> clusters = np.zeros(len(X))
```

With all the components ready, we can train the model iteration by iteration where it first checks convergence before performing *steps 3* and *4*, and then visualizes the latest centroids:

```
>>> from copy import deepcopy
>>> while iter < max_iter and centroids_diff > tol:
...     for i in range(len(X)):
...         clusters[i] = assign_cluster(X[i], centroids)
...     centroids_prev = deepcopy(centroids)
...     update_centroids(X, centroids, clusters)
...     iter += 1
...     centroids_diff = np.linalg.norm(centroids -
                                    centroids_prev)
...     print('Iteration:', str(iter))
...     print('Centroids:\n', centroids)
...     print(f'Centroids move: {centroids_diff:5.4f}')
...     visualize_centroids(X, centroids)
```

Let's look at the following outputs generated from the preceding commands:

- **Iteration 1:** Take a look at the following output of iteration 1:

```
Iteration: 1
Centroids:
[[1.462      0.246     ]
 [5.80285714 2.11142857]
 [4.42307692 1.44153846]]
Centroids move: 0.8274
```

The plot of centroids after iteration 1 is as follows:

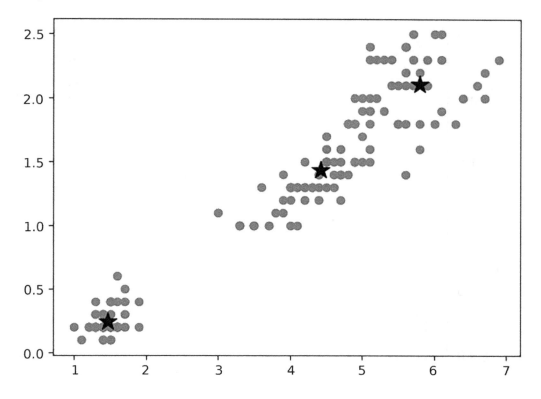

Figure 8.4: k-means clustering result after the first round

- **Iteration 2:** Take a look at the following output of iteration 2:

```
Iteration: 2
Centroids:
[[1.462      0.246     ]
 [5.73333333 2.09487179]
 [4.37704918 1.40819672]]
Centroids move: 0.0913
```

The plot of centroids after iteration 2 is as follows:

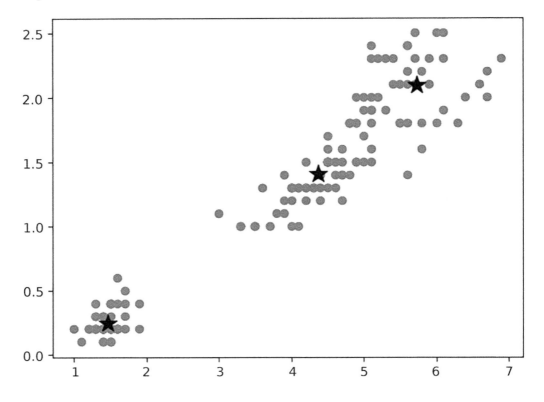

Figure 8.5: k-means clustering result after the second round

- **Iteration 6:** Take a look at the following output of iteration 6 (we herein skip iterations 3 to 5 to avoid tedium):

```
Iteration: 6
Centroids:
[[1.462      0.246     ]
 [5.62608696 2.04782609]
 [4.29259259 1.35925926]]
Centroids move: 0.0225
```

The plot of centroids after iteration 6 is as follows:

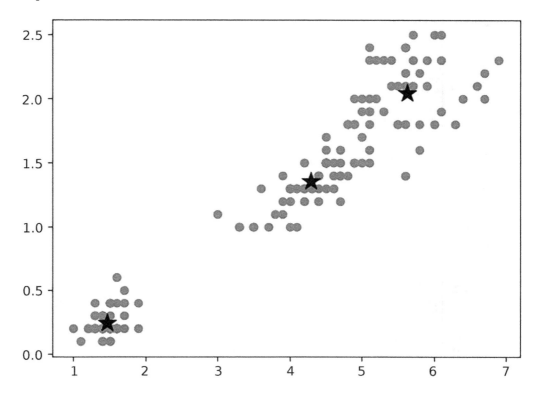

Figure 8.6: k-means clustering result after the sixth round

- **Iteration 7:** Take a look at the following output of iteration 7:

```
Iteration: 7
Centroids:
[[1.462      0.246      ]
 [5.62608696 2.04782609]
 [4.29259259 1.35925926]]
Centroids move: 0.0000
```

The plot of centroids after iteration 7 is as follows:

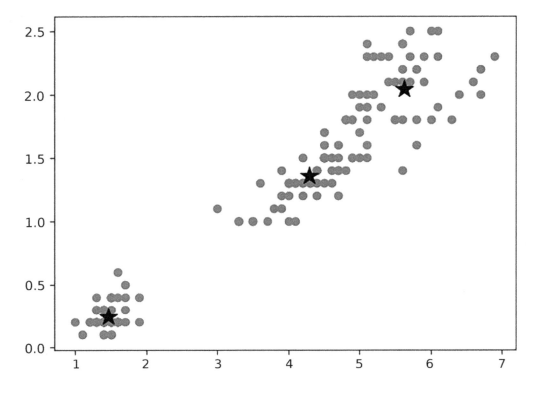

Figure 8.7: k-means clustering result after the seventh round

The model converges after seven iterations. The resulting centroids look promising, and we can also plot the clusters:

```
>>> plt.scatter(X[:, 0], X[:, 1], c=clusters)
>>> plt.scatter(centroids[:, 0], centroids[:, 1], marker='*',
                                      s=200, c='r')
>>> plt.show()
```

Refer to the following screenshot for the end result:

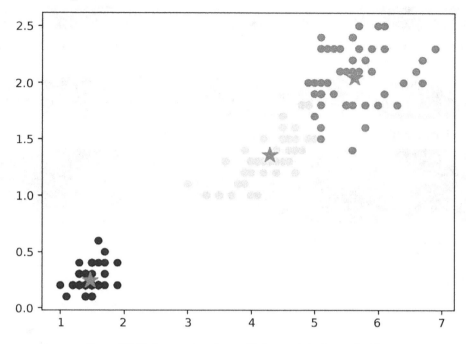

Figure 8.8: Data samples along with learned cluster centroids

As you can see, samples around the same centroid form a cluster. After seven iterations (you might see slightly more or fewer iterations in your case if you change the random seed in `np.random.seed(0)`), the model converges and the centroids will no longer be updated.

Implementing k-means with scikit-learn

Having developed our own k-means clustering model, we will now discuss how to use scikit-learn for a quicker solution by performing the following steps:

1. First, import the `KMeans` class and initialize a model with three clusters, as follows:

```
>>> from sklearn.cluster import KMeans
>>> kmeans_sk = KMeans(n_clusters=3, n_init='auto', random_state=42)
```

The `KMeans` class takes in the following important parameters:

Constructor parameter	Default value	Example values	Description
`n_clusters`	8	3, 5, 10	k clusters
`max_iter`	300	10, 100, 500	Maximum number of iterations
`tol`	1e-4	1e-5, 1e-8	Tolerance to declare convergence
`random_state`	None	0, 42	Random seed for program reproducibility

Table 8.1: Parameters of the KMeans class

2. We then fit the model on the data:

```
>>> kmeans_sk.fit(X)
```

3. After that, we can obtain the clustering results, including the clusters for data samples and centroids of individual clusters:

```
>>> clusters_sk = kmeans_sk.labels_
>>> centroids_sk = kmeans_sk.cluster_centers_
```

4. Similarly, we plot the clusters along with the centroids:

```
>>> plt.scatter(X[:, 0], X[:, 1], c=clusters_sk)
>>> plt.scatter(centroids_sk[:, 0], centroids_sk[:, 1], marker='*',
s=200, c='r')
>>> plt.show()
```

This will result in the following output:

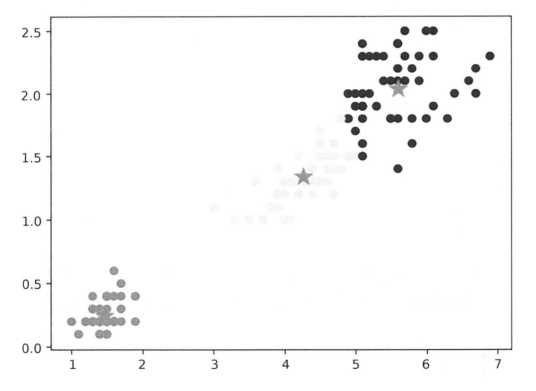

Figure 8.9: Data samples along with learned cluster centroids using scikit-learn

We get a similar result to the previous one using the model we implemented from scratch.

Choosing the value of k

Let's return to our earlier discussion on what the right value for k is. In the preceding example, it is more intuitive to set it to 3 since we know there are three classes in total. However, in most cases, we don't know how many groups are sufficient or efficient, and meanwhile, the algorithm needs a specific value of k to start with. So, how can we choose the value of k? There is a famous heuristic approach called the **elbow method**.

In the elbow method, different values of k are chosen and corresponding models are trained; for each trained model, the **sum of squared errors**, or SSE (also called the **sum of within-cluster distances**), of centroids is calculated and is plotted against k. Note that for one cluster, the squared error (or the within-cluster distance) is computed as the sum of the squared distances from individual samples in the cluster to the centroid. The optimal k is chosen where the marginal drop of SSE starts to decrease dramatically. This means that further clustering does not provide any substantial gain.

Let's apply the elbow method to the example we covered in the previous section (learning by example is what this book is all about). We perform k-means clustering under different values of k on the `iris` data:

```
>>> X = iris.data
>>> y = iris.target
>>> k_list = list(range(1, 7))
>>> sse_list = [0] * len(k_list)
```

We use the whole feature space and k ranges from 1 to 6. Then, we train individual models and record the resulting SSE, respectively:

```
>>> for k_ind, k in enumerate(k_list):
...     kmeans = KMeans(n_clusters=k, n_init='auto', random_state=42)
...     kmeans.fit(X)
...     clusters = kmeans.labels_
...     centroids = kmeans.cluster_centers_
...     sse = 0
...     for i in range(k):
...         cluster_i = np.where(clusters == i)
...         sse += np.linalg.norm(X[cluster_i] - centroids[i])
...     print(f'k={k}, SSE={sse}')
...     sse_list[k_ind] = sse
k=1, SSE=26.103076447039722
k=2, SSE=16.469773740281195
k=3, SSE=15.089477089696558
k=4, SSE=15.0307321707491
k=5, SSE=14.858930749063735
k=6, SSE=14.883090350867239
```

Finally, we plot the SSE versus the various k ranges, as follows:

```
>>> plt.plot(k_list, sse_list)
>>> plt.show()
```

This will result in the following output:

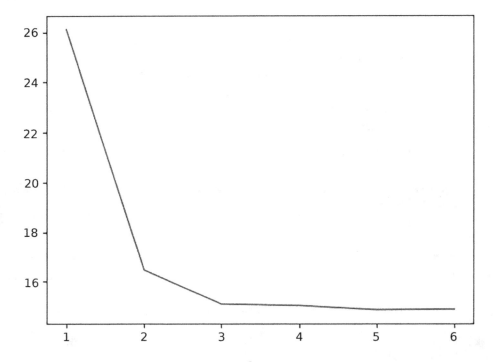

Figure 8.10: k-means elbow – SSE versus k

Best practice

Choosing the right similarity measure for distance calculation in k-means clustering depends on the nature of your data and the specific goals of your analysis. Some common similarity measures include the following:

- **Euclidean distance:** This default measure is suitable for continuous data where the difference between feature values matters.
- **Manhattan distance (also known as L1 norm):** This calculates the sum of the absolute differences between the coordinates of two points. It is suitable for high-dimensional data and when the dimensions are not directly comparable.
- **Cosine similarity:** This is useful for text data or data represented as vectors where the magnitude of the vectors is less important than the orientation.
- **Jaccard similarity:** This measures the similarity between two sets by comparing their intersection to their union. It is commonly used for binary or categorical data.

Apparently, the elbow point is k=3, since the drop in SSE slows down dramatically right after 3. Hence, k=3 is an optimal solution in this case, which is consistent with the fact that there are three classes of flowers.

Clustering newsgroups dataset

You should now be very familiar with k-means clustering. Next, let's see what we are able to mine from the newsgroups dataset using this algorithm. We will use all the data from four categories, 'alt. atheism', 'talk.religion.misc', 'comp.graphics', and 'sci.space', as an example. We will then use ChatGPT to describe the generated newsgroup clusters. ChatGPT can generate natural language descriptions of the clusters formed by k-means clustering. This can help in understanding the characteristics and themes of each cluster.

Clustering newsgroups data using k-means

We first load the data from those newsgroups and preprocess it as we did in *Chapter 7, Mining the 20 Newsgroups Dataset with Text Analysis Techniques*:

```
>>> from sklearn.datasets import fetch_20newsgroups
>>> categories = [
...     'alt.atheism',
...     'talk.religion.misc',
...     'comp.graphics',
...     'sci.space',
... ]
>>> groups = fetch_20newsgroups(subset='all',
                                    categories=categories)
>>> labels = groups.target
>>> label_names = groups.target_names
>>> from nltk.corpus import names
>>> from nltk.stem import WordNetLemmatizer
>>> all_names = set(names.words())
>>> lemmatizer = WordNetLemmatizer()
>>> def get_cleaned_data(groups, lemmatizer, remove_words):
        data_cleaned = []
        for doc in groups.data:
...         doc = doc.lower()
...         doc_cleaned = ' '.join(lemmatizer.lemmatize(word)
                                for word in doc.split()
                                if word.isalpha() and
                                word not in remove_words)
...         data_cleaned.append(doc_cleaned)
...     return data_cleaned
>>> data_cleaned = get_cleaned_data(groups, lemmatizer, all_names)
```

We then convert the cleaned text data into count vectors using `CountVectorizer` from scikit-learn:

```
>>> from sklearn.feature_extraction.text import CountVectorizer
>>> count_vector = CountVectorizer(stop_words="english",
                        max_features=None, max_df=0.5, min_df=2)
>>> data_cv = count_vector.fit_transform(data_cleaned)
```

Note that the vectorizer we use here does not limit the number of features (word tokens), but the minimum and maximum document frequency (`min_df` and `max_df`), which are 2% and 50% of the dataset, respectively. The **document frequency** of a word is measured by the fraction of documents (samples) in the dataset that contain this word. This helps filter out rare or spurious terms that may not be relevant to the analysis.

With the input data ready, we will now try to cluster them into four groups as follows:

```
>>> k = 4
>>> kmeans = KMeans(n_clusters=k, n_init='auto', random_state=42)
>>> kmeans.fit(data_cv)
```

Let's do a quick check on the sizes of the resulting clusters:

```
>>> clusters = kmeans.labels_
>>> from collections import Counter
>>> print(Counter(clusters))
Counter({3: 3360, 0: 17, 1: 7, 2: 3})
```

The clusters don't look absolutely correct, with most samples (3360 samples) congested in one big cluster (cluster 3). What could have gone wrong? It turns out that our count-based features are not sufficiently representative. A better numerical representation for text data is the **term frequency-in-verse document frequency (tf-idf)**. Instead of simply using the token count, or the so-called **term frequency (tf)**, it assigns each term frequency a weighting factor that is inversely proportional to the document frequency. In practice, the **idf** factor of a term t in documents D is calculated as follows:

$$idf(t, D) = log\frac{n_D}{1 + n_t}$$

Here, n_D is the total number of documents, n_t is the number of documents containing the term t, and *1* is added to avoid division by 0.

With the `idf` factor incorporated, the `tf-idf` representation diminishes the weight of common terms (such as *get* and *make*) and emphasizes terms that rarely occur but convey an important meaning.

To use the `tf-idf` representation, we just need to replace `CountVectorizer` with `TfidfVectorizer` from scikit-learn as follows:

```
>>> from sklearn.feature_extraction.text import TfidfVectorizer
>>> tfidf_vector = TfidfVectorizer(stop_words='english',
                        max_features=None, max_df=0.5, min_df=2)
```

The parameter max_df is used to ignore terms that have a document frequency higher than the given threshold. In this case, terms that appear in more than 50% of the documents will be ignored during the vectorization process. min_df specifies the minimum document frequency required for a term to be included in the output. Terms that appear in fewer than two documents will be ignored.

Now, redo feature extraction using the tf-idf vectorizer and the k-means clustering algorithm on the resulting feature space:

```
>>> data_tv = tfidf_vector.fit_transform(data_cleaned)
>>> kmeans.fit(data_tv)
>>> clusters = kmeans.labels_
>>> print(Counter(clusters))
Counter({1: 1478, 2: 797, 3: 601, 0: 511})
```

The clustering result becomes more reasonable.

We also take a closer look at the clusters by examining what they contain and the top 10 terms (the terms with the 10 highest tf-idf scores) representing each cluster:

```
>>> cluster_label = {i: labels[np.where(clusters == i)] for i in
                                                         range(k)}
>>> terms = tfidf_vector.get_feature_names_out()
>>> centroids = kmeans.cluster_centers_
>>> for cluster, index_list in cluster_label.items():
...     counter = Counter(cluster_label[cluster])
...     print(f'cluster_{cluster}: {len(index_list)} samples')
...     for label_index, count in sorted(counter.items(),
                             key=lambda x: x[1], reverse=True):
...         print(f'- {label_names[label_index]}: {count} samples')
...     print('Top 10 terms:')
...     for ind in centroids[cluster].argsort()[-10:]:
...         print(' %s' % terms[ind], end="")
...     print()
cluster_0: 601 samples
- sci.space: 598 samples
- alt.atheism: 1 samples
- talk.religion.misc: 1 samples
- comp.graphics: 1 samples
Top 10 terms: just orbit moon hst nasa mission launch wa shuttle space

cluster_1: 1478 samples
- alt.atheism: 522 samples
```

```
- talk.religion.misc: 387 samples
- sci.space: 338 samples
- comp.graphics: 231 samples
Top 10 terms: say people know like think ha just university wa article


cluster_2: 797 samples
- comp.graphics: 740 samples
- sci.space: 49 samples
- talk.religion.misc: 5 samples
- alt.atheism: 3 samples
Top 10 terms: computer need know looking thanks university program file graphic
image


cluster_3: 511 samples
- alt.atheism: 273 samples
- talk.religion.misc: 235 samples
- sci.space: 2 samples
- comp.graphics: 1 samples
Top 10 terms: doe bible think believe say people christian jesus wa god
```

From what we observe in the preceding results:

- cluster_0 is obviously about space and includes almost all sci.space samples and related terms such as orbit, moon, nasa, launch, shuttle, and space
- cluster_1 is more of a generic topic
- cluster_2 is more about computer graphics and related terms, such as computer, program, file, graphic, and image
- cluster_3 is an interesting one, which successfully brings together two overlapping topics, atheism and religion, with key terms including bible, believe, jesus, christian, and god

Feel free to try different values of k, or use the elbow method to find the optimal one (this is actually an exercise later in this chapter).

It is quite interesting to find key terms for each text group via clustering. It will be more fun if we can describe each cluster based on its key terms. Let's see how we do so with ChatGPT in the next section.

Describing the clusters using GPT

ChatGPT (https://chat.openai.com/) is an AI language model developed by **OpenAI** (https://openai.com/). It is part of the **Generative Pre-trained Transformer (GPT)** family of models, specifically based on GPT-3.5 (GPT-4 is in beta at the time of writing) architecture. ChatGPT is designed to engage in natural language conversations with users and provide human-like responses.

The model is trained on a vast amount of diverse text data from the internet, allowing it to understand and generate human-like text across a wide range of topics and contexts. ChatGPT can comprehend questions, prompts, and instructions given by users and generate coherent responses based on its training.

ChatGPT has been used in various applications, including chatbots, virtual assistants, content generation, language translation, and more. Users interact with ChatGPT through API calls or interactive interfaces, and the model generates responses in real time. However, it is essential to note that while ChatGPT can produce impressive and contextually relevant responses, it may also occasionally generate incorrect or nonsensical answers due to the limitations of current language models. ChatGPT responses should be sense-checked to improve the quality and reliability of the generated text and minimize the risk of misinformation.

We will ask ChatGPT to describe the clusters we just generated in the following steps.

First, we obtain the top 100 terms as follows:

```
>>> keywords = ' '.join(
                   terms[ind] for ind in centroids[0].argsort()[-100:])
>>> print(keywords)
big power vehicle using alaska look mass money marketing company loss pluto
russian scheduled office express probably research software funding billboard
online pat access doe telescope april jet usa digest light want prize forwarded
way large mar project sci center command technology air government commercial
good work servicing know going comet world propulsion people idea design data
university day international use orbital long science need time sky program
thing make spencer new year earth spacecraft flight henry billion rocket think
ha station lunar solar like cost satellite article toronto zoology just orbit
moon hst nasa mission launch wa shuttle space
```

After signing up (or logging in if you have an account) at https://chat.openai.com, we ask ChatGPT to describe the topic based on these keywords using the prompt Describe a common topic based on the following keywords:. Refer to the following screenshot for the entire question and answer:

Default (GPT-3.5)

 Describe a common topic based on the following keywords:
big power vehicle using alaska look mass money marketing company loss pluto russian
scheduled office express probably research software funding billboard online pat access
doe telescope april jet usa digest light want prize forwarded way large mar project sci center
command technology air government commercial good work servicing know going comet
world propulsion people idea design data university day international use orbital long
science need time sky program thing make spencer new year earth spacecraft flight henry
billion rocket think ha station lunar solar like cost satellite article toronto zoology just orbit
moon hst nasa mission launch wa shuttle space

 The common topic based on the provided keywords seems to be related to space
exploration and space technology, particularly focusing on various aspects such as big
power vehicles, research, funding, companies, missions, and celestial bodies like Pluto and
comet.

The topic covers a wide range of subjects, including space agencies like NASA, space
missions, spacecraft, rocket launches, orbital mechanics, celestial bodies, and government
and commercial involvement in space exploration. There are also mentions of technology,
software, and data research associated with space science.

The use of terms like "big power vehicle," "mass money marketing company," "loss,"
"scheduled office," "express," "billboard online," "access," "telescope," "jet," "government,"
"sci center," "technology air," and "commercial good work servicing" indicates a discussion of
the involvement of various organizations and industries in space-related projects and
research.

Some specific events or missions might be mentioned, such as "Pluto Russian," "April jet
USA," "doe telescope," "forwarded way large mar project," and "comet world propulsion."
These could refer to specific missions or projects related to celestial bodies and space
propulsion.

There's also a mention of "Spencer new year" which could be related to an individual's name
or possibly a particular event in space exploration.

Overall, the common topic revolves around the various aspects of space exploration,
research, technology, and missions, with mentions of key players and celestial bodies in the
field.

Figure 8.11: Asking ChatGPT to describe the topic of cluster 0

As ChatGPT pointed out correctly, "the common topic revolves around the various aspects of space exploration, research, technology, and missions, with mentions of key players and celestial bodies in the field." Feel free to repeat the same process for other clusters. You can also achieve the same using the ChatGPT API in Python by following these steps:

1. Install the OpenAI library with `pip`:

```
pip install openai
```

You can also do this with `conda`:

```
conda install openai
```

2. Generate an API key at `https://platform.openai.com/account/api-keys`. Note that you will need to log in or sign up to do this.

3. Import the library and set your API key:

```
>>> import openai
>>> openai.api_key = '<YOUR API KEY>'
```

4. Create a function that allows you to obtain a response from ChatGPT:

```
>>> def get_completion(prompt, model="text-davinci-003"):
        messages = [{"role": "user", "content": prompt}]
        response = openai.ChatCompletion.create(
            model=model,
            messages=messages,
            temperature=0
        )
        return response.choices[0].message["content"]
```

Here, we use the `text-davinci-003` model. Check out the page at `https://platform.openai.com/docs/models` for more information on the various models available.

5. Query the API:

```
>>> response = get_completion(f"Describe a common topic based on the
        following keywords: {keywords}")
>>> print(response)
```

This will yield a response similar to what you previously read in the web interface. Note that the API call is subject to your plan quota.

Up to this point, we have produced topical keywords by first grouping documents into clusters and subsequently extracting the top terms within each cluster. **Topic modeling** is another approach to produce topical keywords but in a much more direct way. It does not simply search for the key terms in individual clusters generated beforehand. What it does is directly extract collections of key terms from documents. You will see how this works in the next section.

Density-Based Spatial Clustering of Applications with Noise (DBSCAN) is another popular clustering algorithm used for identifying clusters in spatial data. Unlike centroid-based algorithms like k-means, DBSCAN does not require specifying the number of clusters in advance and can discover clusters of arbitrary shapes. It works by partitioning the dataset into clusters of contiguous high-density regions, separated by regions of low density, while also identifying outliers as noise.

The algorithm requires two parameters: epsilon (ε), which defines the maximum distance between two samples for them to be considered as part of the same neighborhood, and `min_samples`, which specifies the minimum number of samples required to form a dense region.

DBSCAN starts by randomly selecting a point and expanding its neighborhood to find all reachable points within ε distance. If the number of reachable points exceeds `min_samples`, the point is labeled as a core point and a new cluster is formed. The process is repeated recursively for all core points and their neighborhoods until all points are assigned to a cluster or labeled as noise.

Discovering underlying topics in newsgroups

A **topic model** is a type of statistical model for discovering the probability distributions of words linked to the topic. The topic in topic modeling does not exactly match the dictionary definition but corresponds to a nebulous statistical concept, which is an abstraction that occurs in a collection of documents.

When we read a document, we expect certain words appearing in the title or the body of the text to capture the semantic context of the document. An article about Python programming might have words such as *class* and *function*, while a story about snakes might have words such as *eggs* and *afraid*. Documents usually have multiple topics; for instance, this section is about three things: topic modeling, non-negative matrix factorization, and latent Dirichlet allocation, which we will discuss shortly. We can therefore define an additive model for topics by assigning different weights to topics.

Topic modeling is widely used for mining hidden semantic structures in given text data. There are two popular topic modeling algorithms—**non-negative matrix factorization** (NMF) and **latent Dirichlet allocation** (LDA). We will go through both of these in the next two sections.

Topic modeling using NMF

Non-negative matrix factorization (NMF) is a dimensionality reduction technique used for feature extraction and data representation. It factorizes a non-negative input matrix, V, into a product of two smaller matrices, W and H, in such a way that these three matrices have no negative values. These two lower-dimensional matrices represent features and their associated coefficients. In the context of NLP, these three matrices have the following meanings:

- The input matrix **V** is the term count or tf-idf matrix of size $n * m$, where n is the number of documents or samples, and m is the number of terms.
- The first decomposition output matrix **W** is the feature matrix of size $t * m$, where t is the number of topics specified. Each row of **W** represents a topic with each element in the row representing the rank of a term in the topic.
- The second decomposition output matrix **H** is the coefficient matrix of size $n * t$. Each row of **H** represents a document, with each element in the row representing the weight of a topic within the document.

How to derive the computation of **W** and **H** is beyond the scope of this book. However, you can refer to the following example to get a better sense of how NMF works:

Input matrix V

	Term1	Term2	Term3	Term4	Term5	Term6
Document1	4	2	0	0	3	1
Document2	0	1	1	0	2	0
Document3	1	0	1	4	0	2
Document4	2	0	0	0	0	1

Feature matrix W

	Term1	Term2	Term3	Term4	Term5	Term6
Topic1	0.2	0	0.5	0	0	0
Topic2	0	1	0	0	0.5	0
Topic3	0	0	1	0	0	0.5

Coefficient matrix H

	Topic1	Topic2	Topic3
Document1	1	0	0
Document2	0	0.5	0.5
Document3	0.2	0	0.8
Document4	0	1	0

Figure 8.12: Example of matrix W and matrix H derived from an input matrix V

 If you are interested in reading more about NMF, feel free to check out the original paper *Generalized Nonnegative Matrix Approximations with Bregman Divergences*, by Inderjit S. Dhillon and Suvrit Sra, in NIPS 2005.

Let's now apply NMF to our newsgroups data. Scikit-learn has a nice module for decomposition that includes NMF:

```
>>> from sklearn.decomposition import NMF
>>> t = 20
>>> nmf = NMF(n_components=t, random_state=42)
```

We specify 20 topics (n_components) as an example. Important parameters of the model are included in the following table:

Constructor parameter	Default value	Example values	Description
n_components	None	5, 10, 20	Number of components—in the context of topic modeling, this corresponds to the number of topics. If None, it becomes the number of input features.
max_iter	200	100, 200	Maximum number of iterations.
tol	1e-4	1e-5, 1e-8	Tolerance to declare convergence.

Table 8.2: Parameters of the NMF class

We used the term matrix as input to the NMF model, but you could also use the tf-idf one instead. Now, fit the NMF model, nmf, on the term matrix, data_cv:

```
>>> nmf.fit(data_cv)
```

We can obtain the resulting topic feature rank W after the model is trained:

```
>>> print(nmf.components_)
[[0.00000000e+00 0.00000000e+00 0.00000000e+00 ... 0.00000000e+00
  0.00000000e+00 1.82524532e-04]
 [0.00000000e+00 0.00000000e+00 0.00000000e+00 ... 0.00000000e+00
  7.77697392e-04 3.85995474e-03]
 [0.00000000e+00 0.00000000e+00 0.00000000e+00 ... 0.00000000e+00
  0.00000000e+00 0.00000000e+00]
 ...
 [0.00000000e+00 0.00000000e+00 0.00000000e+00 ... 2.71332203e-02
  0.00000000e+00 0.00000000e+00]
 [0.00000000e+00 0.00000000e+00 0.00000000e+00 ... 0.00000000e+00
  0.00000000e+00 4.31048632e-05]
 [0.00000000e+00 0.00000000e+00 0.00000000e+00 ... 0.00000000e+00
  0.00000000e+00 0.00000000e+00]]
```

For each topic, we display the top 10 terms based on their ranks:

```
>>> terms_cv = count_vector.get_feature_names_out()
>>> for topic_idx, topic in enumerate(nmf.components_):
...         print("Topic {}:" .format(topic_idx))
...         print(" ".join([terms_cv[i] for i in topic.argsort()[-10:]]))
Topic 0:
available quality program free color version gif file image jpeg
Topic 1:
```

ha article make know doe say like just people think
Topic 2:
include available analysis user software ha processing data tool image
Topic 3:
atmosphere kilometer surface ha earth wa planet moon spacecraft solar
Topic 4:
communication technology venture service market ha commercial space satellite
launch
Topic 5:
verse wa jesus father mormon shall unto mcconkie lord god
Topic 6:
format message server object image mail file ray send graphic
Topic 7:
christian people doe atheism believe religion belief religious god atheist
Topic 8:
file graphic grass program ha package ftp available image data
Topic 9:
speed material unified star larson book universe theory physicist physical
Topic 10:
planetary station program group astronaut center mission shuttle nasa space
Topic 11:
infrared high astronomical center acronym observatory satellite national
telescope space
Topic 12:
used occurs true form ha ad premise conclusion argument fallacy
Topic 13:
gospel people day psalm prophecy christian ha matthew wa jesus
Topic 14:
doe word hanging say greek matthew mr act wa juda
Topic 15:
siggraph graphic file information format isbn data image ftp available
Topic 16:
venera mar lunar surface space venus soviet mission wa probe
Topic 17:
april book like year time people new did article wa
Topic 18:
site retrieve ftp software data information client database gopher search
Topic 19:
use look xv color make program correction bit gamma image

There are a number of interesting topics, for instance:

- Computer graphics-related topics, such as 0, 2, 6, and 8
- Space-related ones, such as 3, 4, and 9
- Religion-related ones, such as 5, 7, and 13

Some topics, such as 1 and 12, are hard to interpret. This is totally fine since topic modeling is a kind of free-form learning.

Topic modeling using LDA

Let's explore another popular topic modeling algorithm, **Latent Dirichlet Allocation (LDA)**. LDA is a generative probabilistic graphical model that explains each input document by means of a mixture of topics with certain probabilities. It assumes that each document is a mixture of multiple topics, and each topic is characterized by a specific word probability distribution. The algorithm iteratively assigns words in documents to topics and updates the topic distributions based on the observed word co-occurrences. Again, **topic** in topic modeling means a collection of words with a certain connection. In other words, LDA basically deals with two probability values, $P(term \lor topic)$ and $P(topic \lor document)$. This can be difficult to understand at the beginning. So, let's start from the bottom, the end result of an LDA model.

Let's take a look at the following set of documents:

```
Document 1: This restaurant is famous for fish and chips.
Document 2: I had fish and rice for lunch.
Document 3: My sister bought me a cute kitten.
Document 4: Some research shows eating too much rice is bad.
Document 5: I always forget to feed fish to my cat.
```

Now, let's say we want two topics. The topics derived from these documents may appear as follows:

```
Topic 1: 30% fish, 20% chip, 30% rice, 10% lunch, 10% restaurant (which we can
interpret Topic 1 to be food related)
Topic 2: 40% cute, 40% cat, 10% fish, 10% feed (which we can interpret Topic 1
to be about pet)
```

Therefore, we find how each document is represented by these two topics:

```
Document 1: 85% Topic 1, 15% Topic 2
Document 2: 88% Topic 1, 12% Topic 2
Document 3: 100% Topic 2
Document 4: 100% Topic 1
Document 5: 33% Topic 1, 67% Topic 2
```

After seeing a toy example, we come back to its learning procedure:

1. Specify the number of topics, *T*. Now we have topics 1, 2, ..., and *T*.
2. For each document, randomly assign one of the topics to each term in the document.
3. For each document, calculate *P(topic = t* V *document)*, which is the proportion of terms in the document that are assigned to the topic *t*.
4. For each topic, calculate *P(term = w* V *topic)*, which is the proportion of term *w* among all terms that are assigned to the topic.
5. For each term *w*, reassign its topic based on the latest probabilities *P(topic = t* V *document)* and *P(term = w* V *topic = t)*.
6. Repeat *steps 3* to *5* under the latest topic distributions for each iteration. The training stops if the model converges or reaches the maximum number of iterations.

LDA is trained in a generative manner, where it tries to abstract from the documents a set of hidden topics that are likely to generate a certain collection of words.

With all this in mind, let's see LDA in action. The LDA model is also included in scikit-learn:

```
>>> from sklearn.decomposition import LatentDirichletAllocation
>>> t = 20
>>> lda = LatentDirichletAllocation(n_components=t,
                    learning_method='batch',random_state=42)
```

Again, we specify 20 topics (`n_components`). The key parameters of the model are included in the following table:

Constructor parameter	Default value	Example values	Description
n_components	10	5, 10, 20	Number of components—in the context of topic modeling, this corresponds to the number of topics.
learning_method	"batch"	"online", "batch"	In batch mode, all training data is used for each update. In online mode, a mini-batch of training data is used for each update. In general, if the data size is large, the online mode is faster.
max_iter	10	10, 20	Maximum number of iterations.
randome_state	None	0, 42	Seed used by the random number generator.

Table 8.3: Parameters of the LatentDirichletAllocation class

For the input data to LDA, remember that LDA only takes in term counts as it is a probabilistic graphical model. This is unlike NMF, which can work with both the term count matrix and the tf-idf matrix as long as they are non-negative data. Again, we use the term matrix defined previously as input to the LDA model. Now, we fit the LDA model on the term matrix, data_cv:

```
>>> lda.fit(data_cv)
```

We can obtain the resulting topic term rank after the model is trained:

```
>>> print(lda.components_)
[[0.05       2.05       2.05       ...  0.05       0.05       0.05 ]
 [0.05       0.05       0.05       ...  0.05       0.05       0.05 ]
 [0.05       0.05       0.05       ...  4.0336285  0.05       0.05 ]
 ...
 [0.05       0.05       0.05       ...  0.05       0.05       0.05 ]
 [0.05       0.05       0.05       ...  0.05       0.05       0.05 ]
 [0.05       0.05       0.05       ...  0.05       0.05       3.05 ]]
```

Similarly, for each topic, we display the top 10 terms based on their ranks as follows:

```
>>> for topic_idx, topic in enumerate(lda.components_):
...         print("Topic {}:" .format(topic_idx))
...         print(" ".join([terms_cv[i] for i in
                               topic.argsort()[-10:]]))
Topic 0:
atheist doe ha believe say jesus people christian wa god
Topic 1:
moment just adobe want know ha wa hacker article radius
Topic 2:
center point ha wa available research computer data graphic hst
Topic 3:
objective argument just thing doe people wa think say article
Topic 4:
time like brian ha good life want know just wa
Topic 5:
computer graphic think know need university just article wa like
Topic 6:
free program color doe use version gif jpeg file image
Topic 7:
gamma ray did know university ha just like article wa
Topic 8:
tool ha processing using data software color program bit image
```

```
Topic 9:
apr men know ha think woman just university article wa
Topic 10:
jpl propulsion mission april mar jet command data spacecraft wa
Topic 11:
russian like ha university redesign point option article space station
Topic 12:
ha van book star material physicist universe physical theory wa
Topic 13:
bank doe book law wa article rushdie muslim islam islamic
Topic 14:
think gopher routine point polygon book university article know wa
Topic 15:
ha rocket new lunar mission satellite shuttle nasa launch space
Topic 16:
want right article ha make like just think people wa
Topic 17:
just light space henry wa like zoology sky article toronto
Topic 18:
comet venus solar moon orbit planet earth probe ha wa
Topic 19:
site format image mail program available ftp send file graphic
```

There are a number of interesting topics that we just mined, for instance:

- Computer graphics-related topics, such as 2, 5, 6, 8, and 19
- Space-related ones, such as 10, 11, 12, and 15
- Religion-related ones, such as 0 and 13

There are also topics involving noise, for example, 9 and 16, which may require some imagination to interpret. Once more, this observation is entirely expected, given that LDA or topic modeling, as mentioned before, falls under the category of free-form learning.

Summary

The project in this chapter was about finding hidden similarities underneath newsgroups data, be it semantic groups, themes, or word clouds. We started with what unsupervised learning does and the typical types of unsupervised learning algorithms. We then introduced unsupervised learning clustering and studied a popular clustering algorithm, k-means, in detail. We also explored using ChatGPT to describe the topics of individual clusters based on their keywords.

We also talked about tf-idf as a more efficient feature extraction tool for text data. After that, we performed k-means clustering on the newsgroups data and obtained four meaningful clusters. After examining the key terms in each resulting cluster, we went straight to extracting representative terms among original documents using topic modeling techniques. Two powerful topic modeling approaches, NMF and LDA, were discussed and implemented. Finally, we had some fun interpreting the topics we obtained from both methods.

So far, we have covered all the main categories of unsupervised learning, including dimensionality reduction, clustering, and topic modeling, which is also dimensionality reduction in a way.

In the next chapter, we will talk about **Support Vector Machines (SVMs)** for face recognition. SVM is a popular choice for a wide range of classification and regression tasks, especially when dealing with complex decision boundaries. We will also cover another dimensionality reduction technique called principal component analysis.

Exercises

1. Ask ChatGPT to describe other clusters we generated through k-means clustering. You may experiment with various prompts and discover intriguing information within these clusters.
2. Perform k-means clustering on newsgroups data using different values of k, or use the elbow method to find the optimal one. See if you get better grouping results.
3. Try different numbers of topics in either NMF or LDA and see which one produces more meaningful topics in the end. This should be a fun exercise.
4. Can you experiment with NMF or LDA on the entire 20 groups of newsgroups data? Are the resulting topics full of noise or gems?

Recognizing Faces with Support Vector Machine

In the previous chapter, we discovered underlying topics using clustering and topic modeling techniques. This chapter continues our journey of supervised learning and classification, with a particular emphasis on **Support Vector Machine (SVM)** classifiers.

SVM is one of the most popular algorithms when it comes to high-dimensional spaces. The goal of the algorithm is to find a decision boundary in order to separate data from different classes. We will discuss in detail how that works. Also, we will implement the algorithm with scikit-learn and apply it to solve various real-life problems, including our main project of face recognition. A dimensionality reduction technique called **principal component analysis**, which boosts the performance of the image classifier, will also be covered in this chapter, as will support vector regression.

This chapter explores the following topics:

- Finding the separating boundary with SVM
- Classifying face images with SVM
- Estimating with support vector regression

Finding the separating boundary with SVM

SVM is another great classifier, which is effective in cases with high-dimensional spaces or where the number of dimensions is greater than the number of samples.

In machine learning classification, SVM finds an optimal hyperplane that best segregates observations from different classes.

A **hyperplane** is a plane of $n - 1$ dimensions that separates the n-dimensional feature space of the observations into two spaces. For example, the hyperplane in a two-dimensional feature space is a line, and in a three-dimensional feature space, the hyperplane is a surface. The optimal hyperplane is picked so that the distance from its nearest points in each space to itself is maximized, and these nearest points are the so-called **support vectors**.

The following toy example demonstrates what support vectors and a separating hyperplane (along with the distance margin, which I will explain later) look like in a binary classification case:

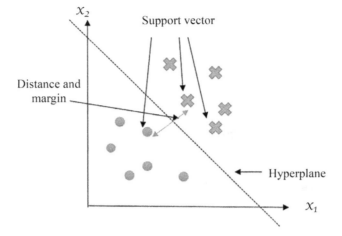

Figure 9.1: Example of support vectors and a hyperplane in binary classification

The ultimate goal of SVM is to find an optimal hyperplane, but the burning question is "How can we find this optimal hyperplane?" You will get the answer as we explore the following scenarios. It's not as hard as you may think. The first thing we will look at is how to find a hyperplane.

Scenario 1 — identifying a separating hyperplane

First, you need to understand what qualifies as a separating hyperplane. In the following example, hyperplane C is the only correct one, as it successfully segregates observations by their labels, while hyperplanes A and B fail:

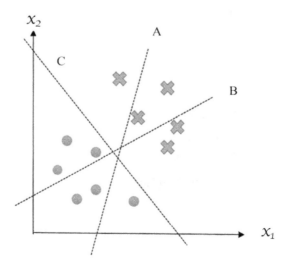

Figure 9.2: Example of qualified and unqualified hyperplanes

This is an easy observation. Let's express a separating hyperplane in a formal or mathematical way next.

In a two-dimensional space, a line can be defined by a slope vector w (represented as a two-dimensional vector), and an intercept b. Similarly, in a space of n dimensions, a hyperplane can be defined by an n-dimensional vector w and an intercept b. Any data point x on the hyperplane satisfies $wx + b = 0$. A hyperplane is a separating hyperplane if the following conditions are satisfied:

- For any data point x from one class, it satisfies $wx + b > 0$
- For any data point x from another class, it satisfies $wx + b < 0$

However, there can be countless possible solutions for w and b. You can move or rotate hyperplane C to a certain extent, and it will still remain a separating hyperplane. Next, you will learn how to identify the best hyperplane among various possible separating hyperplanes.

Scenario 2 – determining the optimal hyperplane

Look at the following example: hyperplane C is preferred, as it enables the maximum sum of the distance between the nearest data point on the positive side and itself, and the distance between the nearest data point on the negative side and itself:

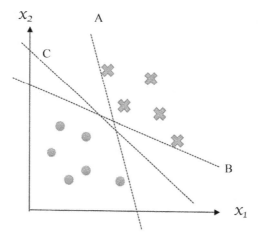

Figure 9.3: An example of optimal and suboptimal hyperplanes

The nearest point(s) on the positive side can constitute a hyperplane parallel to the decision hyperplane, which we call a **positive hyperplane**; conversely, the nearest point(s) on the negative side can constitute the **negative hyperplane**. The perpendicular distance between the positive and negative hyperplanes is called the **margin**, the value of which equates to the sum of the two aforementioned distances. A **decision** hyperplane is deemed **optimal** if the margin is maximized.

The optimal (also called **maximum-margin**) hyperplane and the distance margins for a trained SVM model are illustrated in the following diagram. Again, samples on the margin (two from one class and one from another class, as shown) are the so-called **support vectors**:

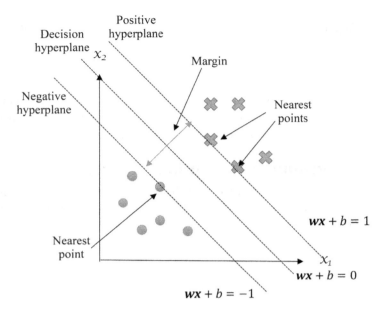

Figure 9.4: An example of an optimal hyperplane and distance margins

We can interpret it mathematically by first describing the positive and negative hyperplanes, as follows:

$$wx^{(p)} + b = 1$$

$$wx^{(n)} + b = -1$$

Here, $x^{(p)}$ is a data point on the positive hyperplane, and $x^{(n)}$ is a data point on the negative hyperplane. The distance between a point $x^{(p)}$ and the decision hyperplane can be calculated as follows:

$$\frac{wx^{(p)} + b}{\|w\|} = \frac{1}{\|w\|}$$

Similarly, the distance between a point $x^{(n)}$ and the decision hyperplane is as follows:

$$\frac{wx^{(n)} + b}{\|w\|} = \frac{1}{\|w\|}$$

So the margin becomes $\frac{2}{\|w\|}$. As a result, we need to minimize $\|w\|$ in order to maximize the margin. Importantly, to comply with the fact that the support vectors on the positive and negative hyperplanes are the nearest data points to the decision hyperplane, we add a condition that no data point falls between the positive and negative hyperplanes:

$$wx^{(i)} + b \geq 1, \quad if \ y^{(i)} = 1$$

$$wx^{(i)} + b \leq 1, \quad if \ y^{(i)} = -1$$

Here, $(x^{(i)}, y^{(i)})$ is an observation. This can be combined further into the following:

$$y^{(i)}(wx^{(i)} + b) \geq 1$$

To summarize, w and b, which determine the SVM decision hyperplane, are trained and solved by the following optimization problem:

- Minimizing $\|w\|$
- Subject to $y^{(i)}(wx^{(i)} + b) \geq 1$, for a training set of $(x^{(1)}, y^{(1)})$, $(x^{(2)}, y^{(2)})$,... $(x^{(i)}, y^{(i)})$..., and $(x^{(m)}, y^{(m)})$

To solve this optimization problem, we need to resort to quadratic programming techniques, which are beyond the scope of our learning journey. Therefore, we will not cover the computation methods in detail and, instead, will implement the classifier using the SVC and LinearSVC modules from scikit-learn, which are respectively based on libsvm (https://www.csie.ntu.edu.tw/~cjlin/libsvm/) and liblinear (https://www.csie.ntu.edu.tw/~cjlin/liblinear/), two popular open-source SVM machine learning libraries. However, it is always valuable to understand the concepts of computing SVM.

 Pegasos: Primal estimated sub-gradient solver for SVM (*Mathematical Programming*, March 2011, volume 127, issue 1, pp. 3–30) by Shai Shalev-Shwartz et al. and *A dual coordinate descent method for large-scale linear SVM* (*Proceedings of the 25th international conference on machine learning*, pp 408–415) by Cho-Jui Hsieh et al. are great learning materials. They cover two modern approaches, sub-gradient descent and coordinate descent.

The learned model parameters w and b are then used to classify a new sample x', based on the following conditions:

$$y' = \begin{cases} 1, & if \ wx' + b > 0 \\ -1, & if \ wx' + b < 0 \end{cases}$$

Moreover, $\|wx' + b\|$ can be portrayed as the distance from the data point x' to the decision hyperplane, and it can also be interpreted as the confidence of prediction: the higher the value, the further away the data point is from the decision boundary, hence the higher prediction certainty.

Although you might be eager to implement the SVM algorithm, let's take a step back and look at a common scenario where data points are not linearly separable in a strict way. Try to find a separating hyperplane in the following example:

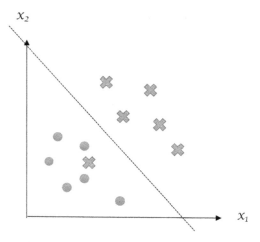

Figure 9.5: An example of data points that are not strictly linearly separable

How can we deal with cases where it is impossible to strictly linearly segregate a set of observations containing outliers? Let's see in the next section.

Scenario 3 – handling outliers

To handle scenarios where we cannot linearly segregate a set of observations containing outliers, we can actually allow the misclassification of outliers and try to minimize the error introduced. The misclassification error $\zeta^{(i)}$ (also called **hinge loss**) for a sample $x^{(i)}$ can be expressed as follows:

$$\zeta^{(i)} \begin{cases} 1 - y^{(i)}(wx^{(i)} + b), & if\ misclassified \\ 0 & , \qquad otherwise \end{cases}$$

Together with the ultimate term $||w||$ that we want to reduce, the final objective value we want to minimize becomes the following:

$$||w|| + C \frac{\sum_{i=1}^{m} \zeta^{(i)}}{m}$$

As regards a training set of m samples $(x^{(1)}, y^{(1)})$, $(x^{(2)}, y^{(2)})$,... $(x^{(i)}, y^{(i)})$..., and $(x^{(m)}, y^{(m)})$, where the hyperparameter C controls the trade-off between the two terms, the following apply:

- If a large value of C is chosen, the penalty for misclassification becomes relatively high. This means the rule of thumb of data segregation becomes stricter and the model might be prone to overfitting, since few mistakes are allowed during training. An SVM model with a large C has a low bias, but it might suffer from high variance.

- Conversely, if the value of C is sufficiently small, the influence of misclassification becomes fairly low. This model allows more misclassified data points than a model with a large C. Thus, data separation becomes less strict. Such a model has low variance, but it might be compromised by high bias.

A comparison between a large and small C is shown in the following diagram:

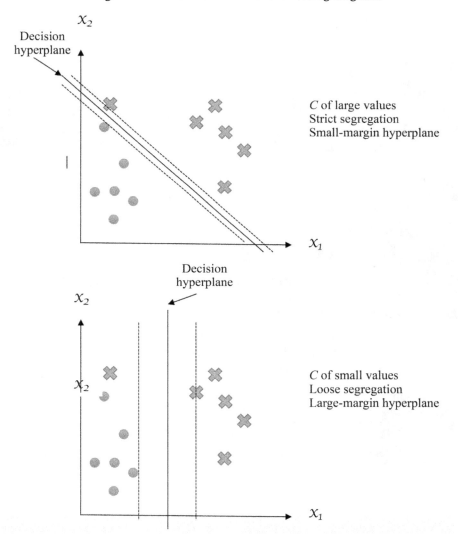

Figure 9.6: How the value of C affects the strictness of segregation and the margin

The parameter C determines the balance between bias and variance. It can be fine-tuned with cross-validation, which we will practice shortly.

Implementing SVM

We have largely covered the fundamentals of the SVM classifier. Now, let's apply it right away to an easy binary classification dataset. We will use the classic breast cancer Wisconsin dataset (https://scikit-learn.org/stable/modules/generated/sklearn.datasets.load_breast_cancer.html) from scikit-learn.

Let's take a look at the following steps:

1. We first load the dataset and do some basic analysis, as follows:

```
>>> from sklearn.datasets import load_breast_cancer
>>> cancer_data = load_breast_cancer()
>>> X = cancer_data.data
>>> Y = cancer_data.target
>>> print('Input data size :', X.shape)
Input data size : (569, 30)
>>> print('Output data size :', Y.shape)
Output data size : (569,)
>>> print('Label names:', cancer_data.target_names)
Label names: ['malignant' 'benign']
>>> n_pos = (Y == 1).sum()
>>> n_neg = (Y == 0).sum()
>>> print(f'{n_pos} positive samples and {n_neg} negative samples.')
357 positive samples and 212 negative samples.
```

As you can see, the dataset has 569 samples with 30 features; its label is binary, and 63% of samples are positive (benign). Again, always check whether classes are imbalanced before trying to solve any classification problem. In this case, they are relatively balanced.

2. Next, we split the data into training and testing sets:

```
>>> from sklearn.model_selection import train_test_split
>>> X_train, X_test, Y_train, Y_test = train_test_split(X, Y, random_
state=42)
```

For reproducibility, don't forget to specify a random seed.

3. We can now apply the SVM classifier to the data. We first initialize an `SVC` model with the kernel parameter set to `linear` (linear kernel refers to the use of a linear decision boundary to separate classes in the input space. I will explain what kernel means in *Scenario 5*) and the penalty hyperparameter `C` set to the default value, `1.0`:

```
>>> from sklearn.svm import SVC
>>> clf = SVC(kernel='linear', C=1.0, random_state=42)
```

4. We then fit our model on the training set, as follows:

```
>>> clf.fit(X_train, Y_train)
```

5. Then, we predict on the testing set with the trained model and obtain the prediction accuracy directly:

```
>>> accuracy = clf.score(X_test, Y_test)
>>> print(f'The accuracy is: {accuracy*100:.1f}%')
The accuracy is: 95.8%
```

Our first SVM model works just great, achieving an accuracy of 95.8%. How about dealing with more than two topics? How does SVM handle multiclass classification?

Scenario 4 – dealing with more than two classes

SVM and many other classifiers can be applied to cases with more than two classes. There are two typical approaches we can take, **one-vs-rest** (also called **one-vs-all**) and **one-vs-one**.

One-vs-rest

In the one-vs-rest setting, for a K-class problem, we construct K different binary SVM classifiers. For the kth classifier, it treats the kth class as the positive case and the remaining K-1 classes as the negative case as a whole; the hyperplane denoted as (w_k, b_k) is trained to separate these two cases. To predict the class of a new sample, x', it compares the resulting predictions $w_k x' + b_k$ from K individual classifiers from 1 to k. As we discussed in the previous section, the larger value of $w_k x' + b_k$ means higher confidence that x' belongs to the positive case. Therefore, it assigns x' to the class i where $w_i x' + b_i$ has the largest value among all prediction results:

$$y' = argmax_i(w_i x' + b_i)$$

The following diagram shows how the one-vs-rest strategy works in a three-class case:

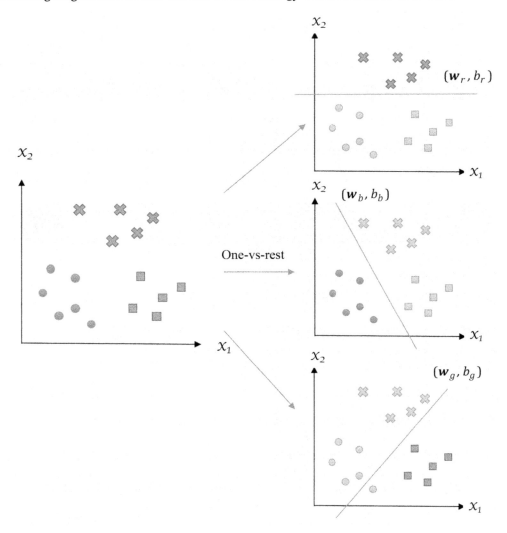

Figure 9.7: An example of three-class classification using the one-vs-rest strategy

For instance, if we have the following (r, b, and g denote the red cross, blue dot, and green square classes, respectively):

$$w_r x'+b_r = 0.78$$

$$w_b x'+b_b = 0.35$$

$$w_g x'+b_g = -0.64$$

we can say x' belongs to the red cross class, since *0.78 > 0.35 > -0.64*.

If we have the following:

$$w_r x'+b_r = -0.78$$

$$w_b x'+b_b = -0.35$$

$$w_g x'+b_g = -0.64$$

then we can determine that x' belongs to the blue dot class regardless of the sign, since *-0.35 > -0.64 > -0.78*.

One-vs-one

In the one-vs-one strategy, we conduct a pairwise comparison by building a set of SVM classifiers that can distinguish data points from each pair of classes. This will result in $\frac{K(K-1)}{2}$ different classifiers.

For a classifier associated with classes i and j, the hyperplane denoted as (w_{ij}, b_{ij}) is trained only on the basis of observations from i (can be viewed as a positive case) and j (can be viewed as a negative case); it then assigns the class, either i or j, to a new sample, x', based on the sign of $w_{ij} x'+b_{ij}$. Finally, the class with the highest number of assignments is considered the predicted result of x'. The winner is the class that gets the most votes.

The following diagram shows how the one-vs-one strategy works in a three-class case:

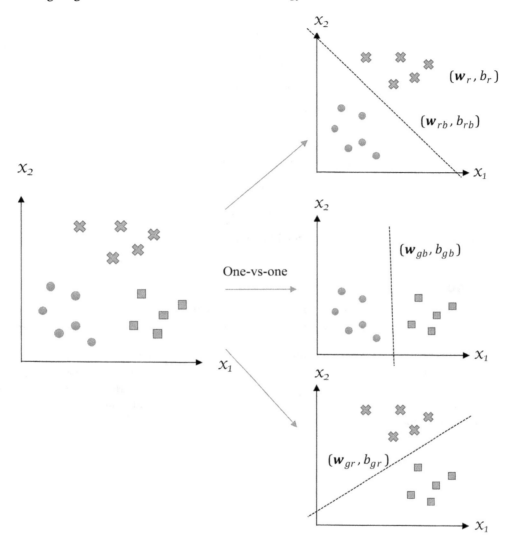

Figure 9.8: An example of three-class classification using the one-vs-one strategy

In general, an SVM classifier with a one-vs-rest setting and a classifier with a one-vs-one setting perform comparably in terms of accuracy. The choice between these two strategies is largely computational.

Although one-vs-one requires more classifiers, $\frac{K(K-1)}{2}$, than one-vs-rest (K), each pairwise classifier only needs to learn on a small subset of data, as opposed to the entire set in the one-vs-rest setting. As a result, training an SVM model in the one-vs-one setting is generally more memory-efficient and less computationally expensive; hence, it is preferable for practical use, as argued in Chih-Wei Hsu and Chih-Jen Lin's *A comparison of methods for multiclass support vector machines* (*IEEE Transactions on Neural Networks*, March 2002, Volume 13, pp. 415–425).

Multiclass cases in scikit-learn

In scikit-learn, classifiers handle multiclass cases internally, and we do not need to explicitly write any additional code to enable this. You can see how simple it is in the wine classification example (https://scikit-learn.org/stable/modules/generated/sklearn.datasets.load_wine.html#sklearn.datasets.load_wine) with three classes, as follows:

1. We first load the dataset and do some basic analysis, as follows:

```
>>> from sklearn.datasets import load_wine
>>> wine_data = load_wine()
>>> X = wine_data.data
>>> Y = wine_data.target
>>> print('Input data size :', X.shape)
Input data size : (178, 13)
>>> print('Output data size :', Y.shape)
Output data size : (178,)
>>> print('Label names:', wine_data.target_names)
Label names: ['class_0' 'class_1' 'class_2']
>>> n_class0 = (Y == 0).sum()
>>> n_class1 = (Y == 1).sum()
>>> n_class2 = (Y == 2).sum()
>>> print(f'{n_class0} class0 samples,\n{n_class1} class1 samples,\n{n_
class2} class2 samples.')
59 class0 samples,
71 class1 samples,
48 class2 samples.
```

As you can see, the dataset has 178 samples with 13 features; its label has three possible values taking up 33%, 40%, and 27%, respectively.

2. Next, we split the data into training and testing sets:

```
>>> X_train, X_test, Y_train, Y_test = train_test_split(X, Y, random_
state=42)
```

3. We can now apply the SVM classifier to the data. We first initialize an SVC model and fit it against the training set:

```
>>> clf = SVC(kernel='linear', C=1.0, random_state=42)
>>> clf.fit(X_train, Y_train)
```

In an SVC model, multiclass support is implicitly handled according to the one-vs-one scheme.

4. Next, we predict on the testing set with the trained model and obtain the prediction accuracy directly:

```
>>> accuracy = clf.score(X_test, Y_test)
>>> print(f'The accuracy is: {accuracy*100:.1f}%')
The accuracy is: 97.8%
```

Our SVM model also works well in this multiclass case, achieving an accuracy of 97.8%.

5. We also check how it performs for individual classes:

```
>>> from sklearn.metrics import classification_report
>>> pred = clf.predict(X_test)
>>> print(classification_report(Y_test, pred))
              precision    recall  f1-score   support
           0       1.00      1.00      1.00        15
           1       1.00      0.94      0.97        18
           2       0.92      1.00      0.96        12
    accuracy                           0.98        45
   macro avg       0.97      0.98      0.98        45
weighted avg       0.98      0.98      0.98        45
```

It looks excellent! Is the example too easy? Maybe. What do we do in tricky cases? Of course, we could tweak the values of the kernel and C hyperparameters. As discussed, factor C controls the strictness of separation, and it can be tuned to achieve the best trade-off between bias and variance. How about the kernel? What does it mean and what are the alternatives to a linear kernel?

In the next section, we will answer those two questions we just raised. You will see how the kernel trick makes SVM so powerful.

Scenario 5 — solving linearly non-separable problems with kernels

The hyperplanes we have found so far are linear, for instance, a line in a two-dimensional feature space, or a surface in a three-dimensional one. However, in the following example, we are not able to find a linear hyperplane that can separate two classes:

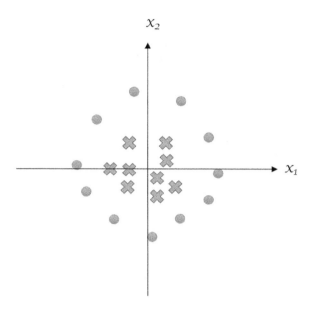

Figure 9.9: The linearly non-separable case

Intuitively, we observe that data points from one class are closer to the origin than those from another class. The distance to the origin provides distinguishable information. So we add a new feature, $z=(x_1^2+x_2^2)^2$, and transform the original two-dimensional space into a three-dimensional one. In the new space, as displayed in the following diagram, we can find a surface hyperplane separating the data (see the bottom left graph in *Figure 9.10*), or a line in the two-dimensional view (see the bottom right graph in *Figure 9.10*). With the additional feature, the dataset becomes linearly separable in the higher dimensional space, (x_1, x_2, z):

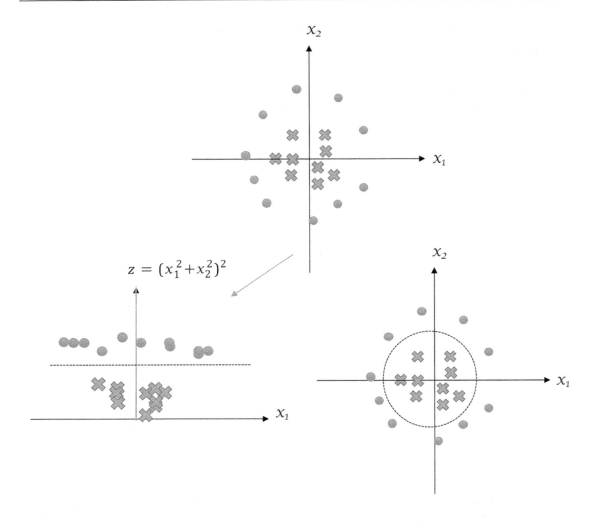

Figure 9.10: Making a non-separable case separable

Based upon similar logic, **SVMs with kernels** were invented to solve non-linear classification problems by converting the original feature space, $x^{(i)}$, to a higher dimensional feature space with a transformation function, Φ, such that the transformed dataset $\Phi(x^{(i)})$ is linearly separable.

A linear hyperplane (w_ϕ, b_ϕ) is then learned, using observations $(\phi(x^{(i)}), y^{(i)})$. For an unknown sample x', it is first transformed into $\phi(x')$; the predicted class is determined by $w_\phi x' + b_\phi$.

An SVM with kernels enables non-linear separation, but it does not explicitly map each original data point to the high-dimensional space and then perform expensive computation in the new space. Instead, it approaches this in a tricky way.

During the course of solving the SVM optimization problems, feature vectors $x^{(1)}, x^{(2)},...,x^{(m)}$ are involved only in the form of a pairwise dot product $x^{(i)} x^{(j)}$, although we will not expand this mathematically in this book. With kernels, the new feature vectors are $\phi(x^{(1)}), \phi(x^{(2)}), ..., \phi(x^{(m)})$, and their pairwise dot products can be expressed as $\phi(x^{(i)}) . \phi(x^{(j)})$. It would be computationally efficient to first implicitly conduct a pairwise operation on two low-dimensional vectors and later map the result to the high-dimensional space. In fact, a function K that satisfies this does exist:

$$K(x^{(i)}, x^{(j)}) = \phi(x^{(i)}) . \phi(x^{(j)})$$

The function K is the so-called **kernel function**. It is the mathematical formula that does the transformation.

 There are different types of kernels, each suited for different kinds of data.

With the kernel function, the transformation Φ becomes implicit, and the non-linear decision boundary can be efficiently learned by simply replacing the term $\phi(x^{(i)}) . \phi(x^{(j)})$ with $K(x^{(i)}, x^{(j)})$.

The data is transformed into a higher-dimensional space. You don't actually need to compute this space explicitly; the kernel function just works with the original data and performs the calculations necessary for the SVM.

The most popular kernel function is probably the **Radial Basis Function (RBF)** kernel (also called the **Gaussian** kernel), which is defined as follows:

$$K(x^{(i)}, x^{(j)}) = exp\left(-\frac{\left\|x^{(i)} - x^{(j)}\right\|}{2\sigma^2}\right) = exp\left(-\gamma\left\|x^{(i)} - x^{(j)}\right\|^2\right)$$

Here, $\gamma = \frac{1}{2\sigma^2}$ In the Gaussian function, the standard deviation σ controls the amount of variation or dispersion allowed: the higher the σ (or the lower the γ), the larger the width of the bell, and the wider the range in which data points are allowed to spread out over. Therefore, γ as the **kernel coefficient** determines how strictly or generally the kernel function fits the observations. A large γ implies a small variance allowed and a relatively exact fit on the training samples, which might lead to overfitting.

Conversely, a small γ implies a high variance allowed and a loose fit on the training samples, which might cause underfitting.

To illustrate this trade-off, let's apply the RBF kernel with different values to a toy dataset:

```
>>> import numpy as np
>>> import matplotlib.pyplot as plt
>>> X = np.c_[# negative class
...           (.3, -.8),
...           (-1.5, -1),
```

```
...              (-1.3, -.8),
...              (-1.1, -1.3),
...              (-1.2, -.3),
...              (-1.3, -.5),
...              (-.6, 1.1),
...              (-1.4, 2.2),
...              (1, 1),
...              # positive class
...              (1.3, .8),
...              (1.2, .5),
...              (.2, -2),
...              (.5, -2.4),
...              (.2, -2.3),
...              (0, -2.7),
...              (1.3, 2.1)].T
>>> Y = [-1] * 8 + [1] * 8
```

Eight data points are from one class, and eight are from another. We take three values, 1, 2, and 4, for kernel coefficient options as an example:

```
>>> gamma_option = [1, 2, 4]
```

Under each kernel coefficient, we fit an individual SVM classifier and visualize the trained decision boundary:

```
>>> for i, gamma in enumerate(gamma_option, 1):
...     svm = SVC(kernel='rbf', gamma=gamma)
...     svm.fit(X, Y)
...     plt.scatter(X[:, 0], X[:, 1], c=['b']*8+['r']*8, zorder=10)
...     plt.axis('tight')
...     XX, YY = np.mgrid[-3:3:200j, -3:3:200j]
...     Z = svm.decision_function(np.c_[XX.ravel(), YY.ravel()])
...     Z = Z.reshape(XX.shape)
...     plt.pcolormesh(XX, YY, Z > 0 , cmap=plt.cm.Paired)
...     plt.contour(XX, YY, Z, colors=['k', 'k', 'k'],
...             linestyles=['--', '-', '--'], levels=[-.5, 0, .5])
...     plt.title('gamma = %d' % gamma)
...     plt.show()
```

Refer to the following screenshot for the end results:

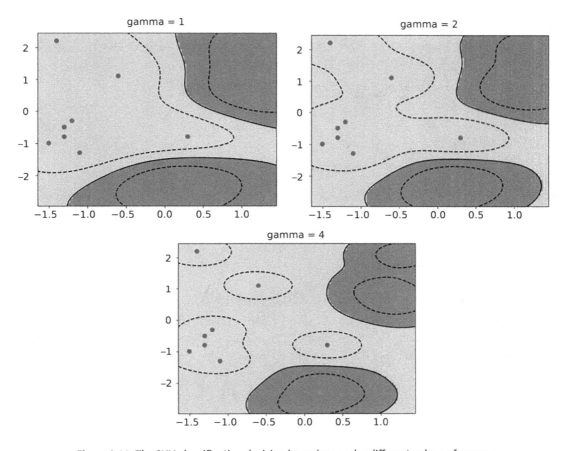

Figure 9.11: The SVM classification decision boundary under different values of gamma

🔍 **Quick tip:** Need to see a high-resolution version of this image? Open this book in the next-gen Packt Reader or view it in the PDF/ePub copy.

🔒 **The next-gen Packt Reader** and a **free PDF/ePub copy** of this book are included with your purchase. Unlock them by scanning the QR code below or visiting https://www.packtpub.com/unlock/9781835085622.

We can observe that a larger γ results in narrow regions, which means a stricter fit on the dataset; a smaller γ results in broad regions, which means a loose fit on the dataset. Of course, γ can be fine-tuned through cross-validation to obtain the best performance.

Some other common kernel functions include the **polynomial** kernel:

$$K(x^{(i)}, x^{(j)}) = (x^{(i)}.x^{(j)} + \gamma)^d$$

and the **sigmoid** kernel:

$$K(x^{(i)}, x^{(j)}) = tanh(x^{(i)}.x^{(j)} + \gamma)$$

In the absence of prior knowledge of the distribution, the RBF kernel is usually preferable in practical usage, as there is an additional parameter to tweak in the polynomial kernel (polynomial degree d), and the empirical sigmoid kernel can perform approximately on par with the RBF, but only under certain parameters. Hence, we arrive at a debate between the linear (also considered no kernel) and the RBF kernel when given a dataset.

Choosing between linear and RBF kernels

Of course, linear separability is the rule of thumb when choosing the right kernel to start with due to its simplicity and efficiency. However, most of the time, this is very difficult to identify, unless you have sufficient prior knowledge of the dataset, or its features are of low dimensions (1 to 3).

 Some general prior knowledge that is commonly known is that text data is often linearly separable, while data generated from the XOR function (https://en.wikipedia.org/wiki/XOR_gate) is not.

Now, let's look at the following three scenarios where the linear kernel is favored over RBF.

Scenario 1: Both the number of features and the number of instances are large (more than 104 or 105). Since the dimension of the feature space is high enough, additional features as a result of RBF transformation will not provide a performance improvement, but this will increase the computational expense. Some examples from the UCI machine learning repository (a collection of databases, and data generators widely used for empirical analysis of ML algorithms) are of this type:

- **URL Reputation Dataset:** https://archive.ics.uci.edu/ml/datasets/URL+Reputation (the number of instances: 2,396,130; the number of features: 3,231,961). This is designed for malicious URL detection based on their lexical and host information.

- **YouTube Multiview Video Games Dataset:** https://archive.ics.uci.edu/ml/datasets/YouTube+Multiview+Video+Games+Dataset (the number of instances: 120,000; the number of features: 1,000,000). This is designed for topic classification.

Scenario 2: The number of features is noticeably large compared to the number of training samples. Apart from the reasons stated in *scenario 1*, the RBF kernel is significantly more prone to overfitting. Such a scenario occurs, for example, in the following examples:

- **Dorothea Dataset:** `https://archive.ics.uci.edu/ml/datasets/Dorothea` (the number of instances: 1,950; the number of features: 100,000). This is designed for drug discovery that classifies chemical compounds as active or inactive, according to their structural molecular features.

- **Arcene Dataset:** `https://archive.ics.uci.edu/ml/datasets/Arcene` (the number of instances: 900; the number of features: 10,000). This represents a mass spectrometry dataset for cancer detection.

Scenario 3: The number of instances is significantly large compared to the number of features. For a dataset of low dimensions, the RBF kernel will, in general, boost the performance by mapping it to a higher-dimensional space. However, due to the training complexity, it usually becomes inefficient on a training set with more than 106 or 107 samples. Example datasets include the following:

- *Heterogeneity Activity Recognition Dataset*: `https://archive.ics.uci.edu/ml/datasets/Het erogeneity+Activity+Recognition` (the number of instances: 43,930,257; the number of features: 16). This is designed for human activity recognition.

- *HIGGS Dataset*: `https://archive.ics.uci.edu/ml/datasets/HIGGS` (the number of instances: 11,000,000; the number of features: 28). This is designed to distinguish between a signal process producing Higgs bosons or a background process.

Aside from these three scenarios, you can consider experimenting with RBF kernels.

The rules for choosing between linear and RBF kernels can be summarized as follows:

Scenario	Linear	RBF
Prior knowledge	If linearly separable	If nonlinearly separable
Visualizable data of 1 to 3 dimension(s)	If linearly separable	If nonlinearly separable
Both the number of features and number of instances are large.	First choice	
Features >> Instances	First choice	
Instances >> Features	First choice	
Others		First choice

Table 9.1: Rules for choosing between linear and RBF kernels

Once again, **first choice** means we can **begin with** this option; it does not mean that this is the only option moving forward.

Next, let's take a look at classifying face images.

Classifying face images with SVM

Finally, it is time to build an SVM-based face image classifier using everything you just learned. We will do so in parts, exploring the image dataset first.

Exploring the face image dataset

We will use the **Labeled Faces in the Wild (LFW)** people dataset (https://scikit-learn.org/stable/modules/generated/sklearn.datasets.fetch_lfw_people.html) from scikit-learn. It consists of more than 13,000 curated face images of more than 5,000 famous people. Each class has various numbers of image samples.

First, we load the face image data as follows:

```
>>> from sklearn.datasets import fetch_lfw_people
Downloading LFW metadata: https://ndownloader.figshare.com/files/5976012
Downloading LFW metadata: https://ndownloader.figshare.com/files/5976009
Downloading LFW metadata: https://ndownloader.figshare.com/files/5976006
Downloading LFW data (~200MB): https://ndownloader.figshare.com/files/5976015
>>> face_data = fetch_lfw_people(min_faces_per_person=80)
```

We only load classes with at least 80 samples so that we will have enough training data. Note that if you run into the problem of `ImportError: The Python Imaging Library (PIL) is required to load data from jpeg files`, please install the `pillow` package in the terminal, as follows:

```
pip install pillow
```

If you encounter an `urlopen` error, you can download the four data files manually from the following URLs:

- `pairsDevTrain.txt`: https://ndownloader.figshare.com/files/5976012
- `pairsDevTest.txt`: https://ndownloader.figshare.com/files/5976009
- `pairs.txt`: https://ndownloader.figshare.com/files/5976006
- `lfw-funneled.tgz`: https://ndownloader.figshare.com/files/5976015

You can then place them in a designated folder, for example, the current path, `./`. Accordingly, the code to load image data becomes the following:

```
>>> face_data = fetch_lfw_people(data_home='./',
                                 min_faces_per_person=80,
                                 download_if_missing=False )
```

Next, we take a look at the data we just loaded:

```
>>> X = face_data.data
>>> Y = face_data.target
>>> print('Input data size :', X.shape)
Input data size : (1140, 2914)
>>> print('Output data size :', Y.shape)
Output data size : (1140,)
>>> print('Label names:', face_data.target_names)
Label names: ['Colin Powell' 'Donald Rumsfeld' 'George W Bush' 'Gerhard
Schroeder' 'Tony Blair']
```

This five-class dataset contains $1,140$ samples and a sample is of $2,914$ dimensions. As a good practice, we analyze the label distribution as follows:

```
>>> for i in range(5):
...     print(f'Class {i} has {(Y == i).sum()} samples.')
Class 0 has 236 samples.
Class 1 has 121 samples.
Class 2 has 530 samples.
Class 3 has 109 samples.
Class 4 has 144 samples.
```

The dataset is rather imbalanced. Let's keep this in mind when we build the model.

Now, let's plot a few face images:

```
>>> fig, ax = plt.subplots(3, 4)
>>> for i, axi in enumerate(ax.flat):
...     axi.imshow(face_data.images[i], cmap='bone')
...     axi.set(xticks=[], yticks=[],
...             xlabel=face_data.target_names[face_data.target[i]])
...
>>> plt.show()
```

You will see the following 12 images with their labels:

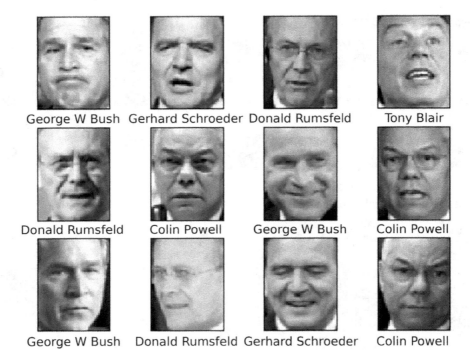

Figure 9.12: Samples from the LFW people dataset

Now that we have covered exploratory data analysis, we will move on to the model development phase in the next section.

Building an SVM-based image classifier

First, we split the data into the training and testing set:

```
>>> X_train, X_test, Y_train, Y_test = train_test_split(X, Y,
                                           random_state=42)
```

In this project, the number of dimensions is greater than the number of samples. This is a classification case that SVM is effective at solving. In our solution, we will tune the hyperparameters, including the penalty C, the kernel (linear or RBF), and γ (for the RBF kernel) through cross-validation.

We then initialize a common SVM model:

```
>>> clf = SVC(class_weight='balanced', random_state=42)
```

The dataset is imbalanced, so we set `class_weight='balanced'` to emphasize the underrepresented classes.

We utilize the GridSearchCV module from scikit-learn to search for the best combination of hyperparameters over several candidates. We will explore the following hyperparameter candidates:

```
>>> parameters = {'C': [10, 100, 300],
...               'gamma': [0.0001,  0.0003, 0.001],
...               'kernel' : ['rbf', 'linear'] }
>>> from sklearn.model_selection import GridSearchCV
>>> grid_search = GridSearchCV(clf, parameters, n_jobs=-1, cv=5)
```

 If you are unsure about the suitable value of gamma to start with for RBF kernel, opting for 1 divided by the feature dimension is consistently a reliable choice. So in this example, 1/2914 = 0.0003.

The `GridSearchCV` model we just initialized will conduct five-fold cross-validation (`cv=5`) and will run in parallel on all available cores (`n_jobs=-1`). We then perform hyperparameter tuning by simply applying the `fit` method:

```
>>> grid_search.fit(X_train, Y_train)
```

We obtain the optimal set of hyperparameters using the following code:

```
>>> print('The best model:\n', grid_search.best_params_)
The best model:
 {'C': 300, 'gamma': 0.001, 'kernel': 'rbf'}
```

Then, we obtain the best five-fold averaged performance under the optimal set of parameters by using the following code:

```
>>> print('The best averaged performance:', grid_search.best_score_)
 The best averaged performance: 0.8456140350877192
```

We then retrieve the SVM model with the optimal set of hyperparameters and apply it to the testing set:

```
>>> clf_best = grid_search.best_estimator_
>>> pred = clf_best.predict(X_test)
```

We then calculate the accuracy and classification report:

```
>>> print(f'The accuracy is: {clf_best.score(X_test,
...        Y_test)*100:.1f}%')
The accuracy is: 89.8%
```

```
>>> from sklearn.metrics import classification_report
>>> print(classification_report(Y_test, pred,
...           target_names=face_data.target_names))
                   precision    recall  f1-score   support
     Colin Powell       0.90      0.88      0.89        64
  Donald Rumsfeld       0.90      0.84      0.87        32
    George W Bush       0.89      0.94      0.92       127
Gerhard Schroeder       0.90      0.90      0.90        29
       Tony Blair       0.90      0.85      0.88        33
         Accuracy                           0.90       285
        macro avg       0.90      0.88      0.89       285
     weighted avg       0.90      0.90      0.90       285
```

It should be noted that we tune the model based on the original training set, which is divided into folds for cross-training and validation internally, and that we apply the optimal model to the original testing set. We examine the classification performance in this manner to measure how well generalized the model is, in order to make correct predictions on a completely new dataset. An accuracy of 89.8% is achieved with the best SVM model.

There is another SVM classifier, LinearSVC (https://scikit-learn.org/stable/modules/generated/sklearn.svm.LinearSVC.html), from scikit-learn. How is it different from SVC? LinearSVC is similar to SVC with linear kernels, but it is implemented based on the liblinear library, which is better optimized than libsvm with the linear kernel, and its penalty function is more flexible.

In general, training with the LinearSVC model is faster than SVC. This is because the liblinear library with high scalability is designed for large datasets, while the libsvm library, with more than quadratic computation complexity, is not able to scale well with more than 10^5 training instances. But again, the LinearSVC model is limited to only linear kernels.

Boosting image classification performance with PCA

We can also improve the image classifier by compressing the input features with **Principal Component Analysis (PCA)** (*A Tutorial on Principal Component Analysis* by Jonathon Shlens). This reduces the dimension of the original feature space and preserves the most important internal relationships among features. In simple terms, PCA projects the original data into a smaller space with the most important directions (coordinates). We hope that in cases where we have more features than training samples, considering fewer features as a result of dimensionality reduction using PCA can prevent overfitting.

Here's how PCA works:

1. **Data Standardization:** Before applying PCA, it is essential to standardize the data by subtracting the mean and dividing it by the standard deviation for each feature. This step ensures that all features are on the same scale and prevents any single feature from dominating the analysis.

2. **Covariance Matrix Calculation:** PCA calculates the covariance matrix of the standardized data. The covariance matrix shows how each pair of features varies together. The diagonal elements of the covariance matrix represent the variance of individual features, while the off-diagonal elements represent the covariance between pairs of features.

3. **Eigendecomposition:** The next step is to perform eigendecomposition on the covariance matrix. Eigendecomposition breaks down the covariance matrix into its eigenvectors and eigenvalues. The eigenvectors represent the principal components, and the corresponding eigenvalues indicate the amount of variance explained by each principal component.

4. **Selecting Principal Components:** The principal components are sorted based on their corresponding eigenvalues in descending order. The first principal component (PC1) explains the highest variance, followed by PC2, PC3, and so on. Typically, you select a subset of principal components that explain a significant portion (e.g., 95% or more) of the total variance.

5. **Projection:** Finally, the data is projected onto the selected principal components to create a lower-dimensional representation of the original data. This lower-dimensional representation captures most of the variance in the data while reducing the number of features.

You can read more about PCA at `https://www.kaggle.com/nirajvermafcb/principal-component-analysis-explained` if you are interested. We will implement PCA with the PCA module (`https://scikit-learn.org/stable/modules/generated/sklearn.decomposition.PCA.html`) from scikit-learn. We will first apply PCA to reduce the dimensionality and train the classifier on the resulting data.

In machine learning, we usually concatenate multiple consecutive steps and treat them as one "model." We call this process **pipelining**. We utilize the `pipeline` API (`https://scikit-learn.org/stable/modules/generated/sklearn.pipeline.Pipeline.html`) from scikit-learn to facilitate this.

Now, let's initialize a PCA model, an SVC model, and a model pipelining these two:

```
>>> from sklearn.decomposition import PCA
>>> pca = PCA(n_components=100, whiten=True, random_state=42)
>>> svc = SVC(class_weight='balanced', kernel='rbf',
...           random_state=42)
>>> from sklearn.pipeline import Pipeline
>>> model = Pipeline([('pca', pca),
...                    ('svc', svc)])
```

The PCA component projects the original data into a 100-dimension space, followed by the SVC classifier with the RBF kernel. We then perform a grid search for the best model among a few options:

```
>>> parameters_pipeline = {'svc__C': [1, 3, 10],
...                         'svc__gamma': [0.01,  0.03, 0.003]}
>>> grid_search = GridSearchCV(model, parameters_pipeline ,
                               n_jobs=-1, cv=5)
>>> grid_search.fit(X_train, Y_train)
```

Best practice

Choosing the initial values of hyperparameters like C and gamma in grid search for SVMs is crucial for efficiently finding optimal values. Here are some best practices:

- **Start with a coarse grid:** Begin with a coarse grid that covers a wide range of values for C and gamma. This allows you to quickly explore the hyperparameter space and identify promising regions.

- **Consider specific knowledge:** Incorporate any prior knowledge or domain expertise about the problem into the selection of initial values. For example, if you know that the dataset is noisy or has outliers, you may want to prioritize larger values of C to allow for more flexibility in the decision boundary.

- **Use cross-validation:** This helps to assess how well the initial values generalize to unseen data and guides the refinement of the grid search.

- **Iteratively refine the grid:** Based on the results of initial cross-validation, iteratively refine the grid around regions that show promising performance. Narrow down the range of values for C and gamma to focus the search on areas where the optimal values are likely to lie.

Finally, we print out the best set of hyperparameters and the classification performance with the best model:

```
>>> print('The best model:\n', grid_search.best_params_)
The best model:
 {'svc__C': 1, 'svc__gamma': 0.01}
>>> print('The best averaged performance:', grid_search.best_score_)
The best averaged performance: 0.8619883040935671
>>> model_best = grid_search.best_estimator_
>>> print(f'The accuracy is: {model_best.score(X_test, Y_test)*100:.1f}%')
The accuracy is: 92.3%
>>> pred = model_best.predict(X_test)
>>> print(classification_report(Y_test, pred,
                          target_names=face_data.target_names))
                   precision    recall  f1-score   support

      Colin Powell      0.94      0.94      0.94        64
   Donald Rumsfeld      0.93      0.84      0.89        32
     George W Bush      0.91      0.97      0.94       127
  Gerhard Schroeder      0.92      0.79      0.85        29
        Tony Blair      0.94      0.91      0.92        33

          accuracy                          0.92       285
         macro avg      0.93      0.89      0.91       285
      weighted avg      0.92      0.92      0.92       285
```

The model composed of a PCA and an SVM classifier achieves an accuracy of 92.3%. PCA boosts the performance of the SVM-based image classifier.

Following the successful application of SVM in image classification, we will look at its application in regression.

Estimating with support vector regression

As the name implies, SVR is part of the support vector family and a sibling of the **Support Vector Machine (SVM)** for classification (or we can just call it SVC).

To recap, SVC seeks an optimal hyperplane that best segregates observations from different classes. In SVR, our goal is to find a decision hyperplane (defined by a slope vector w and intercept b) so that two hyperplanes $wx + b = -\varepsilon$ (negative hyperplane) and $wx + b = \varepsilon$ (positive hyperplane) can cover the ε bands of the optimal hyperplane. Simultaneously, the optimal hyperplane is as flat as possible, which means w is as small as possible, as shown in the following diagram:

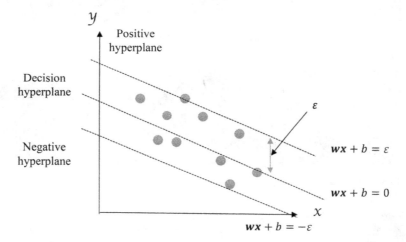

Figure 9.13: Finding the decision hyperplane in SVR

This translates into deriving the optimal w and b by solving the following optimization problem:

- Minimizing $||w||$
- Subject to $|y^{(i)} - (wx^{(i)} + b)| \le \varepsilon$, given a training set of $(x^{(1)}, y^{(1)}), (x^{(2)}, y^{(2)}), \dots (x^{(i)}, y^{(i)})$ ..., $(x^{(m)}, y^{(m)})$

The theory behind SVR is very similar to SVM. In the next section, let's see the implementation of SVR.

Implementing SVR

Again, to solve the preceding optimization problem, we need to resort to quadratic programming techniques, which are beyond the scope of our learning journey. Therefore, we won't cover the computation methods in detail and will implement the regression algorithm using the SVR package (https://scikit-learn.org/stable/modules/generated/sklearn.svm.SVR.html) from scikit-learn.

Important techniques used in SVM, such as penalty as a trade-off between bias and variance, and the kernel (RBF, for example) handling linear non-separation, are transferable to SVR. The SVR package from scikit-learn also supports these techniques.

Let's solve the previous diabetes prediction problem with SVR this time, as we did in *Chapter 5, Predicting Stock Prices with Regression Algorithms*:

1. Initially, we load the dataset and check the data size, as follows:

```
>>> from sklearn import datasets
>>> diabetes = datasets.load_diabetes()
>>> X = diabetes.data
>>> Y = diabetes.target
>>> print('Input data size :', X.shape)
Input data size : (442, 10)
>>> print('Output data size :', Y.shape)
Output data size : (442,)
```

2. Next, we designate the last 30 samples as the testing set, while the remaining samples serve as the training set:

```
>>> num_test = 30
>>> X_train = diabetes.data[:-num_test, :]
>>> y_train = diabetes.target[:-num_test]
>>> X_test = diabetes.data[-num_test:, :]
>>> y_test = diabetes.target[-num_test:]
```

3. We can now apply the SVR regressor to the data. We first initialize an SVC model and fit it against the training set:

```
>>> from sklearn.svm import SVR
>>> regressor = SVR(C=100, kernel='linear')
>>> regressor.fit(X_train, y_train)
```

Here, we start with a linear kernel.

4. We predict on the testing set with the trained model and obtain the prediction performance:

```
>>> from sklearn.metrics import r2_score
>>> predictions = regressor.predict(X_test)
>>> print(r2_score(y_test, predictions))
0.5868189735154503
```

With this simple model, we are able to achieve an R^2 of 0.59.

5. Let's further improve it with a grid search to find the best model from the following options:

```
>>> parameters = {'C': [300, 500, 700],
                  'gamma': [0.3, 0.6, 1],
                  'kernel' : ['rbf', 'linear']}
>>> regressor = SVR()
>>> grid_search = GridSearchCV(regressor, parameters, n_jobs=-1,
                               cv=5)
>>> grid_search.fit(X_train, y_train)
```

6. After searching over 18 sets of hyperparameters, we find the best model with the following combination of hyperparameters:

```
>>> print('The best model:\n', grid_search.best_params_)
The best model: {'C': 300, 'gamma': 1.5, 'kernel': 'rbf'}
```

7. Finally, we use the best model to make predictions and evaluate its performance:

```
>>> model_best = grid_search.best_estimator_
>>> predictions = model_best.predict(X_test)
>>> print(r2_score(Y_test, predictions))
```

We are able to boost the R^2 score to 0.68 after fine-tuning.

Unlike SVM for classification, where the goal is to separate data into distinct classes, SVR focuses on finding a function that best fits the data, by minimizing the prediction error while allowing some tolerance.

Summary

In this chapter, we continued our journey of supervised learning with SVM. You learned about the mechanics of an SVM, kernel techniques, implementations of SVM, and other important concepts of machine learning classification, including multiclass classification strategies and grid search, as well as useful tips to use an SVM (for example, choosing between kernels and tuning parameters). Then, we finally put into practice what you learned in the form of real-world use cases, including face recognition. You also learned about SVM's extension to regression, SVR.

In the next chapter, we will review what you have learned so far in this book and examine the best practices of real-world machine learning. The chapter aims to make your learning foolproof and get you ready for the entire machine learning workflow and productionization. This will be a wrap-up of the general machine learning techniques before we move on to more complex topics in the final three chapters.

Exercises

1. Can you implement SVM using the `LinearSVC` module? What are the hyperparameters that you need to tweak, and what is the best performance of face recognition you can achieve?

2. Can you classify more classes in the image recognition project? As an example, you can set `min_faces_per_person=50`. What is the best performance you can achieve using grid search and cross-validation?

3. Explore stock price prediction using SVR. You can reuse the dataset and feature generation functions from *Chapter 5, Predicting Stock Prices with Regression Algorithms*.

Join our book's Discord space

Join our community's Discord space for discussions with the authors and other readers:

`https://packt.link/yuxi`

10

Machine Learning Best Practices

After working on multiple projects covering important machine learning concepts, techniques, and widely used algorithms, you have a broad picture of the machine learning ecosystem, as well as solid experience in tackling practical problems using machine learning algorithms and Python. However, there will be issues once we start working on projects from scratch in the real world. This chapter aims to get us ready for it with 21 best practices to follow throughout the entire machine learning solution workflow.

We will cover the following topics in this chapter:

- Machine learning solution workflow
- Best practices in the data preparation stage
- Best practices in the training set generation stage
- Best practices in the model training, evaluation, and selection stage
- Best practices in the deployment and monitoring stage

Machine learning solution workflow

In general, the main tasks involved in solving a machine learning problem can be summarized into four areas, as follows:

- Data preparation
- Training set generation
- Model training, evaluation, and selection
- Deployment and monitoring

Starting from data sources and ending with the final machine learning system, a machine learning solution basically follows the paradigm shown here:

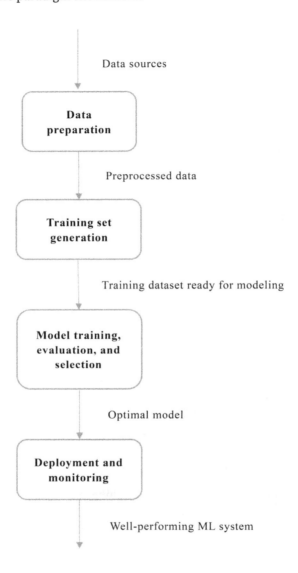

Figure 10.1: The life cycle of a machine learning solution

In the following sections, we will learn about the typical tasks, common challenges, and best practices for each of these four stages.

Best practices in the data preparation stage

No machine learning system can be built without data. Therefore, **data collection** should be our first focus.

Best practice 1 — Completely understanding the project goal

Before starting to collect data, we should make sure that the goal of the project and the business problem are completely understood, as this will guide us on what data sources to look into, and where sufficient domain knowledge and expertise is also required. For example, in a previous chapter, *Chapter 5, Predicting Stock Prices with Regression Algorithms*, our goal was to predict the future prices of the stock index, so we first collected data on its past performance, instead of the past performance of an irrelevant European stock. In *Chapter 3, Predicting Online Ad Click-Through with Tree-Based Algorithms*, for example, the business problem was to optimize advertising, targeting efficiency measured by click-through rate, so we collected the clickstream data of who clicked or did not click on what ad on what page, instead of merely using how many ads were displayed in a web domain.

Best practice 2 — Collecting all fields that are relevant

With a set goal in mind, we can narrow down potential data sources to investigate. Now the question becomes: is it necessary to collect the data of all fields available in a data source, or is a subset of attributes enough? It would be perfect if we knew in advance which attributes were key indicators or key predictive factors. However, it is in fact very difficult to ensure that the attributes hand-picked by a domain expert will yield the best prediction results. Hence, for each data source, it is recommended to collect all of the fields that are related to the project, especially in cases where recollecting the data is time-consuming, or even impossible.

For example, in the stock price prediction example, we collected the data of all fields, including **Open**, **High**, **Low**, and **Volume**, even though we were initially not certain of how useful **high** and **low** predictions would be. Retrieving the stock data is quick and easy, however. In another example, if we ever want to collect data ourselves by scraping online articles for topic classification, we should store as much information as possible. Otherwise, if any piece of information is not collected but is later found to be valuable, such as hyperlinks in an article, the article might already have been removed from the web page; if it still exists, rescraping those pages can be costly.

After collecting the datasets that we think are useful, we need to ensure the data quality by inspecting its **consistency** and **completeness**. Consistency refers to how the distribution of data changes over time. Completeness means how much data is present across fields and samples. They are explained in detail in the following two practices.

Best practice 3 — Maintaining the consistency and normalization of field values

In a dataset that already exists, or in one that we collect from scratch, we often see different values representing the same meaning. For example, we see *American*, *US*, and *U.S.A* in the Country field, and *male* and *M* in the Gender field. It is necessary to unify or standardize the values in a field, otherwise, it will mess up the algorithms in later stages as different feature values will be treated differently even if they have the same meaning. For example, we keep only the three options *M*, *F*, and *gender-diverse* in the Gender field, and replace other alternative values. It is also a great practice to keep track of what values are mapped to the default value of a field.

In addition, the format of values in the same field should also be consistent. For instance, in the *age* field, there could be true age values, such as *21* and *35*, and incorrect age values, such as *1990* and *1978*; in the *rating* field, both cardinal numbers and English numerals could be found, such as *1*, *2*, and *3*, and *one*, *two*, and *three*. Transformation and reformatting should be conducted in order to ensure data consistency.

Best practice 4 – Dealing with missing data

Due to various reasons, datasets in the real world are rarely completely clean and often contain missing or corrupted values. They are usually presented as blanks, *Null, -1, 999999, unknown*, or any other placeholder. Samples with missing data not only provide incomplete predictive information but also confuse the machine learning model as it cannot tell whether *-1* or *unknown* holds a meaning. It is important to pinpoint and deal with missing data in order to avoid jeopardizing the performance of models in the later stages.

Here are three basic strategies that we can use to tackle the missing data issue:

- Discarding samples containing any missing values.
- Discarding fields containing missing values in any sample.
- Inferring the missing values based on the known part of the attribute. This process is called **missing data imputation**. Typical imputation methods include replacing missing values with the mean or median value of the field across all samples, or the most frequent value for categorical data.

The first two strategies are simple to implement; however, they come at the expense of the data lost, especially when the original dataset is not large enough. The third strategy doesn't abandon any data but does try to fill in the blanks.

Let's look at how each strategy is applied in an example where we have a dataset (age, income) consisting of six samples – (30, 100), (20, 50), (35, *unknown*), (25, 80), (30, 70), and (40, 60):

- If we process this dataset using the first strategy, it becomes (30, 100), (20, 50), (25, 80), (30, 70), and (40, 60).
- If we employ the second strategy, the dataset becomes (30), (20), (35), (25), (30), and (40), where only the first field remains.
- If we decide to complete the unknown value instead of skipping it, the sample (35, *unknown*) can be transformed into (35, 72) with the mean of the rest of the values in the second field, or (35, 70), with the median value in the second field.

In scikit-learn, the SimpleImputer class (https://scikit-learn.org/stable/modules/generated/sklearn.impute.SimpleImputer.html) provides a nicely written imputation transformer. We can use it for the following small example:

```
>>> import numpy as np
>>> from sklearn.impute import SimpleImputer
```

Represent the unknown value with `np.nan` in numpy, as detailed in the following:

```
>>> data_origin = [[30, 100],
...                 [20, 50],
...                 [35, np.nan],
...                 [25, 80],
...                 [30, 70],
...                 [40, 60]]
```

Initialize the imputation transformer with the mean value and obtain the mean value from the original data:

```
>>> imp_mean = SimpleImputer(missing_values=np.nan, strategy='mean')
>>> imp_mean.fit(data_origin)
```

Complete the missing value as follows:

```
>>> data_mean_imp = imp_mean.transform(data_origin)
>>> print(data_mean_imp)
[[ 30. 100.]
 [ 20.  50.]
 [ 35.  72.]
 [ 25.  80.]
 [ 30.  70.]
 [ 40.  60.]]
```

Similarly, initialize the imputation transformer with the median value, as detailed in the following:

```
>>> imp_median = SimpleImputer(missing_values=np.nan, strategy='median')
>>> imp_median.fit(data_origin)
>>> data_median_imp = imp_median.transform(data_origin)
>>> print(data_median_imp)
[[ 30. 100.]
 [ 20.  50.]
 [ 35.  70.]
 [ 25.  80.]
 [ 30.  70.]
 [ 40.  60.]]
```

When new samples come in, the missing values (in any attribute) can be imputed using the trained transformer, for example, with the mean value, as shown here:

```
>>> new = [[20, np.nan],
...        [30, np.nan],
...        [np.nan, 70],
...        [np.nan, np.nan]]
```

```
>>> new_mean_imp = imp_mean.transform(new)
>>> print(new_mean_imp)
[[ 20. 72.]
 [ 30. 72.]
 [ 30. 70.]
 [ 30. 72.]]
```

Note that 30 in the age field is the mean of those six age values in the original dataset.

Now that we have seen how imputation works, as well as its implementation, let's explore how the strategy of imputing missing values and discarding missing data affects the prediction results through the following example:

1. First, we load the diabetes dataset, as shown here:

```
>>> from sklearn import datasets
>>> dataset = datasets.load_diabetes()
>>> X_full, y = dataset.data, dataset.target
```

2. Simulate a corrupted dataset by adding 25% missing values:

```
>>> m, n = X_full.shape
>>> m_missing = int(m * 0.25)
>>> print(m, m_missing)
442 110
```

3. Randomly select the m_missing samples, as follows:

```
>>> np.random.seed(42)
>>> missing_samples = np.array([True] * m_missing + [False] * (m - m_
missing))
>>> np.random.shuffle(missing_samples)
```

4. For each missing sample, randomly select 1 out of n features:

```
>>> missing_features = np.random.randint(low=0, high=n, size=m_missing)
```

5. Represent missing values with nan, as shown here:

```
>>> X_missing = X_full.copy()
>>> X_missing[np.where(missing_samples)[0], missing_features] = np.nan
```

6. Then, we deal with this corrupted dataset by discarding the samples containing a missing value:

```
>>> X_rm_missing = X_missing[~missing_samples, :]
>>> y_rm_missing = y[~missing_samples]
```

7. Measure the effects of using this strategy by estimating the averaged regression score R^2, with a regression forest model in a cross-validation manner. Estimate R^2 on the dataset with the missing samples removed, as follows:

```
>>> from sklearn.ensemble import RandomForestRegressor
>>> from sklearn.model_selection import cross_val_score
>>> regressor = RandomForestRegressor(random_state=42,
                             max_depth=10, n_estimators=100)
>>> score_rm_missing = cross_val_score(regressor,X_rm_missing,
                               y_rm_missing).mean()
>>> print(f'Score with the data set with missing samples removed: {score_
rm_missing:.2f}')
Score with the data set with missing samples removed: 0.38
```

8. Now we approach the corrupted dataset differently by imputing missing values with the mean, as shown here:

```
>>> imp_mean = SimpleImputer(missing_values=np.nan, strategy='mean')
>>> X_mean_imp = imp_mean.fit_transform(X_missing)
```

9. Similarly, measure the effects of using this strategy by estimating the averaged R^2, as follows:

```
>>> regressor = RandomForestRegressor(random_state=42,
                             max_depth=10,
                             n_estimators=100)
>>> score_mean_imp = cross_val_score(regressor, X_mean_imp, y).mean()
>>> print(f'Score with the data set with missing values replaced by mean:
{score_mean_imp:.2f}')
Score with the data set with missing values replaced by mean: 0.41
```

10. An imputation strategy works better than discarding in this case. So, how far is the imputed dataset from the original full one? We can check it again by estimating the averaged regression score on the original dataset, as follows:

```
>>> regressor = RandomForestRegressor(random_state=42,
                             max_depth=10,
                             n_estimators=500)
>>> score_full = cross_val_score(regressor, X_full, y).mean()
>>> print(f'Score with the full data set: {score_full:.2f}')
Score with the full data set: 0.42
```

It turns out that little information is compromised in the imputed dataset.

However, there is no guarantee that an imputation strategy always works better, and sometimes, dropping samples with missing values can be more effective. Hence, it is a great practice to compare the performance of different strategies via cross-validation, as we have done previously.

Best practice 5 – Storing large-scale data

With the ever-growing size of data, oftentimes, we can't simply fit the data on our single local machine and need to store it on the cloud or distributed filesystems. As this is mainly a book on machine learning with Python, we will just touch on some basic areas that you can look into. The two main strategies for storing big data are **scale up** and **scale out**:

- A **scale-up** approach increases storage capacity if data exceeds the current system capacity, such as by adding more disks. This is useful in fast-access platforms.

- In a **scale-out** approach, storage capacity grows incrementally with additional nodes in a storage cluster. Hadoop Distributed File System (HDFS) (https://hadoop.apache.org/) and Spark (https://spark.apache.org/) are used to store and process big data in scale-out clusters, where data is spread across hundreds or even thousands of nodes. Also, there are cloud-based distributed file services, such as S3 in Amazon Web Services (https://aws.amazon.com/s3/), Google Cloud Storage in Google Cloud (https://cloud.google.com/storage/), and Storage in Microsoft Azure (https://azure.microsoft.com/en-us/services/storage/). They are massively scalable and are designed for secure and durable storage.

Besides choosing the right storage system to increase capacity, you also need to pay attention to the following practices:

- **Data partitioning**: Divide your data into smaller partitions or shards. This distributes the load across multiple servers or nodes, enabling better parallel processing and retrieval.

- **Data compression and encoding**: Implement data compression techniques to reduce storage space and optimize data retrieval times.

- **Replication and redundancy**: Replicate data across multiple storage nodes or geographical locations to ensure data availability and fault tolerance.

- **Security and access control**: Implement robust access control mechanisms to ensure that only authorized personnel can access sensitive data.

With well-prepared data, it is safe to move on to the training set generation stage. Let's see the next section.

Best practices in the training set generation stage

Typical tasks in this stage can be summarized into two major categories: **data preprocessing** and **feature engineering**.

To begin, data preprocessing usually involves categorical feature encoding, feature scaling, feature selection, and dimensionality reduction.

Best practice 6 – Identifying categorical features with numerical values

In general, categorical features are easy to spot, as they convey qualitative information, such as risk level, occupation, and interests. However, it gets tricky if the feature takes on a discreet and countable (limited) number of numerical values, for instance, 1 to 12 representing months of the year, and 1 and 0 indicating true and false.

The key to identifying whether such a feature is categorical or numerical is whether it provides a mathematical or ranking implication; if it does, it is a numerical feature, such as a product rating from 1 to 5; otherwise, it is categorical, such as the month, or day of the week.

Best practice 7 – Deciding whether to encode categorical features

If a feature is considered categorical, we need to decide whether we should encode it. This depends on what prediction algorithm(s) we will use in later stages. Naïve Bayes and tree-based algorithms can directly work with categorical features, while other algorithms in general cannot, in which case encoding is essential.

As the output of the feature generation stage is the input of the model training stage, *steps taken in the feature generation stage should be compatible with the prediction algorithm*. Therefore, we should look at the two stages of feature generation and predictive model training as a whole, instead of two isolated components. The next two practical tips also reinforce this point.

Best practice 8 – Deciding whether to select features and, if so, how to do so

You have seen, in *Chapter 4, Predicting Online Ad Click-Through with Logistic Regression*, how feature selection can be performed using L1-based regularized logistic regression and random forest. The benefits of feature selection include the following:

- Reducing the training time of prediction models as redundant or irrelevant features are eliminated
- Reducing overfitting for the same preceding reason
- Likely improving performance, as prediction models will learn from data with more significant features

Note that we used the word *likely* because there is no absolute certainty that feature selection will increase prediction accuracy. It is, therefore, good practice to compare the performances of conducting feature selection and not doing so via cross-validation. For example, by executing the following steps, we can measure the effects of feature selection by estimating the averaged classification accuracy with an SVC model in a cross-validation manner:

1. First, we load the handwritten digits dataset from `scikit-learn`, as follows:

```
>>> from sklearn.datasets import load_digits
>>> dataset = load_digits()
>>> X, y = dataset.data, dataset.target
>>> print(X.shape)
(1797, 64)
```

2. Next, estimate the accuracy of the original dataset, which is 64-dimensional, as detailed here:

```
>>> from sklearn.svm import SVC
```

```
>>> from sklearn.model_selection import cross_val_score
>>> classifier = SVC(gamma=0.005, random_state=42)
>>> score = cross_val_score(classifier, X, y).mean()
>>> print(f'Score with the original data set: {score:.2f}')
Score with the original data set: 0.90
```

3. Then, conduct feature selection based on random forest and sort the features based on their importance scores:

```
>>> from sklearn.ensemble import RandomForestClassifier
>>> random_forest = RandomForestClassifier(n_estimators=100,
                                            criterion='gini',
                                            n_jobs=-1,
                                            random_state=42)
>>> random_forest.fit(X, y)
>>> feature_sorted = np.argsort(random_forest.feature_importances_)
```

4. Now select a different number of top features to construct a new dataset, and estimate the accuracy on each dataset, as follows:

```
>>> K = [10, 15, 25, 35, 45]
>>> for k in K:
...     top_K_features = feature_sorted[-k:]
...     X_k_selected = X[:, top_K_features]
...     # Estimate accuracy on the data set with k
          selected features
...     classifier = SVC(gamma=0.005)
...     score_k_features = cross_val_score(classifier,
                            X_k_selected, y).mean()
...     print(f'Score with the dataset of top {k} features:
            {score_k_features:.2f}')
...
Score with the dataset of top 10 features: 0.86
Score with the dataset of top 15 features: 0.92
Score with the dataset of top 25 features: 0.95
Score with the dataset of top 35 features: 0.93
Score with the dataset of top 45 features: 0.90
```

If we use the top 25 features selected by random forest, the SVM classification performance can increase from 0.9 to 0.95.

Best practice 9 – Deciding whether to reduce dimensionality and, if so, how to do so

Feature selection and dimensionality are different in the sense that the former chooses features from the original data space, while the latter does so from a projected space from the original space. Dimensionality reduction has the following advantages that are similar to feature selection:

- Reducing the training time of prediction models, as redundant or correlated features are merged into new ones
- Reducing overfitting for the same reason
- Likely improving performance, as prediction models will learn from data with less redundant or correlated features

Again, it is not guaranteed that dimensionality reduction will yield better prediction results. In order to examine its effects, integrating dimensionality reduction in the model training stage is recommended. Reusing the preceding handwritten digits example, we can measure the effects of **Principal Component Analysis (PCA)**-based dimensionality reduction, where we keep a different number of top components to construct a new dataset, and estimate the accuracy on each dataset:

```
>>> from sklearn.decomposition import PCA
>>> # Keep different number of top components
>>> N = [10, 15, 25, 35, 45]
>>> for n in N:
...     pca = PCA(n_components=n)
...     X_n_kept = pca.fit_transform(X)
...     # Estimate accuracy on the data set with top n components
...     classifier = SVC(gamma=0.005)
...     score_n_components =
...             cross_val_score(classifier, X_n_kept, y).mean()
...     print(f'Score with the dataset of top {n} components:
                            {score_n_components:.2f}')
Score with the dataset of top 10 components: 0.94
Score with the dataset of top 15 components: 0.95
Score with the dataset of top 25 components: 0.93
Score with the dataset of top 35 components: 0.91
Score with the dataset of top 45 components: 0.90
```

If we use the top 15 features generated by PCA, the SVM classification performance can increase from 0.9 to 0.95.

Best practice 10 – Deciding whether to rescale features

As seen in *Chapter 5, Predicting Stock Prices with Regression Algorithms*, and *Chapter 6, Predicting Stock Prices with Artificial Neural Networks*, SGD-based linear regression, SVR, and the neural network model require features to be standardized by removing the mean and scaling to unit variance. So, when is feature scaling needed, and when is it not?

In general, Naïve Bayes and tree-based algorithms are not sensitive to features at different scales, as they look at each feature independently.

In most cases, an algorithm that involves any form of distance (or separation in spaces) of samples in learning requires scaled/standardized inputs, such as SVC, SVR, k-means clustering, and **k-nearest neighbors (KNN)** algorithms. Feature scaling is also a must for any algorithm using SGD for optimization, such as linear or logistic regression with gradient descent, and neural networks.

We have so far covered tips regarding data preprocessing and will next discuss best practices of feature engineering as another major aspect of training set generation. We will do so from two perspectives.

Best practice 11 – Performing feature engineering with domain expertise

If we are lucky enough to possess sufficient domain knowledge, we can apply it in creating domain-specific features; we utilize our business experience and insights to identify what is in the data and formulate new data that correlates to the prediction target. For example, in *Chapter 5, Predicting Stock Prices with Regression Algorithms*, we designed and constructed feature sets for the prediction of stock prices based on factors that investors usually look at when making investment decisions.

While particular domain knowledge is required, sometimes we can still apply some general tips in this category. For example, in fields related to customer analytics, such as marketing and advertising, the time of the day, day of the week, and month are usually important signals. Given a data point with the value *2020/09/01* in the Date column and *14:34:21* in the Time column, we can create new features including *afternoon*, *Tuesday*, and *September*. In retail, information covering a period of time is usually aggregated to provide better insights. The number of times a customer visited a store in the past three months, or the average number of products purchased weekly in the previous year, for instance, can be good predictive indicators for customer behavior prediction.

Best practice 12 – Performing feature engineering without domain expertise

If, unfortunately, we have very little domain knowledge, how can we generate features? Don't panic. There are several generic approaches that you can follow, such as binarization, discretization, interaction, and polynomial transformation.

Binarization and discretization

Binarization is the process of converting a numerical feature to a binary one with a preset threshold. For example, in spam email detection, for the feature (or term) *prize*, we can generate a new feature, `whether_term_prize_occurs`: any term frequency value greater than 1 becomes 1; otherwise, it is 0. The feature *number of visits per week* can be used to produce a new feature, `is_frequent_visitor`, by judging whether the value is greater than or equal to 3. We implement such binarization using scikit-learn, as follows:

```
>>> from sklearn.preprocessing import Binarizer
>>> X = [[4], [1], [3], [0]]
>>> binarizer = Binarizer(threshold=2.9)
>>> X_new = binarizer.fit_transform(X)
>>> print(X_new)
[[1]
 [0]
 [1]
 [0]]
```

Discretization is the process of converting a numerical feature to a categorical feature with limited possible values. Binarization can be viewed as a special case of discretization. For example, we can generate an *age group* feature: "*18-24*" for ages from 18 to 24, "*25-34*" for ages from 25 to 34, "*34-54*", and "*55+*".

Interaction

This includes the sum, multiplication, or any operations of two numerical features, and the joint condition check of two categorical features. For example, *the number of visits per week* and *the number of products purchased per week* can be used to generate *the number of products purchased per visit* feature; *interest and occupation*, such as *sports* and *engineer*, can form *occupation AND interest*, such as *engineer interested in sports*.

Polynomial transformation

This is the process of generating polynomial and interaction features. For two features, a and b, the two degrees of polynomial features generated are a^2, ab, and b^2. In scikit-learn, we can use the `PolynomialFeatures` class (https://scikit-learn.org/stable/modules/generated/sklearn.preprocessing.PolynomialFeatures.html) to perform polynomial transformation, as follows:

```
>>> from sklearn.preprocessing import PolynomialFeatures
>>> X = [[2, 4],
...      [1, 3],
...      [3, 2],
...      [0, 3]]
>>> poly = PolynomialFeatures(degree=2)
```

```
>>> X_new = poly.fit_transform(X)
>>> print(X_new)
[[ 1.  2.  4.  4.  8. 16.]
 [ 1.  1.  3.  1.  3.  9.]
 [ 1.  3.  2.  9.  6.  4.]
 [ 1.  0.  3.  0.  0.  9.]]
```

Note the resulting new features consist of 1 (bias, intercept), a, b, a^2, ab, and b^2.

Best practice 13 — Documenting how each feature is generated

We have covered the rules of feature engineering with domain knowledge, and in general, there is one more thing worth noting: documenting how each feature is generated. It sounds trivial, but oftentimes we just forget about how a feature is obtained or created. We usually need to go back to this stage after some failed trials in the model training stage and attempt to create more features with the hope of improving performance. We have to be clear on what and how features are generated, in order to remove those that do not quite work out, and to add new ones that have more potential.

Best practice 14 — Extracting features from text data

We will start with a traditional approach to extract features from text, tf, and tf-idf. Then, we will continue with a modern approach: word embedding. Specifically, we will look at word embedding using Word2Vec models, and embedding layers in neural network models.

tf and tf-idf

We have worked intensively with text data in *Chapter 7, Mining the 20 Newsgroups Dataset with Text Analysis Techniques*, and *Chapter 8, Discovering Underlying Topics in the Newsgroups Dataset with Clustering and Topic Modeling*, where we extracted features from text based on **term frequency (tf)** and **term frequency-inverse document frequency (tf-idf)**. Both methods consider each document of words (terms) a collection of words, or a **bag of words (BoW)**, disregarding the order of the words but keeping multiplicity. A tf approach simply uses the counts of tokens, while tf-idf extends tf by assigning each tf a weighting factor that is inversely proportional to the document frequency. With the idf factor incorporated, tf-idf diminishes the weight of common terms (such as "get" and "make") that occur frequently, and emphasizes terms that rarely occur but convey important meaning. Hence, oftentimes, features extracted from tf-idf are more representative than those from tf.

As you may remember, a document is represented by a very sparse vector where only present terms have non-zero values. The vector's dimensionality is usually high, which is determined by the size of the vocabulary and the number of unique terms. Also, such a one-hot encoding approach treats each term as an independent item and does not consider the relationship across words (referred to as "context" in linguistics).

Word embedding

On the contrary, another approach, called **word embedding**, is able to capture the meanings of words and their context. In this approach, a word is represented by a vector of float numbers. Its dimensionality is a lot lower than the size of the vocabulary and is usually several hundred only.

The embedding vectors are of real values, where each dimension encodes an aspect of meaning for the words in the vocabulary. This helps preserve the semantic information of the words, as opposed to discarding it as in the one-hot encoding approach using tf or tf-idf. An interesting phenomenon is that vectors from semantically similar words are proximate to each other in geometric space. For example, both the words *clustering and grouping* refer to unsupervised clustering in the context of machine learning, hence their embedding vectors are close together.

Here are some popular ways to obtain word embeddings:

- **Word2Vec:** Train your own Word2Vec embeddings on your specific corpus using the Skip-gram or Continuous Bag of Words (CBOW) models. We covered this in *Chapter 7, Mining the 20 Newsgroups Dataset with Text Analysis Techniques*. Libraries like Gensim in Python provide easy-to-use interfaces for training Word2Vec embeddings. We will present a simple example shortly.
- **Pre-trained embeddings:** Use pre-trained word embeddings that are trained on large corpora. We also talked about this in *Chapter 7*. Popular examples include:
  - FastText
  - **GloVe (Global Vectors for Word Representation)**
  - **BERT (Bidirectional Encoder Representations from Transformers)**
  - **GPT (Generative Pre-trained Transformer)**
  - **USE (Universal Sentence Encoder)** embeddings
- **Training custom models with an embedding layer:** If you have a specific domain or dataset, you can train your own word embeddings using custom neural network models.

Word2Vec embedding

Prior to delving into training a custom model for word embeddings, let's begin with the following example of training a basic Word2Vec model using gensim:

We first import the gensim module:

```
>>> from gensim.models import Word2Vec
```

We define some sample sentences for training:

```
>>> sentences = [
    ["i", "love", "machine", "learning", "by", "example"],
    ["machine", "learning", "and", "deep", "learning", "are", "fascinating"],
    ["word", "embedding", "is", "essential", "for", "many", "nlp", "tasks"],
    ["word2vec", "produces", "word", "embeddings"]
]
```

In practice, you will need to format the sentences in plain text into a list of word lists just like the sentences object.

We then create a Word2Vec model with various parameters, such as `vector_size` (embedding dimension), `window` (context window size), `min_count` (minimum frequency of words), and `sg` (training algorithm – 0 for CBOW, 1 for Skip-gram):

```
>>> model = Word2Vec(sentences=sentences, vector_size=100, window=5,
                     min_count=1, sg=0)
```

> 💡 **Quick tip:** Enhance your coding experience with the **AI Code Explainer** and **Quick Copy** features. Open this book in the next-gen Packt Reader. Click the **Copy** button (1) to quickly copy code into your coding environment, or click the **Explain** button (2) to get the AI assistant to explain a block of code to you.
>
> Copy   Explain
>
> ```
> function calculate(a, b) {
>   return {sum: a + b};
> };
> ```
> ①   ②

🔒 **The next-gen Packt Reader** is included for free with the purchase of this book. Unlock it by scanning the QR code below or visiting
`https://www.packtpub.com/unlock/9781835085622`.

After training, we access word vectors using the model's `wv` property. Here, we display the embedding vector for the word *machine:*

```
>>> vector = model.wv["machine"]
>>> print("Vector for 'machine':", vector)
Vector for 'machine': [ 9.2815855e-05  3.0779743e-03 -6.8117767e-03
 -1.3753572e-03 7.6693585e-03  7.3465472e-03 -3.6724545e-03  2.6435424e-
03 -8.3174659e-03  6.2051434e-03 -4.6373457e-03 -3.1652437e-03 9.3113342e-
03  8.7273103e-04  7.4911476e-03 -6.0739564e-03 5.1591368e-03  9.9220201e-03
 -8.4587047e-03 -5.1362212e-03 -7.0644980e-03 -4.8613679e-03 -3.7768795e-03
 -8.5355258e-03 7.9550967e-03 -4.8430962e-03  8.4243221e-03  5.2609886e-03
 -6.5501807e-03  3.9575580e-03  5.4708594e-03 -7.4282014e-03 -7.4055856e-
03 -2.4756377e-03 -8.6252270e-03 -1.5801827e-03 -4.0236043e-04  3.3001360e-
03  1.4415972e-03 -8.8241365e-04 -5.5940133e-03  1.7302597e-03 -8.9826871e-
04  6.7939684e-03 3.9741215e-03  4.5290575e-03  1.4341431e-03 -2.6994087e-03
```

```
-4.3666936e-03 -1.0321270e-03  1.4369689e-03 -2.6467817e-03 -7.0735654e-03
-7.8056543e-03 -9.1217076e-03 -5.9348154e-03 -1.8470082e-03 -4.3242811e-
03 -6.4605214e-03 -3.7180765e-03 4.2892280e-03 -3.7388816e-03  8.3797537e-
03  1.5337169e-03 -7.2427099e-03  9.4338059e-03  7.6304432e-03  5.4950463e-03
-6.8496312e-03  5.8225882e-03  4.0093577e-03  5.1861661e-03 4.2569390e-
03  1.9407619e-03 -3.1710821e-03  8.3530620e-03 9.6114948e-03  3.7916750e-
03 -2.8375010e-03  6.6632601e-06 1.2186278e-03 -8.4594022e-03 -8.2233679e-03
-2.3177716e-04 1.2370384e-03 -5.7435711e-03 -4.7256653e-03 -7.3463405e-03
8.3279097e-03  1.2112247e-04 -4.5090448e-03  5.7024667e-03 9.1806483e-03
-4.0998533e-03  7.9661217e-03  5.3769764e-03 5.8786790e-03  5.1239668e-04
8.2131373e-03 -7.0198057e-03]
```

Keep in mind that this is a basic example. In practice, you might need to preprocess your data more thoroughly, adjust hyperparameters, and train on a larger corpus for better embeddings.

Embedding layers in custom neural networks

In a complete deep neural network for NLP tasks, we would typically combine an embedding layer with other layers, like fully connected (dense) layers, or recurrent layers (we will talk about recurrent layers in *Chapter 12, Making Predictions with Sequences Using Recurrent Neural Networks*) to build a more sophisticated model. The embedding layer allows the network to learn meaningful representations for words in the input data.

Let's look at a simplified example of using an embedding layer for word embeddings. In PyTorch, we use the nn.Embedding module (https://pytorch.org/docs/stable/generated/torch.nn.Embedding.html) for embedding layers:

```
>>> import torch
>>> import torch.nn as nn
>>> input_data = torch.LongTensor([[1, 2, 3, 4], [5, 1, 6, 3]])
>>> # Define the embedding layer
>>> vocab_size = 10  # Total number of unique words
>>> embedding_dim = 3  # Dimensionality of the embeddings
>>> embedding_layer = nn.Embedding(vocab_size, embedding_dim)
...
>>> embedded_data = embedding_layer(input_data)
>>> print("Embedded Data:\n", embedded_data)
Embedded Data:
tensor([[[-1.2462,  0.4035,  0.4463],
         [-0.5218,  0.8302, -0.6920],
         [-0.4720, -1.2894,  1.0763],
         [-2.2879, -0.4834,  0.3416]],
        [[ 1.5886, -0.3489, -0.4579],
         [-1.2462,  0.4035,  0.4463],
         [-1.2322, -0.5981, -0.1349],
         [-0.4720, -1.2894,  1.0763]]], grad_fn=<EmbeddingBackward0>)
```

In this example, we first import the necessary modules from PyTorch. We define some sample input data containing word indices (for example, 1 represents *I*, 2 represents *love*, 3 represents *machine*, and 4 represents *learning*). Then, we define the embedding layer using nn.Embedding with vocab_size as the total number of unique words in the vocabulary, and embedding_dim as the desired dimensionality of the embeddings. The embedding layer is usually the first layer of a neural network model after the input layer. Upon completion of network training, when we pass the input data through the embedding layer, it returns embedded vectors for each input word index. The shape of output embedded_data will be (sample size, sequence length, embedding_dim), which is (2, 4, 3) in our case.

Once again, this is a simplified example. In practice, the embedding layers are involved in more complex architectures with additional layers, in order to process and interpret the embeddings for specific tasks, such as classification, sentiment analysis, or sequence generation. Keep an eye out for the upcoming *Chapter 12*.

Curious about the choice between tf-idf and word embeddings? In conventional NLP applications, such as simple text classification and topic modeling, tf, or tf-idf, remains an exceptional method for feature extraction. In more complicated areas, such as text summarization, machine translation, named entity resolution, question answering, and information retrieval, word embeddings are extensively utilized and yield significantly enhanced features compared to conventional methods.

Now that you have reviewed the best practices for data and feature generation, let's look at model training next.

Best practices in the model training, evaluation, and selection stage

Given a supervised machine learning problem, the first question many people ask is usually *What is the best classification or regression algorithm to solve it?* However, there is no one-size-fits-all solution and no free lunch. No one could know which algorithm will work best before trying multiple ones and fine-tuning the optimal one. We will be looking into best practices around this in this section.

Best practice 15 – Choosing the right algorithm(s) to start with

Due to the fact that there are several parameters to tune for an algorithm, exhausting all algorithms and fine-tuning each one can be extremely time-consuming and computationally expensive. We should instead shortlist one to three algorithms to start with using the general guidelines that follow (note we herein focus on classification, but the theory transcends to regression, and there is usually a counterpart algorithm in regression).

There are several things we need to be clear about before shortlisting potential algorithms, as described in the following:

- The size of the training dataset
- The dimensionality of the dataset
- Whether the data is linearly separable
- Whether features are independent

- Tolerance and trade-off of bias and variance
- Whether online learning is required

Now, let's look at how we choose the right algorithm to start with, taking into account the aforementioned perspectives.

Naïve Bayes

This is a very simple algorithm. For a relatively small training dataset, if features are independent, Naïve Bayes will usually perform well. For a large dataset, Naïve Bayes will still work well as feature independence can be assumed in this case, regardless of the truth. The training of Naïve Bayes is usually faster than any other algorithm due to its computational simplicity. However, this may lead to a high bias (but low variance).

Logistic regression

This is probably the most widely used classification algorithm, and the first algorithm that a machine learning practitioner usually tries when given a classification problem. It performs well when data is linearly separable or approximately **linearly separable**. Even if it is not linearly separable, it might be possible to convert the linearly non-separable features into separable ones and apply logistic regression afterward.

In the following instance, data in the original space is not linearly separable, but it becomes separable in a transformed space created from the interaction of two features:

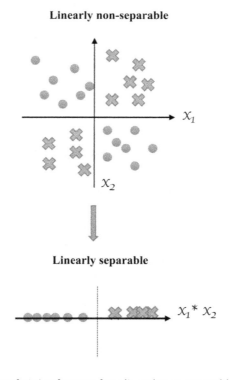

Figure 10.2: Transforming features from linearly non-separable to separable

Also, logistic regression is extremely scalable to large datasets with SGD optimization, which makes it efficient in solving big data problems. Plus, it makes online learning feasible. Although logistic regression is a low-bias, high-variance algorithm, we overcome the potential overfitting by adding L1, L2, or a mix of the two regularizations.

SVM

This is versatile enough to adapt to the linear separability of data. For a separable dataset, SVM with a linear kernel performs comparably to logistic regression. Beyond this, SVM also works well for a non-separable dataset if equipped with a non-linear kernel, such as RBF. Logistic regression may face challenges in high-dimensional datasets, while SVM still performs well. A good example of this can be in news classification, where the feature dimensionality is in the tens of thousands. In general, very high accuracy can be achieved by SVM with the right kernel and parameters. However, this might be at the expense of intense computation and high memory consumption.

Random forest (or decision tree)

The linear separability of the data does not matter to this algorithm, and it works directly with categorical features without encoding, which provides great ease of use. Also, the trained model is very easy to interpret and explain to non-machine learning practitioners, which cannot be achieved with most other algorithms. Additionally, random forest boosts the decision tree algorithm, which can reduce overfitting by ensembling a collection of separate trees. Its performance is comparable to SVM, while fine-tuning a random forest model is less difficult compared to SVM and neural networks.

Neural networks

These are extremely powerful, especially with the development of deep learning. However, finding the right topology (layers, nodes, activation functions, and so on) is not easy, not to mention the time-consuming model of training and tuning. Hence, they are not recommended as an algorithm to start with for general machine learning problems. However, for computer vision and many NLP tasks, the neural network is still the go-to model. In summary, here are some scenarios where using neural networks is particularly beneficial:

- **Complex patterns:** When the task involves learning complex patterns or relationships within the data that may be difficult for traditional algorithms to capture.

- **Large amounts of data:** Neural networks tend to perform well when you have a substantial amount of data available for training, as they are capable of learning from large datasets.

- **Unstructured data:** Neural networks excel in handling unstructured data types like images, audio, and text, where traditional methods might struggle to extract meaningful features. For NLP tasks like sentiment analysis, machine translation, named entity recognition, and text generation, neural networks, especially recurrent and transformer models, have shown remarkable performance. In image classification, object detection, segmentation, and image generation tasks, deep neural networks have revolutionized computer vision.

Best practice 16 — Reducing overfitting

We touched on ways to avoid overfitting when discussing the pros and cons of algorithms in the last practice. We herein formally summarize them, as follows:

- **More data, if possible:** Increase the size of your training dataset. More data can help the model learn relevant patterns and reduce its tendency to memorize noise.

- **Simplification, if possible:** The more complex the model is, the higher the chance of overfitting. Complex models include a tree or forest with excessive depth, a linear regression with a high degree of polynomial transformation, and an SVM with a complicated kernel.

- **Cross-validation:** A good habit that we have built over all of the chapters in this book.

- **Regularization:** This adds penalty terms to reduce the error caused by fitting the model perfectly on the given training set.

- **Early stopping:** Monitor the model's performance on a validation set during training. Stop training when the performance on the validation set starts to degrade, indicating that the model is starting to overfit.

- **Dropout:** In neural networks, apply dropout layers during training. Dropout randomly drops out a fraction of neurons during each forward pass, preventing reliance on specific neurons.

- **Feature selection:** Select a subset of relevant features. Removing irrelevant or redundant features can prevent the model from fitting noise.

- **Ensemble learning:** This involves combining a collection of weak models to form a stronger one.

So, how can we tell whether a model suffers from overfitting, or the other extreme, underfitting? Let's see the next section.

Best practice 17 — Diagnosing overfitting and underfitting

A **learning curve** is usually used to evaluate the bias and variance of a model. A learning curve is a graph that compares the cross-validated training and testing scores over a given number of training samples.

For a model that fits well on the training samples, the performance of the training samples should be beyond what's desired. Ideally, as the number of training samples increases, the model performance on the testing samples will improve; eventually, the performance on the testing samples will become close to that of the training samples.

When the performance on the testing samples converges at a value much lower than that of the training performance, overfitting can be concluded. In this case, the model fails to generalize to instances that have not been seen.

For a model that does not even fit well on the training samples, underfitting is easily spotted: both performances on the training and testing samples are below the desired performance in the learning curve.

Here is an example of the learning curve in an ideal case:

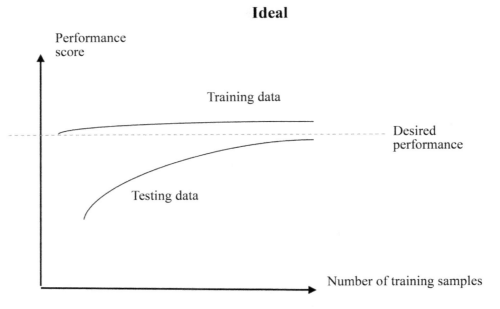

Figure 10.3: Ideal learning curve

An example of the learning curve for an overfitted model is shown in the following diagram:

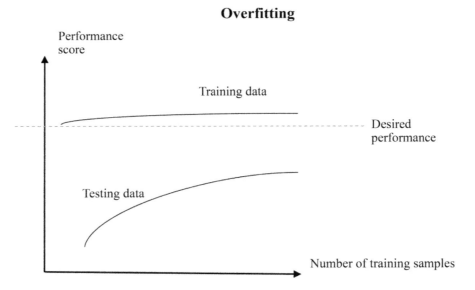

Figure 10.4: Overfitting learning curve

The learning curve for an underfitted model may look like the following diagram:

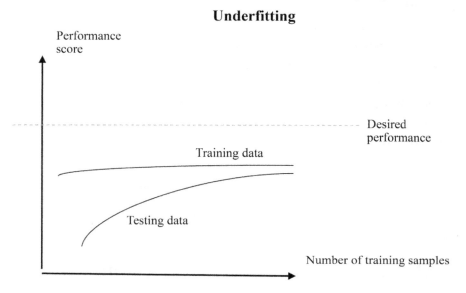

Figure 10.5: Underfitting learning curve

To generate the learning curve, you can utilize the learning_curve module (https://scikit-learn. org/stable/modules/generated/sklearn.model_selection.learning_curve.html#sklearn.model_ selection.learning_curve) from scikit-learn, and the plot_learning_curve function defined at https://scikit-learn.org/stable/auto_examples/model_selection/plot_learning_curve.html.

Best practice 18 — Modeling on large-scale datasets

We gained experience working with large datasets in *Chapter 4, Predicting Online Ad Click-Through with Logistic Regression*. There are a few tips that can help you model on large-scale data more efficiently.

First, start with a small subset, for instance, a subset that can fit on your local machine. This can help speed up early experimentation. Obviously, you don't want to train on the entire dataset just to find out whether SVM or random forest works better. Instead, you can randomly sample data points and quickly run a few models on the selected set.

The second tip is choosing scalable algorithms, such as logistic regression, linear SVM, and SGD-based optimization. This is quite intuitive.

Here are other best practices for modeling on large-scale datasets:

- **Sampling and subset selection:** When starting model development, work with smaller subsets of your data to iterate and experiment quickly. Once your model architecture and parameters are tuned, scale up to the full dataset.
- **Distributed computing:** Utilize distributed computing frameworks like Apache Spark to handle large-scale data processing and model training across multiple nodes or clusters.

- **Feature engineering:** Focus on relevant features and avoid unnecessary dimensions. Use dimensionality reduction techniques like PCA or t-SNE to reduce feature space if needed.
- **Parallelization:** Explore techniques to parallelize training, like data parallelism or model parallelism, to leverage multiple GPUs or distributed systems.
- **Memory management:** Optimize memory usage by using data generators, streaming data from storage, and releasing memory when no longer needed.
- **Optimized libraries:** Choose libraries and frameworks that are optimized for large-scale data, such as TensorFlow, PyTorch, scikit-learn, and XGBoost.
- **Incremental learning:** For streaming data or dynamic datasets, consider incremental learning techniques that update the model as new data arrives.

Last but not least, don't forget to save the trained model. Training on a large dataset takes a long time, which you would want to avoid redoing, if possible. We will explore saving and loading models in detail in *Best practice 19 – Saving, loading, and reusing models*, which is a part of the deployment and monitoring stage.

Best practices in the deployment and monitoring stage

After performing all processes in the previous three stages, we now have a well-established data pre-processing pipeline and a correctly trained prediction model. The last stage of a machine learning system involves saving those resulting models from previous stages and deploying them on new data, as well as monitoring their performance and updating the prediction models regularly. We also need to implement monitoring and logging to track model performance, training progress, and potential issues during training.

Best practice 19 — Saving, loading, and reusing models

When machine learning is deployed, new data should go through the same data preprocessing procedures (scaling, feature engineering, feature selection, dimensionality reduction, and so on) as in the previous stages. The preprocessed data is then fed into the trained model. We simply cannot rerun the entire process and retrain the model every time new data comes in. Instead, we should save the established preprocessing models and trained prediction models after the corresponding stages have been completed. In deployment mode, these models are loaded in advance and are used to produce prediction results from the new data. Let's explore methods for saving and loading models using pickle, TensorFlow, and PyTorch below.

Saving and restoring models using pickle

We start with using `pickle`. This can be illustrated via the diabetes example, where we standardize the data and employ an SVR model, as follows:

```
>>> dataset = datasets.load_diabetes()
>>> X, y = dataset.data, dataset.target
>>> num_new = 30 # the last 30 samples as new data set
```

```
>>> X_train = X[:-num_new, :]
>>> y_train = y[:-num_new]
>>> X_new = X[-num_new:, :]
>>> y_new = y[-num_new:]
```

Preprocess the training data with scaling, as shown in the following commands:

```
>>> from sklearn.preprocessing import StandardScaler
>>> scaler = StandardScaler()
>>> scaler.fit(X_train)
```

Now save the established standardizer, the `scaler` object with `pickle`, as follows:

```
>>> import pickle
>>> pickle.dump(scaler, open("scaler.p", "wb" ))
```

This generates a `scaler.p` file.

Move on to training an SVR model on the scaled data, as follows:

```
>>> X_scaled_train = scaler.transform(X_train)
>>> from sklearn.svm import SVR
>>> regressor = SVR(C=20)
>>> regressor.fit(X_scaled_train, y_train)
```

Save the trained `regressor` object with `pickle`, as follows:

```
>>> pickle.dump(regressor, open("regressor.p", "wb"))
```

This generates a `regressor.p` file.

In the deployment stage, we first load the saved standardizer and the `regressor` object from the preceding two files, as follows:

```
>>> my_scaler = pickle.load(open("scaler.p", "rb" ))
>>> my_regressor = pickle.load(open("regressor.p", "rb"))
```

Then, we preprocess the new data using the standardizer and make a prediction with the `regressor` object just loaded, as follows:

```
>>> X_scaled_new = my_scaler.transform(X_new)
>>> predictions = my_regressor.predict(X_scaled_new)
```

Saving and restoring models in TensorFlow

I will also demonstrate how to save and restore models in TensorFlow. As an example, we will train a simple logistic regression model on the cancer dataset, save the trained model, and reload it in the following steps:

1. Import the necessary TensorFlow modules and load the cancer dataset from `scikit-learn` and rescale the data:

```
>>> import tensorflow as tf
>>> from tensorflow import keras
>>> from sklearn import datasets
>>> cancer_data = datasets.load_breast_cancer()
>>> X = cancer_data.data
>>> X = scaler.fit_transform(X)
>>> y = cancer_data.target
```

2. Build a simple logistic regression model using the Keras Sequential API, along with several specified parameters:

```
>>> learning_rate = 0.005
>>> n_iter = 10
>>> tf.random.set_seed(42)
>>> model = keras.Sequential([
...     keras.layers.Dense(units=1, activation='sigmoid')
... ])
>>> model.compile(loss='binary_crossentropy',
...               optimizer=tf.keras.optimizers.Adam(learning_rate))
```

3. Train the TensorFlow model against the data:

```
>>> model.fit(X, y, epochs=n_iter)
Epoch 1/10
18/18 [==============================] - 0s 943us/step - loss: 0.2288
Epoch 2/10
18/18 [==============================] - 0s 914us/step - loss: 0.1591
Epoch 3/10
18/18 [==============================] - 0s 825us/step - loss: 0.1303
Epoch 4/10
18/18 [==============================] - 0s 865us/step - loss: 0.1147
Epoch 5/10
18/18 [==============================] - 0s 795us/step - loss: 0.1042
Epoch 6/10
18/18 [==============================] - 0s 796us/step - loss: 0.0971
Epoch 7/10
```

```
18/18 [==============================] - 0s 862us/step - loss: 0.0917
Epoch 8/10
18/18 [==============================] - 0s 913us/step - loss: 0.0871
Epoch 9/10
18/18 [==============================] - 0s 795us/step - loss: 0.0835
Epoch 10/10
18/18 [==============================] - 0s 767us/step - loss: 0.0806
```

4. Display the model's architecture:

```
>>> model.summary()
Model: "sequential"

_____
Layer (type)                 Output Shape              Param #
=================================================================
dense (Dense)                multiple                  31
=================================================================
Total params: 31
Trainable params: 31
Non-trainable params: 0
_____
```

We will see if we can retrieve the same model later.

5. Hopefully, the previous steps look familiar to you. If not, feel free to review our TensorFlow implementation. Now we save the model to a path:

```
>>> path = './model_tf'
>>> model.save(path)
```

After this, you will see that a folder called model_tf is created. The folder contains the trained model's architecture, weights, and training configuration.

6. Finally, we load the model from the previous path and display the loaded model's path:

```
>>> new_model = tf.keras.models.load_model(path)
>>> new_model.summary()
Model: "sequential"

_____
Layer (type)                 Output Shape        Param #
=======================================================
dense (Dense)                multiple            31
=======================================================
Total params: 31
```

```
Trainable params: 31
Non-trainable params: 0
```

We just loaded back the exact same model.

Saving and restoring models in PyTorch

Finally, let's see how to save and restore models in PyTorch. Similarly, we will train a simple logistic regression model on the same cancer dataset, save the trained model, and reload it in the following steps:

1. Convert the data `torch` tensors used for modeling:

```
>>> X_torch = torch.FloatTensor(X)
>>> y_torch = torch.FloatTensor(y.reshape(y.shape[0], 1))
```

2. Build a simple logistic regression model using the `nn.sequential` module, along with the loss function and optimizer:

```
>>> torch.manual_seed(42)
>>> model = nn.Sequential(nn.Linear(X.shape[1], 1),
                          nn.Sigmoid())
>>> loss_function = nn.BCELoss()
>>> optimizer = torch.optim.Adam(model.parameters(), lr=learning_rate)
```

3. Reuse the `train_step` function we developed previously in *Chapter 6, Predicting Stock Prices with Artificial Neural Networks*, and train the `PyTorch` model against the data for 10 iterations:

```
>>> def train_step(model, X_train, y_train, loss_function, optimizer):
        pred_train = model(X_train)
        loss = loss_function(pred_train, y_train)
        model.zero_grad()
        loss.backward()
        optimizer.step()
    return loss.item()
>>> for epoch in range(n_iter):
        loss = train_step(model, X_torch, y_torch, loss_function, optimizer)
```

```
          print(f"Epoch {epoch} - loss: {loss}")

Epoch 0 - loss: 0.8387020826339722
Epoch 1 - loss: 0.7999904751777649
Epoch 2 - loss: 0.76298588514328
Epoch 3 - loss: 0.7277476787567139
Epoch 4 - loss: 0.6943162679672241
Epoch 5 - loss: 0.6627081036567688
Epoch 6 - loss: 0.6329135298728943
Epoch 7 - loss: 0.6048969030380249
Epoch 8 - loss: 0.5786024332046509
Epoch 9 - loss: 0.5539639592170715
```

4. Display the model's architecture:

```
>>> print(model)
Sequential(
  (0): Linear(in_features=30, out_features=1, bias=True)
  (1): Sigmoid()
)
```

We will see if we can retrieve the same model later.

5. Hopefully, the previous steps look familiar to you. If not, feel free to review our PyTorch implementation. Now we save the model to a path:

```
>>> path = './model.pth '
>>> torch.save(model, path)
```

After this, you will see that a folder called model.pth is created. The folder contains the entire trained model's architecture, weights, and training configuration.

6. Finally, we load the model from the previous path and display the loaded model's path:

```
>>> new_model = torch.load(path)
>>> print(new_model)
Sequential(
  (0): Linear(in_features=30, out_features=1, bias=True)
  (1): Sigmoid()
)
```

We just loaded back the exact same model.

Best practice 20 – Monitoring model performance

The machine learning system is now up and running. To make sure everything is on the right track, we need to conduct performance checks on a regular basis. To do so, besides making a prediction in real time, we should also record the ground truth at the same time. Here are some best practices for monitoring model performance:

- **Define evaluation metrics**: Choose appropriate evaluation metrics that align with your problem's goals. Accuracy, precision, recall, F1-score, AUC-ROC, R^2, and mean squared error are some common metrics.

- **Baseline performance**: Establish a baseline model or a simple rule-based approach to compare your model's performance. This provides context for understanding whether your model is adding value.

- **Learning curves**: Plot learning curves showing training and validation loss or evaluation metrics over epochs. This helps identify overfitting or underfitting issues, as mentioned in *Best practice 17 – Diagnosing overfitting and underfitting*.

Continuing with the diabetes example from earlier in the chapter, we conduct a performance check as follows:

```
>>> from sklearn.metrics import r2_score
>>> print(f'Health check on the model, R^2: {r2_score(y_new,
            predictions):.3f}')
Health check on the model, R^2: 0.613
```

We should log the performance and set up an alert for any decayed performance.

Best practice 21 – Updating models regularly

If the performance is getting worse, chances are that the pattern of data has changed. We can work around this by updating the model. Depending on whether online learning is feasible or not with the model, the model can be modernized with the new set of data (online updating) or retrained completely with the most recent data. Here are some best practices for the last section of the chapter:

- **Monitor model performance**: Continuously monitor model performance metrics. If there's a significant drop, it's a sign that the model needs updating.

- **Scheduled updates**: Implement a schedule for model updates based on the frequency of data changes and business needs. This ensures that the model remains relevant, without unnecessary updates.

- **Online updating**: For models that support online learning, update the model incrementally with new data. This applies to models based on gradient descent algorithms, or Naïve Bayes. Online updating minimizes the need for retraining the entire model and adapts it to changing patterns over time.

- **Version control:** Maintain version control of models and datasets to track changes and facilitate rollback if necessary. This helps in comparing model performance over time and reverting to previous versions if updates lead to performance degradation.

- **Regular auditing:** Periodically review model performance, reevaluate business goals, and update your evaluation metrics if needed.

Remember that monitoring should be an ongoing process, starting from model development through deployment and maintenance. It ensures that your machine learning models remain effective, trustworthy, and aligned with your business objectives.

Summary

The purpose of this chapter is to prepare you for real-world machine learning problems. We started with the general workflow that a machine learning solution follows: data preparation, training set generation, algorithm training, evaluation and selection, and finally, system deployment and monitoring. We then went, in depth, through the typical tasks, common challenges, and best practices for each of these four stages.

Practice makes perfect. The most important best practice is practice itself. Get started with a real-world project to deepen your understanding and apply what you have learned so far.

In the next chapter, we will start our deep learning journey by categorizing clothing images using convolutional neural networks.

Exercises

1. Can you use word embedding to extract text features and develop a multiclass classifier to classify the newsgroup data? (Note that you might not be able to get better results with word embedding than tf-idf, but it is good practice.)

2. Can you find several challenges in Kaggle (`www.kaggle.com`) and practice what you have learned throughout the entire book?

Unlock this book's exclusive benefits now

This book comes with additional benefits designed to elevate your learning experience.

Note: Have your purchase invoice ready before you begin. `https://www.packtpub.com/unlock/9781835085622`

11

Categorizing Images of Clothing with Convolutional Neural Networks

The previous chapter wrapped up our coverage of the best practices for general machine learning. Starting from this chapter, we will dive into the more advanced topics of deep learning and reinforcement learning.

When we deal with image classification, we usually flatten the images, get vectors of pixels, and feed them to a neural network (or another model). Although this might do the job, we lose critical spatial information. In this chapter, we will use **Convolutional Neural Networks (CNNs)** to extract rich and distinguishable representations from images. You will see how CNN representations make a "9" a "9", a "4" a "4", a cat a cat, or a dog a dog.

We will start by exploring individual building blocks in the CNN architecture. Then, we will develop a CNN classifier in PyTorch to categorize clothing images and demystify the convolutional mechanism. Finally, we will introduce data augmentation to boost the performance of CNN models.

We will cover the following topics in this chapter:

- Getting started with CNN building blocks
- Architecting a CNN for classification
- Exploring the clothing image dataset
- Classifying clothing images with CNNs
- Boosting the CNN classifier with data augmentation
- Advancing the CNN classifier with transfer learning

Getting started with CNN building blocks

Although regular hidden layers (the fully connected layers we have seen so far) do a good job of extracting features from data at certain levels, these representations might not be useful in differentiating images of different classes. CNNs can be used to extract richer, more distinguishable representations that, for example, make a car a car, a plane a plane, or the handwritten letters "y" and "z" recognizably a "y" and a "z," and so on. CNNs are a type of neural network that is biologically inspired by the human visual cortex. To demystify CNNs, I will start by introducing the components of a typical CNN, including the convolutional layer, the non-linear layer, and the pooling layer.

The convolutional layer

The **convolutional layer** is the first layer in a CNN, or the first few layers in a CNN if it has multiple convolutional layers.

CNNs, specifically their convolutional layers, mimic the way our visual cells work, as follows:

- Our visual cortex has a set of complex neuronal cells that are sensitive to specific sub-regions of the visual field and that are called **receptive fields**. For instance, some cells only respond in the presence of vertical edges; some cells fire only when they are exposed to horizontal edges; some react more strongly when they are shown edges of a certain orientation. These cells are organized together to produce the entire visual perception, with each cell being specialized in a specific component. A convolutional layer in a CNN is composed of a set of filters that act like those cells in humans' visual cortexes.

- A simple cell only responds when edge-like patterns are presented within its receptive sub-regions. A more complex cell is sensitive to larger sub-regions, and as a result, can respond to edge-like patterns across the entire visual field. A stack of convolutional layers is a bunch of complex cells that can detect patterns in a bigger scope.

The convolutional layer processes input images or matrices and mimics how neural cells react to the particular regions they are attuned to by employing a convolutional operation on the input. Mathematically, it computes the **dot product** between the nodes of the convolutional layer and individual small regions in the input layer. The small region is the receptive field, and the nodes of the convolutional layer can be viewed as the values on a filter. As the filter moves along on the input layer, the dot product between the filter and the current receptive field (sub-region) is computed. A new layer called the **feature map** is obtained after the filter has convolved over all the sub-regions. Let's look at a simple example, as follows:

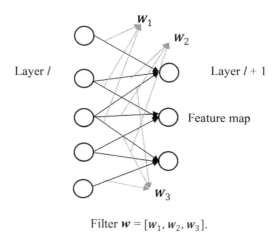

Filter $w = [w_1, w_2, w_3]$.

Figure 11.1: How a feature map is generated

In this example, layer l has 5 nodes and the filter is composed of 3 nodes $[w_1, w_2, w_3]$. We first compute the dot product between the filter and the first three nodes in layer l and obtain the first node in the output feature map; then, we compute the dot product between the filter and the middle three nodes and generate the second node in the output feature map; finally, the third node is generated from the convolution on the last three nodes in layer l.

Now, we'll take a closer look at how convolution works in the following example:

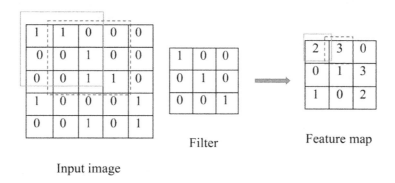

Input image   Filter   Feature map

Figure 11.2: How convolution works

In this example, a 3*3 filter is sliding around a 5*5 input matrix from the top-left sub-region to the bottom-right sub-region. For each sub-region, the dot product is computed using the filter. Take the top-left sub-region (in the orange rectangle) as an example: we have $1 * 1 + 1 * 0 + 1 * 1 = 2$, therefore the top-left node (in the upper-left orange rectangle) in the feature map is of value 2. For the next leftmost sub-region (in the blue dash rectangle), we calculate the convolution as $1 * 1 + 1 * 1 + 1 * 1 = 3$, so the value of the next node (in the upper-middle blue dash rectangle) in the resulting feature map becomes 3. At the end, a 3*3 feature map is generated as a result.

So what do we use convolutional layers for? They are actually used to extract features such as edges and curves. The pixel in the output feature map will be of high value if the corresponding receptive field contains an edge or curve that is recognized by the filter. For instance, in the preceding example, the filter portrays a backslash-shape "\" diagonal edge; the receptive field in the blue dash rectangle contains a similar curve and hence the highest intensity 3 is created. However, the receptive field in the top-right corner does not contain such a backslash shape, hence it results in a pixel of value 0 in the output feature map. The convolutional layer acts as a curve detector or a shape detector.

Also, a convolutional layer usually has multiple filters detecting different curves and shapes. In the simple preceding example, we only apply one filter and generate one feature map, which indicates how well the shape in the input image resembles the curve represented in the filter. In order to detect more patterns from the input data, we can employ more filters, such as horizontal, vertical curve, 30-degree, and right-angle shape.

Additionally, we can stack several convolutional layers to produce higher-level representations such as the overall shape and contour. Chaining more layers will result in larger receptive fields that are able to capture more global patterns.

Right after each convolutional layer, we often apply a non-linear layer.

The non-linear layer

The non-linear layer is basically the activation layer we saw in *Chapter 6, Predicting Stock Prices with Artificial Neural Networks*. It is used to introduce non-linearity, obviously. Recall that in the convolutional layer, we only perform linear operations (multiplication and addition). And no matter how many linear hidden layers a neural network has, it will just behave as a single-layer perceptron. Hence, we need a non-linear activation right after the convolutional layer. Again, ReLU is the most popular candidate for the non-linear layer in deep neural networks.

The pooling layer

Normally after one or more convolutional layers (along with non-linear activation), we can directly use the derived features for classification. For example, we can apply a softmax layer in the multiclass classification case. But let's do some math first.

Given 28 * 28 input images, supposing that we apply 20 5 * 5 filters in the first convolutional layer, we will obtain 20 output feature maps and each feature map layer will be of size (28 – 5 + 1) * (28 – 5 + 1) = 24 * 24 = 576. This means that the number of features as inputs for the next layer increases to 11,520 (20 * 576) from 784 (28 * 28). We then apply 50 5 * 5 filters in the second convolutional layer. The size of the output grows to 50 * 20 * (24 – 5 + 1) * (24 – 5 + 1) = 400,000. This is a lot higher than our initial size of 784. We can see that the dimensionality increases dramatically with every convolutional layer before the final softmax layer. This can be problematic as it leads to overfitting easily, not to mention the cost of training such a large number of weights.

To address the issue of drastically growing dimensionality, we often employ a **pooling layer** after the convolutional and non-linear layers. The pooling layer is also called the **downsampling layer**. As you can imagine, it reduces the dimensions of the feature maps. This is done by aggregating the statistics of features over sub-regions. Typical pooling methods include:

- Max pooling, which takes the max values over all non-overlapping sub-regions
- Mean pooling, which takes the mean values over all non-overlapping sub-regions

In the following example, we apply a 2 * 2 max-pooling filter on a 4 * 4 feature map and output a 2 * 2 one:

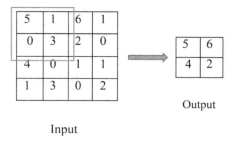

Input   Output

Figure 11.3: How max pooling works

Besides dimensionality reduction, the pooling layer has another advantage: translation invariance. This means that its output doesn't change even if the input matrix undergoes a small amount of translation. For example, if we shift the input image a couple of pixels to the left or right, as long as the highest pixels remain the same in the sub-regions, the output of the max-pooling layer will still be the same. In other words, the prediction becomes less position-sensitive with pooling layers. The following example illustrates how max pooling achieves translation invariance.

Here is the 4 * 4 original image, along with the output from max pooling with a 2 * 2 filter:

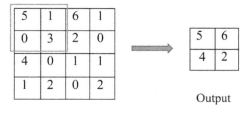

Original image   Output

Figure 11.4: The original image and the output from max pooling

And if we shift the image 1 pixel to the right, we have the following shifted image and the corresponding output:

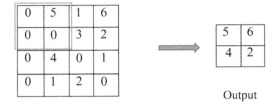

Output

Image shifted 1 pixel right

Figure 11.5: The shifted image and the output

We have the same output even if we horizontally move the input image. Pooling layers increase the robustness of image translation.

You've now learned about all of the components of a CNN. It was easier than you thought, right? Let's see how they compose a CNN next.

Architecting a CNN for classification

Putting the three types of convolutional-related layers together, along with the fully connected layer(s), we can structure the CNN model for classification as follows:

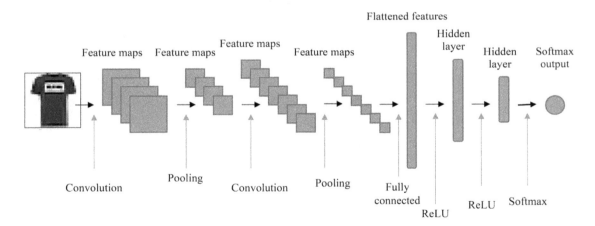

Figure 11.6: CNN architecture

In this example, the input images are first fed into a convolutional layer (with ReLU activation) composed of a bunch of filters. The coefficients of the convolutional filters are trainable. A well-trained initial convolutional layer is able to derive good low-level representations of the input images, which will be critical to downstream convolutional layers if there are any, and also downstream classification tasks. Each resulting feature map is then downsampled by the pooling layer.

Next, the aggregated feature maps are fed into the second convolutional layer. Similarly, the second pooling layer reduces the size of the output feature maps. You can chain as many pairs of convolutional and pooling layers as you want. The second (or more, if any) convolutional layer tries to compose high-level representations, such as the overall shape and contour, through a series of low-level representations derived from previous layers.

Up until this point, the feature maps are matrices. We need to flatten them into a vector before performing any downstream classification. The flattened features are just treated as the input to one or more fully connected hidden layers. We can think of a CNN as a hierarchical feature extractor on top of a regular neural network. CNNs are well suited to exploiting strong and unique features that differentiate images.

The network ends up with a logistic function if we deal with a binary classification problem, a softmax function for a multiclass case, or a set of logistic functions for multi-label cases.

By now, you should have a good understanding of CNNs and should be ready to solve the clothing image classification problem. Let's start by exploring the dataset.

Exploring the clothing image dataset

The clothing dataset Fashion-MNIST (`https://github.com/zalandoresearch/fashion-mnist`) is a dataset of images from Zalando (Europe's biggest online fashion retailer). It consists of 60,000 training samples and 10,000 test samples. Each sample is a 28 * 28 grayscale image, associated with a label from the following 10 classes, each representing articles of clothing:

- 0: T-shirt/top
- 1: Trouser
- 2: Pullover
- 3: Dress
- 4: Coat
- 5: Sandal
- 6: Shirt
- 7: Sneaker
- 8: Bag
- 9: Ankle boot

Zalando aims to make the dataset as popular as the handwritten digits MNIST dataset for benchmarking algorithms and hence calls it Fashion-MNIST.

You can download the dataset from the direct links in the *Get the data* section using the GitHub link or simply import it from PyTorch, which already includes the dataset and its data loader API. We will take the latter approach, as follows:

```
>>> import torch, torchvision
>>> from torchvision import transforms
```

```
>>> image_path = './'
>>> transform = transforms.Compose([transforms.ToTensor()])
>>> train_dataset = torchvision.datasets.FashionMNIST(root=image_path,
                                                       train=True,
                                                       transform=transform,
                                                       download=True)

>>> test_dataset = torchvision.datasets.FashionMNIST(root=image_path,
                                                      train=False,
                                                      transform=transform,
                                                      download=False)
```

We just import `torchvision`, a package in PyTorch that provides access to datasets, model architectures, and various image transformation utilities for computer vision tasks.

The `torchvision` library includes the following key components:

- **Datasets and data loaders:** `torchvision.datasets` provides access to standard datasets for tasks like image classification, object detection, semantic segmentation, etc. Examples include MNIST, CIFAR-10, ImageNet, `FashionMNIST`, etc. `torch.utils.data.DataLoader` helps in creating data loaders to efficiently load and preprocess batches of data from datasets.

- **Transformations:** `torchvision.transforms` offers a variety of image transformations for data augmentation, normalization, and preprocessing. Common transformations include resizing, cropping, normalization, and more.

- **Model architectures:** `torchvision.models` provides pre-trained model architectures for various computer vision tasks.

- **Utilities:** `torchvision.utils` includes utility functions for visualizing images, converting images into different formats, and more.

The Fashion-MNIST dataset we just loaded comes with a pre-specified training and test dataset partitioning scheme. The training set is stored at `image_path`. Then we convert them into Tensor format. Output these two dataset objects to obtain additional details:

```
>>> print(train_dataset)
Dataset FashionMNIST
    Number of datapoints: 60000
    Root location: ./
    Split: Train
    StandardTransform
```

```
Transform: Compose(
    ToTensor()
    )
>>> print(test_dataset)
Dataset FashionMNIST
    Number of datapoints: 10000
    Root location: ./
    Split: Test
    StandardTransform
Transform: Compose(
            ToTensor()
        )
```

As you can see, there are 60,000 training samples and 10,000 test samples.

Next, we load the training set into batches of 64 samples, as follows:

```
>>> from torch.utils.data import DataLoader
>>> batch_size = 64
>>> torch.manual_seed(42)
>>> train_dl = DataLoader(train_dataset, batch_size, shuffle=True)
```

In PyTorch, `DataLoader` is a utility that provides an efficient way to load and preprocess data from a dataset during training or evaluation of machine learning models. It essentially wraps around a dataset and provides methods to iterate over batches of data. This is particularly useful when working with large datasets that do not fit entirely in memory.

Key features of DataLoader:

- **Batching:** It automatically divides the dataset into batches of specified size, allowing you to work with mini-batches of data during training.
- **Shuffling:** You can set the `shuffle` parameter to `True` to shuffle the data before each epoch, which helps in reducing bias and improving convergence.

Feel free to inspect the image samples and their labels from the first batch, for example:

```
>>> data_iter = iter(train_dl)
>>> images, labels = next(data_iter)
>>> print(labels)
tensor([5, 7, 4, 7, 3, 8, 9, 5, 3, 1, 2, 3, 2, 3, 3, 7, 9, 9, 3, 2, 4, 6, 3, 5,
5, 3, 2, 0, 0, 8, 4, 2, 8, 5, 9, 2, 4, 9, 4, 4, 3, 4, 9, 7, 2, 0, 4, 5, 4, 8,
2, 6, 7, 0, 2, 0, 6, 3, 3, 5, 6, 0, 0, 8])
```

The label arrays do not include class names. Hence, we define them as follows and will use them for plotting later on:

```
>>> class_names = ['T-shirt/top', 'Trouser', 'Pullover', 'Dress', 'Coat',
'Sandal', 'Shirt', 'Sneaker', 'Bag', 'Ankle boot']
```

Take a look at the format of the image data as follows:

```
>>> print(images[0].shape)
torch.Size([1, 28, 28])
>>> print(torch.max(images), torch.min(images))
tensor(1.) tensor(0.)
```

Each image is represented as 28 * 28 pixels, whose values are in the range [0, 1].

Let's now display an image as follows:

```
>>> import numpy as np
>>> import matplotlib.pyplot as plt
>>> npimg = images[1].numpy()
>>> plt.imshow(np.transpose(npimg, (1, 2, 0)))
>>> plt.colorbar()
>>> plt.title(class_names[labels[1]])
>>> plt.show()
```

In PyTorch, `np.transpose(npimg, (1, 2, 0))` is used when visualizing images using `matplotlib`. `(1, 2, 0)` is a tuple representing the new order of dimensions. In PyTorch, images are represented in the format (`channels`, `height`, `width`). However, `matplotlib` expects images to be in the format (`height`, `width`, `channels`). `np.transpose(npimg, (1, 2, 0))` is used to rearrange the dimensions of the image array to match the format that `matplotlib` expects.

Refer to the following sneaker image – the end result:

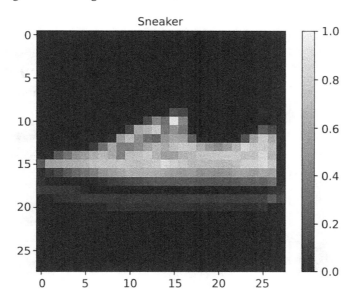

Figure 11.7: A training sample from Fashion-MNIST

Similarly, we display the first 16 training samples, as follows:

```
>>> for i in range(16):
...     plt.subplot(4, 4, i + 1)
...     plt.subplots_adjust(hspace=.3)
...     plt.xticks([])
...     plt.yticks([])
...     npimg = images[i].numpy()
...     plt.imshow(np.transpose(npimg, (1, 2, 0)), cmap="Greys")
...     plt.title(class_names[labels[i]])
... plt.show()
```

Refer to the following image for the result:

Figure 11.8: 16 training samples from Fashion-MNIST

🔍 **Quick tip:** Need to see a high-resolution version of this image? Open this book in the next-gen Packt Reader or view it in the PDF/ePub copy.

📖 **The next-gen Packt Reader** and a **free PDF/ePub copy** of this book are included with your purchase. Unlock them by scanning the QR code below or visiting `https://www.packtpub.com/unlock/9781835085622`.

In the next section, we will be building our CNN model to classify these clothing images.

Classifying clothing images with CNNs

As mentioned, the CNN model has two main components: the feature extractor composed of a set of convolutional and pooling layers, and the classifier backend, similar to a regular neural network.

Let's start this project by architecting the CNN model.

Architecting the CNN model

We import the necessary module and initialize a Sequential-based model:

```
>>> import torch.nn as nn
>>> model = nn.Sequential()
```

For the convolutional extractor, we are going to use three convolutional layers. We start with the first convolutional layer with 32 small-sized 3 * 3 filters. This is implemented with the following code:

```
>>> model.add_module('conv1',
                     nn.Conv2d(in_channels=1,
                               out_channels=32,
                               kernel_size=3)
                     )
>>> model.add_module('relu1', nn.ReLU())
```

Note that we use ReLU as the activation function.

The convolutional layer is followed by a max-pooling layer with a 2 * 2 filter:

```
>>> model.add_module('pool1', nn.MaxPool2d(kernel_size=2))
```

Here comes the second convolutional layer. It has 64 3 * 3 filters and comes with a ReLU activation function as well:

```
>>> model.add_module('conv2',
                     nn.Conv2d(in_channels=32,
                               out_channels=64,
                               kernel_size=3)
                     )
>>> model.add_module('relu2', nn.ReLU())
```

The second convolutional layer is followed by another max-pooling layer with a 2 * 2 filter:

```
>>> model.add_module('pool2', nn.MaxPool2d(kernel_size=2))
```

We continue adding the third convolutional layer. It has 128 3 * 3 filters at this time:

```
>>> model.add_module('conv3',
                     nn.Conv2d(in_channels=64,
                               out_channels=128,
                               kernel_size=3)
                     )
>>> model.add_module('relu3', nn.ReLU())
```

Let's take a pause here and see what the resulting filter maps are. We feed a random batch (of 64 samples) into the model we have built so far:

```
>>> x = torch.rand((64, 1, 28, 28))
>>> print(model(x).shape)
torch.Size([64, 128, 3, 3])
```

By providing the input shape as (64, 1, 28, 28), which means 64 images within the batch, and image size 28 * 28, the output has a shape of (64, 128, 3, 3), indicating feature maps with 128 channels and a spatial size of 3 * 3.

Next, we need to flatten these small 128 * 3 * 3 spatial representations to provide features to the downstream classifier backend:

```
>>> model.add_module('flatten', nn.Flatten())
```

As a result, we have a flattened output of shape (64, 1152), as computed by the following code:

```
>>> print(model(x).shape)
torch.Size([64, 1152])
```

For the classifier backend, we just use one hidden layer with 64 nodes:

```
>>> model.add_module('fc1', nn.Linear(1152, 64))
>>> model.add_module('relu4', nn.ReLU())
```

The hidden layer here is the regular fully connected dense layer, with ReLU as the activation function.

Finally, the output layer has 10 nodes representing 10 different classes in our case, along with a softmax activation:

```
>>> model.add_module('fc2', nn.Linear(64, 10))
>>> model.add_module('output', nn.Softmax(dim = 1))
```

Let's take a look at the model architecture, as follows:

```
>>> print(model)
Sequential(
  (conv1): Conv2d(1, 32, kernel_size=(3, 3), stride=(1, 1))
  (relu1): ReLU()
  (pool1): MaxPool2d(kernel_size=2, stride=2, padding=0, dilation=1, ceil_
mode=False)
  (conv2): Conv2d(32, 64, kernel_size=(3, 3), stride=(1, 1))
  (relu2): ReLU()
  (pool2): MaxPool2d(kernel_size=2, stride=2, padding=0, dilation=1, ceil_
mode=False)
  (conv3): Conv2d(64, 128, kernel_size=(3, 3), stride=(1, 1))
```

```
  (relu3): ReLU()
  (flatten): Flatten(start_dim=1, end_dim=-1)
  (fc1): Linear(in_features=1152, out_features=64, bias=True)
  (relu4): ReLU()
  (fc2): Linear(in_features=64, out_features=10, bias=True)
  (output): Softmax(dim=1)
)
```

If you want to display each layer in detail, including the shape of its output, and the number of its trainable parameters, you can use the torchsummary library. You can install it via pip and use it as follows:

```
>>> pip install torchsummary
>>> from torchsummary import summary
>>> summary(model, input_size=(1, 28, 28), batch_size=-1, device="cpu")
----------------------------------------------------------------
        Layer (type)               Output Shape         Param #
================================================================
            Conv2d-1          [-1, 32, 26, 26]             320
              ReLU-2          [-1, 32, 26, 26]               0
         MaxPool2d-3          [-1, 32, 13, 13]               0
            Conv2d-4          [-1, 64, 11, 11]          18,496
              ReLU-5          [-1, 64, 11, 11]               0
         MaxPool2d-6            [-1, 64, 5, 5]               0
            Conv2d-7           [-1, 128, 3, 3]          73,856
              ReLU-8           [-1, 128, 3, 3]               0
         Flatten-9                 [-1, 1152]               0
          Linear-10                   [-1, 64]          73,792
            ReLU-11                   [-1, 64]               0
          Linear-12                   [-1, 10]             650
         Softmax-13                   [-1, 10]               0
================================================================
Total params: 167,114
Trainable params: 167,114
Non-trainable params: 0
----------------------------------------------------------------
Input size (MB): 0.00
Forward/backward pass size (MB): 0.53
Params size (MB): 0.64
Estimated Total Size (MB): 1.17
----------------------------------------------------------------
```

As you may notice, the output from a convolutional layer is three-dimensional, where the first two are the dimensions of the feature maps and the third is the number of filters used in the convolutional layer. The size (the first two dimensions) of the max-pooling output is half of its input feature map in the example. Feature maps are downsampled by the pooling layer. You may want to see how many parameters there would be to be trained if you took out all the pooling layers. Actually, it is 4,058,314! So, the benefits of applying pooling are obvious: avoiding overfitting and reducing training costs.

You may wonder why the number of convolutional filters keeps increasing over the layers. Recall that each convolutional layer attempts to capture patterns of a specific hierarchy. The first convolutional layer captures low-level patterns, such as edges, dots, and curves. Then, the subsequent layers combine those patterns extracted in previous layers to form high-level patterns, such as shapes and contours. As we move forward in these convolutional layers, there are more and more combinations of patterns to capture in most cases. As a result, we need to keep increasing (or at least not decreasing) the number of filters in the convolutional layers.

Fitting the CNN model

Now it's time to train the model we just built.

First, we compile the model with Adam as the optimizer, cross-entropy as the loss function, and classification accuracy as the metric:

```
>>> device = torch.device("cuda:0")
# device = torch.device("cpu")
>>> model = model.to(device)
>>> loss_fn = nn.CrossEntropyLoss()
>>> optimizer = torch.optim.Adam(model.parameters(), lr=0.001)
```

Here, we employ the GPU for training, so we run `torch.device("cuda:0")` to specify the GPU device (the first device, with index 0) and allocate tensors on it. Opting for the CPU is a working, but comparatively slower option.

Next, we train the model by defining the following function:

```
>>> def train(model, optimizer, num_epochs, train_dl):
        for epoch in range(num_epochs):
            loss_train = 0
            accuracy_train = 0
            for x_batch, y_batch in train_dl:
                x_batch = x_batch.to(device)
                y_batch = y_batch.to(device)
                pred = model(x_batch)
                loss = loss_fn(pred, y_batch)
                loss.backward()
                optimizer.step()
```

```
                optimizer.zero_grad()
                loss_train += loss.item() * y_batch.size(0)
                is_correct = (torch.argmax(pred, dim=1) ==
                                                y_batch).float()
                accuracy_train += is_correct.sum().cpu()

            loss_train /= len(train_dl.dataset)
            accuracy_train /= len(train_dl.dataset)

            print(f'Epoch {epoch+1} - loss: {loss_train:.4f} - accuracy:
                    {accuracy_train:.4f}')
```

We will train the CNN model for 30 iterations and monitor the learning progress:

```
>>> num_epochs = 30
>>> train(model, optimizer, num_epochs, train_dl)
Epoch 1 - loss: 1.7253 - accuracy: 0.7385
Epoch 2 - loss: 1.6333 - accuracy: 0.8287
...
Epoch 10 - loss: 1.5572 - accuracy: 0.9041
...
Epoch 20 - loss: 1.5344 - accuracy: 0.9270
...
Epoch 29 - loss: 1.5249 - accuracy: 0.9362
Epoch 30 - loss: 1.5249 - accuracy: 0.9363
```

We are able to achieve an accuracy of around 94% on the training set. If you want to check the performance on the test set, you can do the following:

```
>>> test_dl = DataLoader(test_dataset, batch_size, shuffle=False)
>>> def evaluate_model(model, test_dl):
        accuracy_test = 0
        with torch.no_grad():
            for x_batch, y_batch in test_dl:
                pred = model.cpu()(x_batch)
                is_correct = torch.argmax(pred, dim=1) == y_batch
                accuracy_test += is_correct.float().sum().item()

        print(f'Accuracy on test set: {100 * accuracy_test / 10000} %')

>>> evaluate_model(model, test_dl)
Accuracy on test set: 90.25 %
```

The model achieves an accuracy of 90% on the test dataset. Note that this result may vary due to factors like differences in hidden layer initializations, or non-deterministic operations in GPUs.

Best practice

Operations that are better suited to execution on a GPU compared to a CPU typically involve parallelizable tasks that benefit from the massive parallelism and computational power offered by GPU architectures. Here are some examples:

- Matrix and convolutional operations
- Processing large batches of data simultaneously. Tasks that involve batch processing, such as training and inference on mini-batches in machine learning models, benefit from the parallel processing capabilities of GPUs.
- Forward and backward propagation in neural networks, which are typically faster on GPUs due to hardware acceleration.

Best practice

Operations that are better suited to execution on a CPU compared to a GPU typically involve less parallelizable tasks and require more sequential processing or small data sizes. Here are some examples:

- Preprocessing such as data loading, feature extraction, and data augmentation.
- Inference on small models. For small models or inference tasks with low computational requirements, performing operations on the CPU can be more cost-effective.
- Control flow operations. Operations involving conditional statements or loops are generally more efficient on the CPU due to its sequential processing nature.

You have seen how the trained model performs, and you may wonder what the convolutional filters look like. You will find out in the next section.

Visualizing the convolutional filters

We extract the convolutional filters from the trained model and visualize them with the following steps.

From the model summary, we know that the layers of conv1, conv2, and conv3 in the model are convolutional layers. Using the third convolutional layer as an example, we obtain its filters as follows:

```
>>> conv3_weight = model.conv3.weight.data
>>> print(conv3_weight.shape)
torch.Size([128, 64, 3, 3])
```

It's apparent that there are 128 filters, where each filter possesses dimensions of 3x3 and contains 64 channels.

Next, for simplification, we visualize only the first channel from the first 16 filters in four rows and four columns:

```
>>> n_filters = 16
>>> for i in range(n_filters):
...     weight = conv3_weight[i].cpu().numpy()
...     plt.subplot(4, 4, i+1)
...     plt.xticks([])
...     plt.yticks([])
...     plt.imshow(weight[0], cmap='gray')
... plt.show()
```

Refer to the following screenshot for the end result:

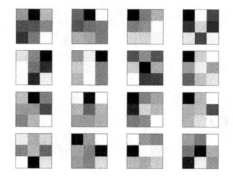

Figure 11.9: Trained convolutional filters

In a convolutional filter, the dark squares represent small weights and the white squares indicate large weights. Based on this intuition, we can see that the second filter in the second row detects the vertical line in a receptive field, while the third filter in the first row detects a gradient from light at the bottom right to dark at the top left.

In the previous example, we trained the clothing image classifier with 60,000 labeled samples. However, it is not easy to gather such a big labeled dataset in reality. Specifically, image labeling is expensive and time-consuming. How can we effectively train an image classifier with a limited number of samples? One solution is data augmentation.

Boosting the CNN classifier with data augmentation

Data augmentation means expanding the size of an existing training dataset in order to improve the generalization performance. It overcomes the cost involved in collecting and labeling more data. In PyTorch, we use the torchvision.transforms module to implement image augmentation in real time.

Flipping for data augmentation

There are many ways to augment image data. The simplest one is probably flipping an image horizontally or vertically. For instance, we will have a new image if we flip an existing image horizontally. To create a horizontally flipped image, we utilize `transforms.functional.hflip`, as follows:

```
>>> image = images[1]
>>> img_flipped = transforms.functional.hflip(image)
```

Let's take a look at the flipped image:

```
>>> def display_image_greys(image):
        npimg = image.numpy()
        plt.imshow(np.transpose(npimg, (1, 2, 0)), cmap="Greys")
        plt.xticks([])
        plt.yticks([])

>>> plt.figure(figsize=(8, 8))
>>> plt.subplot(1, 2, 1)
>>> display_image_greys(image)
>>> plt.subplot(1, 2, 2)
>>> display_image_greys(img_flipped)
>>> plt.show()
```

Refer to the following screenshot for the end result:

Figure 11.10: Horizontally flipped image for data augmentation

In training using data augmentation, we will create manipulated images using a random generator. For horizontal flipping, we will use transforms.RandomHorizontalFlip, which randomly flips images horizontally with a 50% chance, effectively augmenting the dataset. Let's see three output samples:

```
>>> torch.manual_seed(42)
>>>  flip_transform =
             transforms.Compose([transforms.RandomHorizontalFlip()])
>>> plt.figure(figsize=(10, 10))
>>> plt.subplot(1, 4, 1)
>>> display_image_greys(image)
>>> for i in range(3):
        plt.subplot(1, 4, i+2)
        img_flip = flip_transform(image)
        display_image_greys(img_flip)
```

Refer to the following screenshot for the end result:

Figure 11.11: Randomly horizontally flipped images for data augmentation

As you can see, the generated images are either horizontally flipped or not flipped.

In general, the horizontally flipped images convey the same message as the original ones. Vertically flipped images are not frequently seen, although you can generate them using transforms.RandomVerticalFlip. It is also worth noting that flipping only works in orientation-insensitive cases, such as classifying cats and dogs or recognizing parts of cars. On the contrary, it is dangerous to do so in cases where orientation matters, such as classifying between right and left turn signs.

Rotation for data augmentation

Instead of rotating every 90 degrees as in horizontal or vertical flipping, a small-to-medium degree rotation can also be applied in image data augmentation. Let's look at random rotation using transforms. We use RandomRotation in the following example:

```
>>> torch.manual_seed(42)
>>> rotate_transform =
          transforms.Compose([transforms. RandomRotation(20)])
>>> plt.figure(figsize=(10, 10))
>>> plt.subplot(1, 4, 1)
```

```
>>> display_image_greys(image)
>>> for i in range(3):
        plt.subplot(1, 4, i+2)
        img_rotate = rotate_transform(image)
    display_image_greys(img_rotate)
```

Refer to the following screenshot for the end result:

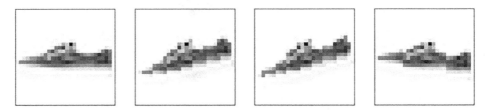

Figure 11.12: Rotated images for data augmentation

In the preceding example, the image is rotated by any degree ranging from -20 (counterclockwise) to 20 (clockwise).

Cropping for data augmentation

Cropping is another commonly used augmentation method. It generates new images by selecting a segment of the original image. Typically, this process is accompanied by resizing the cropped area to a predetermined output size to ensure uniform dimensions.

Now, let's explore how to utilize `transforms.RandomResizedCrop` to randomly select the aspect ratio of the cropped section and subsequently resize the result to match the original dimensions:

```
>>> torch.manual_seed(42)
>>> crop_transform = transforms.Compose([
        transforms.RandomResizedCrop(size=(28, 28), scale=(0.7, 1))])
>>> plt.figure(figsize=(10, 10))
>>> plt.subplot(1, 4, 1)
>>> display_image_greys(image)
>>> for i in range(3):
        plt.subplot(1, 4, i+2)
        img_crop = crop_transform(image)
        display_image_greys(img_crop)
```

Here, `size` specifies the size of the output image after cropping and resizing; `scale` defines the range of scaling for cropping. If set to (`min_scale`, `max_scale`), the crop area's size will be randomly chosen to be between `min_scale` and `max_scale` times the original image's size.

Refer to the following screenshot for the end result:

Figure 11.13: Cropped images for data augmentation

As you can see, `scale=(0.7, 1.0)` indicates that the crop area's size can vary between 70% and 100% of the original image's size.

Improving the clothing image classifier with data augmentation

Armed with several common augmentation methods, we will now apply them to train our image classifier on a small dataset in the following steps:

1. We start by constructing the transform function by combining all the data augmentation techniques we just discussed:

```
>>> torch.manual_seed(42)
>>> transform_train = transforms.Compose([
                    transforms.RandomHorizontalFlip(),
                    transforms.RandomRotation(10),
                    transforms.RandomResizedCrop(size=(28, 28),
                                                  scale=(0.9, 1)),
                    transforms.ToTensor(),
    ])
```

Here, we employ horizontal flip, rotation of up to 10 degrees, and cropping, with dimensions ranging from 90% to 100% of the original size.

2. We reload the training dataset with this transform function and only use 500 samples for training:

```
>>> train_dataset_aug = torchvision.datasets.FashionMNIST(
                                        root=image_path,
                                        train=True,
                                        transform=transform_train,
                                        download=False)
>>> from torch.utils.data import Subset
>>> train_dataset_aug_small = Subset(train_dataset_aug, torch.
arange(500))
```

We will see how data augmentation improves generalization and performance with a very small training set available.

3. Load this small but augmented training set into batches of 64 samples as we did previously:

```
>>> train_dl_aug_small = DataLoader(train_dataset_aug_small,
                                    batch_size,
                                    shuffle=True)
```

Note that even for the same original image, iterating using this data loader will produce different augmented images, which could be flipped, rotated, or cropped within the specified ranges.

4. Next, we initialize the CNN model using the same architecture we used previously and the optimizer accordingly:

```
>>> model = nn.Sequential()
>>> ...(here we skip repeating the same code)
>>> model = model.to(device)
>>> optimizer = torch.optim.Adam(model.parameters(), lr=0.001)
```

5. Now we train the model on the augmented small dataset:

```
>>> train(model, optimizer, 1000, train_dl_aug_small)
Epoch 1 - loss: 2.3013 - accuracy: 0.1400
...
Epoch 301 - loss: 1.6817 - accuracy: 0.7760
...
Epoch 601 - loss: 1.5006 - accuracy: 0.9620
...
Epoch 1000 - loss: 1.4904 - accuracy: 0.9720
```

We train the model for 1,000 iterations.

6. Let's see how it performs on the test set:

```
>>> evaluate_model(model, test_dl)
Accuracy on test set: 79.24%
```

The model with data augmentation has a classification accuracy of 79.24% on the test set. Note that this result may vary.

We also experimented with training without data augmentation, resulting in a test set accuracy of approximately 76%. When employing data augmentation, the accuracy improved to 79%. As always, feel free to fine-tune the hyperparameters as we did in *Chapter 6, Predicting Stock Prices with Artificial Neural Networks*, and see if you can further improve the classification performance.

Transfer learning is an alternative method to enhance the performance of a CNN classifier. Let's proceed to the following section.

Advancing the CNN classifier with transfer learning

Transfer learning is a machine learning technique where a model trained on one task is adapted or fine-tuned for a second related task. In transfer learning, the knowledge acquired during the training of the first task (source task) is leveraged to improve the learning of the second task (target task). This can be particularly useful when you have limited data for the target task because it allows you to transfer knowledge from a larger or more diverse dataset.

The typical workflow of transfer learning involves:

1. **Pretrained model:** Start with a pretrained model that has already been trained on a large and relevant dataset for a different but related task. This model is often a deep neural network, such as a CNN model for image tasks.

2. **Feature Extraction:** Use the pretrained model as a feature extractor. Remove the final classification layers (if they exist) and use the output of one of the intermediate layers as a feature representation for your data. These features can capture high-level patterns and information from the source task.

3. **Fine-Tuning:** Add new layers to the feature extractor. These new layers are specific to your target task and are typically randomly initialized. You then train the entire model, including the feature extractor and the new layers, on your target dataset. Fine-tuning allows the model to adapt to the specifics of the target task.

Prior to implementing transfer learning for our clothing image classification task, let's begin by exploring the evolution of CNN architectures and pretrained models. Even the early CNN architecture is still actively used today! The key point here is that all of these architectures are valuable tools in the modern DL toolbox, particularly when employed for transfer learning tasks.

Development of CNN architectures and pretrained models

The concept of CNNs for image processing dates back to the 1990s. Early architectures like **LeNet-5** (1998) demonstrated the potential of deep neural networks for image classification. LeNet-5 consists of two sets of convolutional layers, followed by two fully connected layers and one output layer. Each convolutional layer uses 5x5 kernels. LeNet-5 played a significant role in demonstrating the effectiveness of deep learning for image classification tasks. It was able to achieve high accuracy on the MNIST dataset, a widely used benchmark dataset for handwritten digit recognition.

The achievements of LeNet-5 paved the way for the creation of more complex architectures, such as **AlexNet** (2012). It consists of eight layers – five sets of convolutional layers followed by three fully connected layers. It used a ReLU activation function for the first time in a deep CNN and utilized dropout in the fully connected layers to prevent overfitting. Data augmentation techniques, such as random cropping and horizontal flipping, were employed to improve the model's generalization. The success of AlexNet triggered a renewed interest in neural networks and led to the development of even deeper and more complex architectures, including VGGNet, GoogLeNet, and ResNet, which have become foundational in computer vision.

VGGNet was introduced by the Visual Geometry Group at the University of Oxford in 2014. VGGNet follows a straightforward and uniform architecture. It consists of a series of convolutional layers, followed by max-pooling layers, with a stack of fully connected layers at the end. It primarily uses 3x3 convolutional filters, allowing the network to capture fine-grained spatial information. The most commonly used versions are VGG16 and VGG19, which have 16 and 19 layers in the network. They are often used as a starting point for transfer learning in various computer vision tasks.

In the same year, **GoogLeNet,** better known as **Inception**, was developed by Google. The hallmark of GoogLeNet is the inception module. Instead of using a single convolutional layer with a fixed filter size, inception modules use multiple filter sizes (1x1, 3x3, 5x5) in parallel. These parallel operations capture features at different scales and provide richer representations. Similar to VGGNet, pretrained GoogLeNet comes in different versions, such as InceptionV1, InceptionV2, InceptionV3, and Inception V4, each with variations and improvements in architecture.

ResNet, short for **Residual Network,** was introduced by Kaiming He et al. in 2015, to address the vanishing gradient problem – gradients of the loss function becoming extremely small in CNNs. Its core innovation is the use of residual connections. These blocks allow the network to skip certain layers during training. Instead of directly learning the desired mapping from input to output, residual blocks learn a residual mapping, which is added to the original input. Deeper networks were made possible this way. Its pretrained models come in various versions, such as ResNet-18, ResNet-34, ResNet-50, ResNet-101, and the extremely deep ResNet-152. Again, the numbers denote the depth of the network.

The development of CNN architectures and pretrained models continues with innovations like EfficientNet, MobileNet, and custom architectures for specific tasks. For instance, **MobileNet** models are designed to be highly efficient in terms of computational resources and memory usage. They are tailored for deployment on devices with limited hardware capabilities, such as smartphones, IoT devices, and edge devices.

 You can see all available pretrained models in PyTorch on this page: `https://pytorch.org/vision/stable/models.html#classification`.

The evolution of CNN architectures and the availability of pretrained models have revolutionized computer vision tasks. They have significantly improved the state of the art in image classification, object detection, segmentation, and many other applications.

Now, let's explore using a pretrained model to enhance our clothing image classifier.

Improving the clothing image classifier by fine-tuning ResNets

We will use the pre-trained ResNets, ResNet-18 to be specific, for transfer learning in the following steps:

1. We start by importing the pretrained ResNet-18 model from `torchvision`:

```
>>> from torchvision.models import resnet18
>>> my_resnet = resnet18(weights='IMAGENET1K_V1')
```

Here, IMAGENET1K refers to the pretrained model that was trained on the ImageNet-1K dataset (https://www.image-net.org/download.php) and V1 refers to version 1 of the pretrained model.

This is the pretrained model step.

2. Since the ImageNet-1K dataset comprises RGB images, the first convolutional layer in the original ResNet is designed for three-dimensional inputs. However, our FashionMNIST dataset contains grayscale images, so we need to modify it to accept one-dimensional inputs:

```
>>> my_resnet.conv1 = nn.Conv2d(1, 64, kernel_size=7, stride=2,
padding=3, bias=False)
```

We just change the first argument, the input dimension, from 3 to 1 in the original definition of the first convolutional:

```
self.conv1 = nn.Conv2d(3, 64, kernel_size=7, stride=2, padding=3,
bias=False)
```

3. Change the output layer to output 10 classes from 1,000 classes:

```
>>> num_ftrs = my_resnet.fc.in_features
>>> my_resnet.fc = nn.Linear(num_ftrs, 10)
```

Here, we only update the output size of the output layer.

Steps 2 and 3 prepare for the **fine-tuning** process.

4. Finally, we **fine-tune** the adapted pretrained model by training it on the full training set:

```
>>> my_resnet = my_resnet.to(device)
>>> optimizer = torch.optim.Adam(my_resnet.parameters(), lr=0.001)
>>> train(my_resnet, optimizer, 10, train_dl)
Epoch 1 - loss: 0.4797 - accuracy: 0.8256
Epoch 2 - loss: 0.3377 - accuracy: 0.8791
Epoch 3 - loss: 0.2921 - accuracy: 0.8944
Epoch 4 - loss: 0.2629 - accuracy: 0.9047
Epoch 5 - loss: 0.2336 - accuracy: 0.9157
Epoch 6 - loss: 0.2138 - accuracy: 0.9221
Epoch 7 - loss: 0.1911 - accuracy: 0.9301
Epoch 8 - loss: 0.1854 - accuracy: 0.9312
Epoch 9 - loss: 0.1662 - accuracy: 0.9385
Epoch 10 - loss: 0.1575 - accuracy: 0.9427
```

After only 10 iterations, an accuracy of 94% is achieved with the fine-tuned ResNet model.

5. How about its performance on the test set? Let's see the following:

```
>>> evaluate_model(my_resnet, test_dl)
Accuracy on test set: 91.01 %
```

We are able to boost the accuracy on the test set from 90% to 91%, with only 10 training iterations.

Transfer learning with CNNs is a powerful technique that allows you to leverage pretrained models and adapt them for your specific image classification tasks.

Summary

In this chapter, we worked on classifying clothing images using CNNs. We started with a detailed explanation of the individual components of a CNN model and learned how CNNs are inspired by the way our visual cells work. We then developed a CNN model to categorize fashion-MNIST clothing images from Zalando. We also talked about data augmentation and several popular image augmentation methods. We practiced transfer learning with ResNets, after discussing the evolution of CNN architectures and pretrained models.

In the next chapter, we will focus on another type of deep learning network: **Recurrent Neural Networks (RNNs)**. CNNs and RNNs are the two most powerful deep neural networks that make deep learning so popular nowadays.

Exercises

1. As mentioned before, can you try to fine-tune the CNN image classifier and see if you can beat what we have achieved?
2. Can you also employ the dropout technique to improve the CNN model?
3. Can you experiment with using the pretrained Vision Transformer model: https://huggingface.co/google/vit-base-patch16-224?

Join our book's Discord space

Join our community's Discord space for discussions with the authors and other readers:

https://packt.link/yuxi

12

Making Predictions with Sequences Using Recurrent Neural Networks

In the previous chapter, we focused on **Convolutional Neural Networks (CNNs)** and used them to deal with image-related tasks. In this chapter, we will explore **Recurrent Neural Networks (RNNs)**, which are suitable for sequential data and time-dependent data, such as daily temperature, DNA sequences, and customers' shopping transactions over time. You will learn how the recurrent architecture works and see variants of the model. We will then work on their applications, including sentiment analysis, time series prediction, and text generation.

We will cover the following topics in this chapter:

- Tracking sequential learning
- Learning the RNN architecture by example
- Training an RNN model
- Overcoming long-term dependencies with **Long Short-Term Memory (LSTM)**
- Analyzing movie review sentiment with RNNs
- Revisiting stock price forecasting with LSTM
- Writing your own War and Peace with LSTM

Introducing sequential learning

The machine learning problems we have solved so far in this book have been time independent. For example, ad click-through doesn't depend on the user's historical ad clicks under our previous approach; in face classification, the model only takes in the current face image, not previous ones. However, there are many cases in life that depend on time. For example, in financial fraud detection, we can't just look at the present transaction; we should also consider previous transactions so that we can model based on their discrepancy. Another example is **Part-of-Speech** (PoS) tagging, where we assign a PoS (verb, noun, adverb, and so on) to a word. Instead of solely focusing on the given word, we must look at some previous words, and sometimes the next words too.

In time-dependent cases like those just mentioned, the current output is dependent on not only the current input but also the previous inputs; note that the length of the previous inputs is not fixed. Using machine learning to solve such problems is called **sequence learning** or **sequence modeling**. Obviously, the time-dependent event is called a **sequence**. Besides events that occur in disjointed time intervals (such as financial transactions and phone calls), text, speech, and video are also sequential data.

You may be wondering why we can't just regularly model the sequential data by feeding in the entire sequence. This can be quite limiting as we have to fix the input size. One problem is that we will lose information if an important event lies outside of the fixed window. But can we just use a very large time window? Note that the feature space grows along with the window size. The feature space will become excessive if we want to cover enough events in a certain time window. Hence, overfitting can be another problem.

I hope you now see why we need to model sequential data in a different way. In the next section, we will talk about an example of a modeling technique used for modern sequence learning: RNNs.

Learning the RNN architecture by example

As you can imagine, RNNs stand out because of their recurrent mechanism. We will start with a detailed explanation of this in the next section. We will talk about different types of RNNs after that, along with some typical applications.

Recurrent mechanism

Recall that in feedforward networks (such as vanilla neural networks and CNNs), data moves one way, from the input layer to the output layer. In RNNs, the recurrent architecture allows data to circle back to the input layer. This means that data is not limited to a feedforward direction. Specifically, in a hidden layer of an RNN, the output from the previous time point will become part of the input for the current time point. The following diagram illustrates how data flows in an RNN in general:

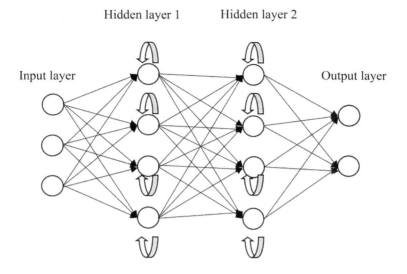

General Form of RNNs

Figure 12.1: The general form of an RNN

Such a recurrent architecture makes RNNs work well with sequential data, including time series (such as daily temperatures, daily product sales, and clinical EEG recordings) and general consecutive data with order (such as words in a sentence and DNA sequences). Take a financial fraud detector as an example; the output features from the previous transaction go into the training for the current transaction. In the end, the prediction for one transaction depends on all of its previous transactions. Let me explain the recurrent mechanism in a mathematical and visual way.

Suppose we have some inputs, x_t. Here, t represents a time step or a sequential order. In a feedforward neural network, we simply assume that inputs at different t are independent of each other. We denote the output of a hidden layer at a time step, t, as $h_t = f(x_t)$, where f is the abstract of the hidden layer.

This is depicted in the following diagram:

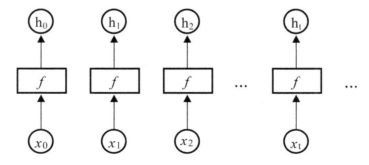

Feedforward NN

Figure 12.2: General form of a feedforward neural network

On the contrary, the feedback loop in an RNN feeds the information of the previous state to the current state. The output of a hidden layer of an RNN at a time step, t, can be expressed as $h_t = f(h_{t-1}, x_t)$. This is depicted in the following diagram:

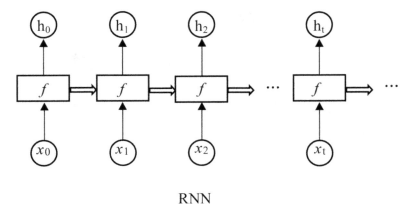

RNN

Figure 12.3: Unfolded recurrent layer over time steps

The same task, f, is performed on each element of the sequence, and the output, h_t, is dependent on the output that's generated from previous computations, h_{t-1}. The chain-like architecture captures the "memory" that has been calculated so far. This is what makes RNNs so successful in dealing with sequential data.

Moreover, thanks to the recurrent architecture, RNNs also have great flexibility in dealing with different combinations of input sequences and/or output sequences. In the next section, we will talk about different categories of RNNs based on input and output, including the following:

- Many to one
- One to many
- Many to many (synced)
- Many to many (unsynced)

We will start by looking at many-to-one RNNs.

Many-to-one RNNs

The most intuitive type of RNN is probably **many to one**. A many-to-one RNN can have input sequences with as many time steps as you want, but it only produces one output after going through the entire sequence. The following diagram depicts the general structure of a many-to-one RNN:

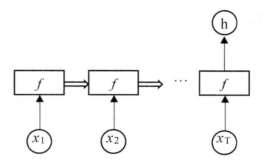

Many-to-one RNN

Figure 12.4: General form of a many-to-one RNN

Here, f represents one or more recurrent hidden layers, where an individual layer takes in its own output from the previous time step. Here is an example of three hidden layers stacking up:

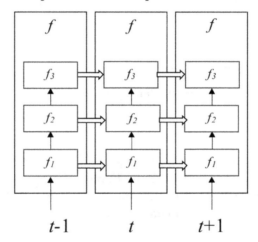

Figure 12.5: Example of three recurrent layers stacking up

Many-to-one RNNs are widely used for classifying sequential data. Sentiment analysis is a good example of this and is where the RNN reads the entire customer review, for instance, and assigns a sentiment score (positive, neutral, or negative sentiment). Similarly, we can also use RNNs of this kind in the topic classification of news articles. Identifying the genre of a song is another application as the model can read the entire audio stream. We can also use many-to-one RNNs to determine whether a patient is having a seizure based on an EEG trace.

One-to-many RNNs

One-to-many RNNs are the exact opposite of many-to-one RNNs. They take in only one input (not a sequence) and generate a sequence of outputs. A typical one-to-many RNN is presented in the following diagram:

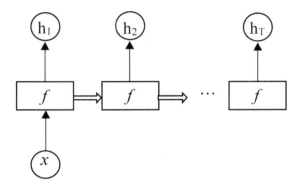

One-to-many RNN

Figure 12.6: General form of a one-to-many RNN

Again, f represents one or more recurrent hidden layers.

Note that "one" here refers to a single time step or a non-sequential input, rather than the number of input features.

One-to-many RNNs are commonly used as sequence generators. For example, we can generate a piece of music given a starting note and/or a genre. Similarly, we can write a movie script like a professional screenwriter using one-to-many RNNs with a starting word we specify. Image captioning is another interesting application: the RNN takes in an image and outputs the description (a sentence of words) of the image.

Many-to-many (synced) RNNs

The third type of RNN, many to many (synced), allows each element in the input sequence to have an output. Let's look at how data flows in the following many-to-many (synced) RNN:

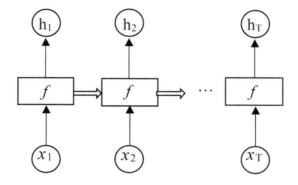

Many-to-many (synced) RNN

Figure 12.7: General form of a many-to-many (synced) RNN

As you can see, each output is calculated based on its corresponding input and all the previous outputs.

One common use case for this type of RNN is time series forecasting, where we want to perform rolling prediction at every time step based on the current and previously observed data. Here are some examples of time series forecasting where we can leverage many-to-many (synced) RNNs:

- Product sales each day for a store
- The daily closing price of a stock
- Power consumption of a factory each hour

They are also widely used in solving NLP problems, including PoS tagging, named-entity recognition, and real-time speech recognition.

Many-to-many (unsynced) RNNs

Sometimes, we only want to generate the output sequence *after* we've processed the entire input sequence. This is the **unsynced** version of a many-to-many RNN.

Refer to the following diagram for the general structure of a many-to-many (unsynced) RNN:

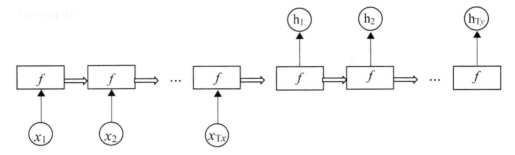

Many-to-many (unsynced) RNN

Figure 12.8: General form of a many-to-many (unsynced) RNN

Note that the length of the output sequence (Ty in the preceding diagram) can be different from that of the input sequence (Tx in the preceding diagram). This provides us with some flexibility.

This type of RNN is the go-to model for machine translation. In French-English translation, for example, the model first reads a complete sentence in French and then produces a translated sentence in English. Multi-step ahead forecasting is another popular example: sometimes, we are asked to predict sales for multiple days in the future when given data from the past month.

You have now learned about four types of RNN based on the model's input and output.

Wait, what about one-to-one RNNs? There is no such thing. One-to-one is just a regular feedforward model.

We will be applying some of these types of RNN to solve projects, including sentiment analysis and word generation, later in this chapter. Now, let's figure out how an RNN model is trained.

Training an RNN model

To explain how we optimize the weights (parameters) of an RNN, we first annotate the weights and the data on the network, as follows:

- U denotes the weights connecting the input layer and the hidden layer.
- V denotes the weights between the hidden layer and the output layer. Note here that we use only one recurrent layer for simplicity.
- W denotes the weights of the recurrent layer; that is, the feedback layer.
- x_t denotes the inputs at time step t.
- s_t denotes the hidden state at time step t.
- h_t denotes the outputs at time step t.

Next, we unfold the simple RNN model over three time steps: $t-1$, t, and $t+1$, as follows:

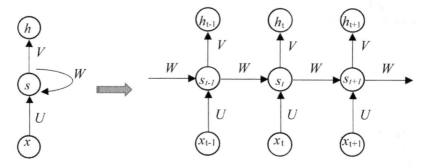

A basic one-layered RNN   Unfolded version

Figure 12.9: Unfolding a recurrent layer

We describe the mathematical relationships between the layers as follows:

*   We let a denote the activation function for the hidden layer. In RNNs, we usually choose tanh or ReLU as the activation function for the hidden layers.
*   Given the current input, x_t, and the previous hidden state, s_{t-1}, we compute the current hidden state, s_t, by $s_t = a(Ux_t + Ws_{t-1})$.

 Feel free to read *Chapter 6, Predicting Stock Prices with Artificial Neural Networks*, again to brush up on your knowledge of neural networks.

*   In a similar manner, we compute s_{t-1} based on

$$s_{t-2}:s_{t-1}=a(Ux_{t-1}+Ws_{t-2})$$

*   We repeat this until s_1, which depends on:

$$s_0:s_1=a(Ux_1+Ws_0)$$

We usually set s_0 to all zeros.

*   We let g denote the activation function for the output layer. It can be a sigmoid function if we want to perform binary classification, a softmax function for multi-class classification, and a simple linear function (that is, no activation) for regression.
*   Finally, we compute the output at time step t:

$$h_t:h_t=g(Vs_t)$$

With the dependency in hidden states over time steps (that is, s_t depends on s_{t-1}, s_{t-1} depends on s_{t-2}, and so on), the recurrent layer brings memory to the network, which captures and retains information from all the previous time steps.

As we did for traditional neural networks, we apply the backpropagation algorithm to optimize all the weights, *U*, *V*, and *W*, in RNNs. However, as you may have noticed, the output at a time step is indirectly dependent on all the previous time steps (h^t depends on s_t, while s_t depends on all the previous ones). Hence, we need to compute the loss over all previous *t*-1 time steps, besides the current time step. Consequently, the gradients of the weights are calculated this way. For example, if we want to compute the gradients at time step *t* = 4, we need to backpropagate the previous four time steps (*t* = 3, *t* = 2, *t* = 1, *t* = 0) and sum up the gradients over these five time steps. This version of the backpropagation algorithm is called **Back Propagation Through Time** (BPTT).

The recurrent architecture enables RNNs to capture information from the very beginning of the input sequence. This advances the predictive capability of sequence learning. You may be wondering whether vanilla RNNs can handle long sequences. They can in theory, but not in practice due to the **vanishing gradient** problem. A vanishing gradient means the gradient will become vanishingly small over long time steps, which prevents the weight from updating. I will explain this in detail in the next section, as well as introduce a variant architecture, LSTM, that helps solve this issue.

Overcoming long-term dependencies with LSTM

Let's start with the vanishing gradient issue in vanilla RNNs. Where does it come from? Recall that during backpropagation, the gradient decays along with each time step in the RNN (that is, $s_t=a(Ux_t+Ws_{t-1})$; early elements in a long input sequence will have little contribution to the computation of the current gradient. This means that vanilla RNNs can only capture the temporal dependencies within a short time window. However, dependencies between time steps that are far away are sometimes critical signals to the prediction. RNN variants, including LSTM and **gated recurrent units** (GRUs), are specifically designed to solve problems that require learning long-term dependencies.

 We will be focusing on LSTM in this book as it is a lot more popular than GRU. LSTM was introduced a decade earlier and is more mature than GRU. If you are interested in learning more about GRU and its applications, feel free to check out *Hands-On Deep Learning Architectures with Python* by Yuxi Hayden Liu (Packt Publishing).

In LSTM, we use a grating mechanism to handle long-term dependencies. Its magic comes from a memory unit and three information gates built on top of the recurrent cell. The word "gate" is taken from the logic gate in a circuit (https://en.wikipedia.org/wiki/Logic_gate). It is basically a sigmoid function whose output value ranges from 0 to 1. 0 represents the "off" logic, while 1 represents the "on" logic.

The LSTM version of the recurrent cell is depicted in the following diagram, right after the vanilla version for comparison:

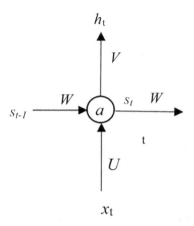

Recurrent cell of a vanilla RNN

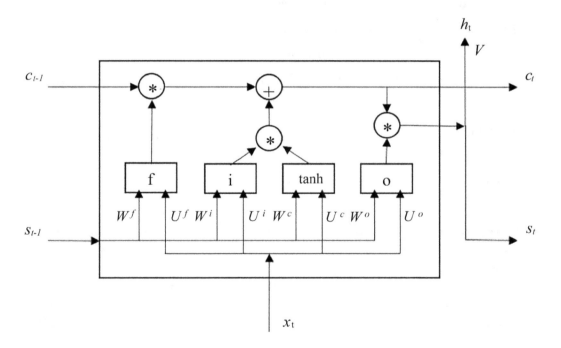

Recurrent cell of an LSTM RNN

Figure 12.10: Recurrent cell in vanilla RNNs versus LSTM RNNs

Let's look at the LSTM recurrent cell in detail from left to right:

- c_t is the **memory unit**. It memorizes information from the very beginning of the input sequence.
- *f* stands for the **forget gate**. It determines how much information from the previous memory state, c_{t-1}, to forget, or, in other words, how much information to pass forward. Let W^f denote the weights between the forget gate and the previous hidden state, s_{t-1}, and U^f denote the weights between the forget gate and the current input, x_t.
- *i* represents the **input gate**. It controls how much information from the current input to put through. W^i and U^i are the weights connecting the input gate to the previous hidden state, s_{t-1}, and the current input, x_t, respectively.
- *tanh* is simply the activation function for the hidden state. It acts as the *a* in the vanilla RNN. Its output is computed based on the current input, x_t, along with the associated weights, U^c, the previous hidden state, s_{t-1}, and the corresponding weights, W^c.
- o serves as the **output gate**. It defines how much information is extracted from the internal memory for the output of the entire recurrent cell. As always, W^o and U^o are the associated weights for the previous hidden state and current input, respectively.

We describe the relationship between these components as follows:

- The output of the forget gate, *f*, at time step *t* is computed as

$$f=sigmoid(W^f s_{t-1}+U^f x_t)$$

- The output of the input gate, *i*, at time step *t* is computed as

$$i=sigmoid(W^i s_{t-1}+U^i x_t)$$

- The output of the tanh activation, *c'*, at time step *t* is computed as

$$c'=tanh(W^c s_{t-1}+U^c x_t)$$

- The output of the output gate, *o*, at time step *t* is computed as

$$o=sigmoid(W^o s_{t-1}+U^o x_t)$$

- The memory unit, c_t, at time step *t* is updated using $c_t=f.*c_{t-1}+i.*c'$ (here, the operator *.** denotes element-wise multiplication). Again, the output of a sigmoid function has a value from 0 to 1. Hence, the forget gate, *f*, and input gate, *i*, control how much of the previous memory, c_{t-1}, and the current memory input, *c'*, to carry forward, respectively.
- Finally, we update the hidden state, s_t, at time step *t* by $s_t=o.*c_t$. Here, the output gate, *o*, governs how much of the updated memory unit, c_t, will be used as the output of the entire cell.

Best practice

LSTM is often considered the default choice for RNN models in practice due to its ability to effectively capture long-term dependencies in sequential data while mitigating the vanishing gradient problem. However, GRUs are also commonly used depending on the specific task and dataset characteristics. The choice between LSTM and GRU depends on the following factors:

- **Model complexity:** LSTMs typically have more parameters than GRUs due to their additional gating mechanisms. If you have limited computational resources or are working with smaller datasets, GRUs may be more suitable due to their simpler architecture.

- **Training speed:** GRUs are generally faster to train than LSTMs. If training time is a concern, GRUs might be a better choice.

- **Performance:** LSTMs tend to have better performance on tasks that require modeling long-term dependencies in sequential data. If your task involves capturing complex temporal patterns and you're concerned about overfitting, LSTMs might be preferable.

As always, we apply the **BPTT** algorithm to train all the weights in LSTM RNNs, including four sets each of weights, U and W, associated with three gates and the tanh activation function. By learning these weights, the LSTM network explicitly models long-term dependencies in an efficient way. Hence, LSTM is the go-to or default RNN model in practice.

Next, you will learn how to use LSTM RNNs to solve real-world problems. We will start by categorizing movie review sentiment.

Analyzing movie review sentiment with RNNs

So, here comes our first RNN project: movie review sentiment. We'll use the IMDb (https://www.imdb.com/) movie review dataset (https://ai.stanford.edu/~amaas/data/sentiment/) as an example. It contains 25,000 highly popular movie reviews for training and another 25,000 for testing. Each review is labeled as 1 (positive) or 0 (negative). We'll build our RNN-based movie sentiment classifier in the following three sections: *Analyzing and preprocessing the movie review data, Developing a simple LSTM network,* and *Boosting the performance with multiple LSTM layers.*

Analyzing and preprocessing the data

We'll start with data analysis and preprocessing, as follows:

1. PyTorch's `torchtext` has a built-in IMDb dataset, so first, we load the dataset:

    ```
    >>> from torchtext.datasets import IMDB
    >>> train_dataset = list(IMDB(split='train'))
    >>> test_dataset = list(IMDB(split='test'))
    >>> print(len(train_dataset), len(test_dataset))
    25000 25000
    ```

    > 💡 **Quick tip:** Enhance your coding experience with the **AI Code Explainer** and **Quick Copy** features. Open this book in the next-gen Packt Reader. Click the **Copy** button (**1**) to quickly copy code into your coding environment, or click the **Explain** button (**2**) to get the AI assistant to explain a block of code to you.

     📱 **The next-gen Packt Reader** is included for free with the purchase of this book. Unlock it by scanning the QR code below or visiting
    `https://www.packtpub.com/unlock/9781835085622`.

    We just load 25,000 training samples and 25,000 test samples.

    If you encounter any errors while running the code, consider installing the `torchtext` and `portalocker` packages. You could use the following commands for installation via conda:

    ```
    conda install -c torchtext
    conda install -c conda-forge portalocker
    ```

Or, via `pip`:

```
pip install portalocker
```

2. Now, let's explore the vocabulary within the training set:

```
>>> import re
>>> from collections import Counter, OrderedDict
>>> def tokenizer(text):
        text = re.sub('<[^>]*>', '', text)
        emoticons = re.findall('(?::|;|=)(?:-)?(?:\)|\(|D|P)', text.
lower())
        text = re.sub('[\W]+', ' ', text.lower()) +\
                        ' '.join(emoticons).replace('-', '')
        tokenized = text.split()
        return tokenized
>>>
>>> token_counts = Counter()
>>> train_labels = []
>>> for label, line in train_dataset:
        train_labels.append(label)
        tokens = tokenizer(line)
        token_counts.update(tokens)
>>> print('Vocab-size:', len(token_counts))
Vocab-size: 75977
>>> print(Counter(train_labels))
Counter({1: 12500, 2: 12500})
```

Here, we define a function to extract tokens (words, in our case) from a given document (movie review, in our case). It first removes HTML-like tags, then extracts and standardizes emoticons, removes non-alphanumeric characters, and tokenizes the text into a list of words for further processing. We store the tokens and their occurrences in the `Counter` object `token_counts`.

As evident, the training set comprises approximately 76,000 unique words, and it exhibits a perfect balance with an equal count of positive (labeled as "2") and negative (labeled as "1") samples.

3. We will feed the word tokens into an embedding layer, nn.Embedding. The embedding layer requires integer input because it's specifically designed to handle discrete categorical data, such as word indices, and transform them into continuous representations that a neural network can work with and learn from. Therefore, we need to first encode each token into a unique integer as follows:

```
>>> from torchtext.vocab import vocab

>>> sorted_by_freq_tuples = sorted(token_counts.items(), key=lambda x:
x[1],
                                    reverse=True)
>>> ordered_dict = OrderedDict(sorted_by_freq_tuples)
>>> vocab_mapping = vocab(ordered_dict)
```

We use the vocab module in PyTorch to create a vocabulary (token mapping) based on the frequency of words in the corpus. But this vocabulary is not complete yet. Let's see why in the next two steps.

4. When examining the document lengths within the training set, you'll notice that they range from 10 to 2,498 words. It's common practice to apply padding to sequences to ensure uniform length during batch processing. So, we insert the special token, "<pad>", representing padding into the vocabulary mapping at index 0 as a **placeholder**:

```
>>> vocab_mapping.insert_token("<pad>", 0)
```

5. We also need to handle unseen words during inference. Similar to the previous step, we insert the special token "<unk>" (short for "unknown") into the vocabulary mapping at index 1. The token represents out-of-vocabulary words or tokens that are not found in the training data:

```
>>> vocab_mapping.insert_token("<unk>", 1)
>>> vocab_mapping.set_default_index(1)
```

We also set the default vocabulary mapping to 1. This means "<unk>" (index 1) is used as the default index for unseen or out-of-vocabulary words.

Let's take a look at the following examples showing the mappings of given words, including an unseen one:

```
>>> print([vocab_mapping[token] for token in ['this', 'is', 'an',
'example']])
[11, 7, 35, 462]
>>> print([vocab_mapping[token] for token in ['this', 'is', 'example2']])
[11, 7, 1]
```

By now, we have the complete vocabulary mapping.

Best practice

Using special tokens like `<pad>` and `<unk>` in RNNs is a common practice for handling variable-length sequences and out-of-vocabulary words. Here are some best practices for their usage:

- Use `<pad>` tokens to pad sequences to a fixed length. This ensures that all input sequences have the same length, which is necessary for efficient batch processing in neural networks. Pad sequences at the end rather than the beginning to preserve the order of the input data. When tokenizing text data, assign a unique integer index to the `<pad>` token and ensure that it corresponds to a vector of zeros in the embedding matrix.

- Use `<unk>` tokens to represent out-of-vocabulary words that are not present in the vocabulary of the model. During inference, replace any words that are not present in the vocabulary with the `<unk>` token to ensure that the model can process the input.

- Exclude `<pad>` tokens from contributing to the loss during training to avoid skewing the learning process.

- Monitor the distribution of `<unk>` tokens in the dataset to assess the prevalence of out-of-vocabulary words and adjust the vocabulary size accordingly.

6. Next, we define the function defining how batches of samples should be collated:

```
>>> import torch
>>> import torch.nn as nn
>>> device = torch.device("cuda" if torch.cuda.is_available() else "cpu")
>>> text_transform = lambda x: [vocab[token] for token in tokenizer(x)]
>>> def collate_batch(batch):
        label_list, text_list, lengths = [], [], []
        for _label, _text in batch:
            label_list.append(1. if _label == 2 else 0.)
            processed_text = [vocab_mapping[token] for token in
tokenizer(_text)]
            text_list.append(torch.tensor(processed_text, dtype=torch.
int64))
            lengths.append(len(processed_text))
        label_list = torch.tensor(label_list)
        lengths = torch.tensor(lengths)
        padded_text_list = nn.utils.rnn.pad_sequence(
                            text_list, batch_first=True)
        return padded_text_list.to(device), label_list.to(device),
                                            lengths.to(device)
```

Besides generating inputs and label outputs as we used to do, we also generate the length of individual samples in a given batch. Note that we convert the positive label from the raw 2 to 1 here, for label standardization and loss function compatibility for binary classification. The length information is used for handling variable-length sequences efficiently. Take a small batch of four samples and examine the processed batch:

```
>>> from torch.utils.data import DataLoader
>>> torch.manual_seed(0)
>>> dataloader = DataLoader(train_dataset, batch_size=4, shuffle=True,
                            collate_fn=collate_batch)
>>> text_batch, label_batch, length_batch = next(iter(dataloader))
>>> print(text_batch)
tensor([[   46,    8,  287,   21,   16,    2,   76, 3987,    3,
          226,  10,  381,    2,  461,   14,   65,    9, 1208,   17,    8,
           13,  856,    2,  156,   70,  398,   50,   32, 2338,   67,  103,
            6,  110,   19,    9,    2,  130,    2,  153,   12,  14,   65,
         1002,   14,    4, 1143,  226,    6, 1061,   31,  2, 1317,  293,
           10,   61,  542, 1459,   24,    6,  105,
         ...
         ...
         ...

            0,    0,    0,    0,    0,    0,    0,    0,    0,    0,  0,
            0,    0,    0,    0,    0,    0,    0,    0,    0, 0,    0,
            0,    0,    0,    0,    0]], device='cuda:0')
>>> print(label_batch)
>>> tensor([0., 1., 1., 0.], device='cuda:0')
>>> print(length_batch)
tensor([106,  76, 247, 158], device='cuda:0')
>>> print(text_batch.shape)
torch.Size([4, 247])
```

You can see the processed text sequences have been standardized to a length of 247 tokens, with the first, second, and fourth samples padded with 0s.

7. Finally, we batch the training and testing set:

```
>>> batch_size = 32
>>> train_dl = DataLoader(train_dataset, batch_size=batch_size,
                          shuffle=True, collate_fn=collate_batch)
>>> test_dl = DataLoader(test_dataset, batch_size=batch_size,
                         shuffle=False, collate_fn=collate_batch)
```

The generated data loaders are ready to use for sentiment prediction.

Let's move on to building an LSTM network.

Building a simple LSTM network

Now that the training and testing data loaders are ready, we can build our first RNN model with an embedding layer that encodes the input word tokens, and an LSTM layer followed by a fully connected layer:

1. First, we define the network hyperparameters, including the input dimension and the embedding dimension of the embedding layer:

```
>>> vocab_size = len(vocab_mapping)
>>> embed_dim = 32
```

We also define the number of hidden nodes in the LSTM layer and the fully connected layer:

```
>>> rnn_hidden_dim = 50
>>> fc_hidden_dim = 32
```

2. Next, we build our RNN model class:

```
>>> class RNN(nn.Module):
        def __init__(self, vocab_size, embed_dim, rnn_hidden_dim, fc_
    hidden_dim):
            super().__init__()
            self.embedding = nn.Embedding(vocab_size,
                                          embed_dim,
                                          padding_idx=0)
            self.rnn = nn.LSTM(embed_dim, rnn_hidden_dim,
                             batch_first=True)
            self.fc1 = nn.Linear(rnn_hidden_dim, fc_hidden_dim)
            self.relu = nn.ReLU()
            self.fc2 = nn.Linear(fc_hidden_dim, 1)
            self.sigmoid = nn.Sigmoid()

        def forward(self, text, lengths):
            out = self.embedding(text)
            out = nn.utils.rnn.pack_padded_sequence(
                                      out, lengths.cpu().numpy(),
                                      enforce_sorted=False,
                                      batch_first=True)
```

```
        out, (hidden, cell) = self.rnn(out)
        out = hidden[-1, :, :]
        out = self.fc1(out)
        out = self.relu(out)
        out = self.fc2(out)
        out = self.sigmoid(out)
        return out
```

The nn.Embedding layer is used to convert input word indices into dense word embeddings. The padding_idx parameter is set to 0, indicating that padding tokens should be ignored during embedding.

The recurrent layer, nn.LSTM, takes the embedded input sequence and processes it sequentially. batch_first=True means that the input has a batch size as the first dimension.

The fully connected hidden layer, fc1, follows the LSTM layer, and the ReLU activation is applied to the output of the fully connected layer.

The final layer has a single output because this model is used for binary classification (sentiment analysis).

In the forward pass method, pack_padded_sequence is used to pack and pad sequences for efficient processing in the LSTM layer. The packed sequence is passed through the LSTM layer, and the final hidden state (hidden[-1, :, :]) is extracted.

3. We then create an instance of the LSTM model with the specific hyperparameters we defined earlier. We also ensure that the model is placed on the specified computing device (GPU if available) for training and inference:

```
>>> model = RNN(vocab_size, embed_dim, rnn_hidden_dim, fc_hidden_dim)
>>> model = model.to(device)
```

4. As for the loss function, we use nn.BCELoss() since it is a binary classification problem. We also set the corresponding optimizer and try with a learning rate of 0.003 as follows:

```
>>> loss_fn = nn.BCELoss()
>>> optimizer = torch.optim.Adam(model.parameters(), lr=0.003)
```

5. Now, we define a training function responsible for training the model for one iteration:

```
>>> def train(model, dataloader, optimizer):
        model.train()
        total_acc, total_loss = 0, 0
        for text_batch, label_batch, length_batch in dataloader:
            optimizer.zero_grad()
            pred = model(text_batch, length_batch)[:, 0]
            loss = loss_fn(pred, label_batch)
```

```
            loss.backward()
            optimizer.step()
            total_acc += ((pred>=0.5).float() == label_batch)
                                        .float().sum().item()
            total_loss += loss.item()*label_batch.size(0)

        total_loss /= len(dataloader.dataset)
        total_acc /= len(train_dl.dataset)
        print(f'Epoch {epoch+1} - loss: {total_loss:.4f} - accuracy:
            {total_acc:.4f}')
```

It also displays the training loss and accuracy at the end of an iteration.

6. We then train the model for 10 iterations:

```
>>> torch.manual_seed(0)
>>> num_epochs = 10
>>> for epoch in range(num_epochs):
        train(model, train_dl, optimizer)
Epoch 1 - loss: 0.5899 - accuracy: 0.6707
Epoch 2 - loss: 0.4354 - accuracy: 0.8019
Epoch 3 - loss: 0.2762 - accuracy: 0.8888
Epoch 4 - loss: 0.1766 - accuracy: 0.9341
Epoch 5 - loss: 0.1215 - accuracy: 0.9563
Epoch 6 - loss: 0.0716 - accuracy: 0.9761
Epoch 7 - loss: 0.0417 - accuracy: 0.9868
Epoch 8 - loss: 0.0269 - accuracy: 0.9912
Epoch 9 - loss: 0.0183 - accuracy: 0.9943
Epoch 10 - loss: 0.0240 - accuracy: 0.9918
```

The training accuracy is close to 100% after 10 iterations.

7. Finally, we evaluate the performance on the test set:

```
>>> def evaluate(model, dataloader):
        model.eval()
        total_acc = 0
        with torch.no_grad():
        for text_batch, label_batch, lengths in dataloader:
            pred = model(text_batch, lengths)[:, 0]
            total_acc += ((pred>=0.5).float() == label_batch)
                                        .float().sum().item()
```

```
print(f'Accuracy on test set: {100 * total_acc/len(dataloader.dataset)}
%')

>>> evaluate(model, test_dl)
Accuracy on test set: 86.1 %
```

We obtained a test accuracy of 86%.

Stacking multiple LSTM layers

We can also stack two (or more) recurrent layers. The following diagram shows how two recurrent layers can be stacked:

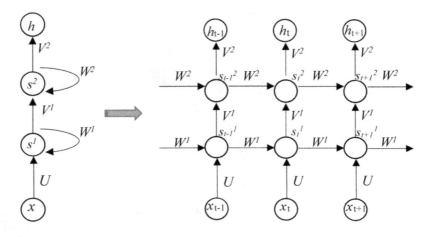

Two-layered RNN   Unfolded version

Figure 12.11: Unfolding two stacked recurrent layers

In PyTorch, stacking multiple RNN layers is simple. Using LSTM as an example once more, it suffices to specify the number of LSTM layers in the `num_layers` argument:

```
>>> nn.LSTM(embed_dim, rnn_hidden_dim, num_layers=2, batch_first=True)
```

In this example, we stack two LSTM layers. Feel free to experiment with a multi-layer RNN model and see whether you can beat the previous single-layered model.

With that, we've just finished the review sentiment classification project using RNNs. In the next project, we will revisit stock price prediction and solve it using RNNs.

Revisiting stock price forecasting with LSTM

Recall in *Chapter 6, Predicting Stock Prices with Artificial Neural Networks*, we derived features from past prices and performance within a specific time step and then trained a standard neural network. In this instance, we will utilize RNNs as the sequential model and harness features from five consecutive time steps rather than just one. Let's examine the process in the following steps:

1. Initially, we load the stock data, create the features and labels, and then split it into training and test sets, mirroring our approach in *Chapter 6*:

```
>>> data_raw = pd.read_csv('19900101_20230630.csv', index_col='Date')
>>> data = generate_features(data_raw)
>>> start_train = '1990-01-01'
>>> end_train = '2022-12-31'
>>> start_test = '2023-01-01'
>>> end_test = '2023-06-30'
>>> data_train = data.loc[start_train:end_train]
>>> X_train = data_train.drop('close', axis=1).values
>>> y_train = data_train['close'].values
>>> data_test = data.loc[start_test:end_test]
>>> X_test = data_test.drop('close', axis=1).values
>>> y_test = data_test['close'].values
```

Here, we reuse the feature and label generation function, `generate_features`, defined in *Chapter 6*. Similarly, we scale the feature space using `StandardScaler` and covert data into `FloatTensor`:

```
>>> from sklearn.preprocessing import StandardScaler
>>> scaler = StandardScaler()
>>> X_scaled_train = torch.FloatTensor(scaler.fit_transform(X_train))
>>> X_scaled_test = torch.FloatTensor(scaler.transform(X_test))
>>> y_train_torch = torch.FloatTensor(y_train)
>>> y_test_torch = torch.FloatTensor(y_test)
```

2. Next, we define a function to create sequences:

```
>>> def create_sequences(data, labels, seq_length):
        sequences = []
        for i in range(len(data) - seq_length):
            seq = data[i:i+seq_length]
            label = labels[i+seq_length-1]
            sequences.append((seq, label))
        return sequences

>>> seq_length = 5
```

```
>>> sequence_train = create_sequences(X_scaled_train, y_train_torch, seq_
length)
>>> sequence_test = create_sequences(X_scaled_test, y_test_torch, seq_
length)
```

Here, every generated sequence comprises two components: the input sequence, which encompasses features from five successive days, and the label, representing the price of the last day in that five-day period. We generate sequences for the training and test sets respectively.

3. Subsequently, we establish a data loader for the training sequences in preparation for model construction and training:

```
>>> batch_size = 128
>>> train_dl = DataLoader(sequence_train, batch_size=batch_size,
                          shuffle=True)
```

We set 128 as the batch size in this project.

4. Now, we define an RNN model with a two-layered LSTM followed by a fully connected layer and an output layer for regression:

```
>>> class RNN(nn.Module):
        def __init__(self, input_dim, rnn_hidden_dim, fc_hidden_dim):
            super().__init__()
            self.rnn = nn.LSTM(input_dim, rnn_hidden_dim, 2,
                               batch_first=True)
            self.fc1 = nn.Linear(rnn_hidden_dim, fc_hidden_dim)
            self.relu = nn.ReLU()
            self.fc2 = nn.Linear(fc_hidden_dim, 1)

        def forward(self, x):
            out, (hidden, cell) = self.rnn(x)
            out = hidden[-1, :, :]
            out = self.fc1(out)
            out = self.relu(out)
            out = self.fc2(out)
            return out
```

The LSTM layer captures sequential dependencies in the input data, and the fully connected layers perform the final regression.

5. Next, we initiate a model after specifying the input dimension and hidden layer dimensions, and use MSE as the loss function:

```
>>> rnn_hidden_dim = 16
>>> fc_hidden_dim = 16
>>> model = RNN(X_train.shape[1], rnn_hidden_dim, fc_hidden_dim)
```

```
>>> device = torch.device("cuda" if torch.cuda.is_available() else "cpu")
>>> model = model.to(device)
>>> loss_fn = nn.MSELoss()
>>> optimizer = torch.optim.Adam(model.parameters(), lr=0.01)
```

Small to medium values (like 16) are often used as starting points for hidden dimensions in RNNs, for computational efficiency. The chosen optimizer (Adam) and learning rate (0.01) are hyperparameters that can be tuned for better performance.

6. Next, we train the model for 1000 iterations as follows:

```
>>> def train(model, dataloader, optimizer):
        model.train()
        total_loss = 0
        for seq, label in dataloader:
            optimizer.zero_grad()
            pred = model(seq.to(device))[:, 0]
            loss = loss_fn(pred, label.to(device))
            loss.backward()
            optimizer.step()
            total_loss += loss.item()*label.size(0)
        return total_loss/len(dataloader.dataset)

>>> num_epochs = 1000
>>> for epoch in range(num_epochs):
>>>     loss = train(model, train_dl, optimizer)
>>>     if epoch % 100 == 0:
>>>         print(f'Epoch {epoch+1} - loss: {loss:.4f}')
Epoch 1 - loss: 24611083.8868
Epoch 101 - loss: 5483.5394
Epoch 201 - loss: 11613.8535
Epoch 301 - loss: 4459.1431
Epoch 401 - loss: 4646.8745
Epoch 501 - loss: 4046.1726
Epoch 601 - loss: 3583.5710
Epoch 701 - loss: 2846.1768
Epoch 801 - loss: 2417.1702
Epoch 901 - loss: 2814.3970
```

The MSE during training is displayed every 100 iterations.

7. Finally, we apply the trained model on the test set and evaluate the performance:

```
>>> predictions, y = [], []
>>> for seq, label in sequence_test:
```

```
         with torch.no_grad():
              pred = model.cpu()(seq.view(1, seq_length, X_test.shape[1]))
   [:, 0]

              predictions.append(pred)
              y.append(label)
>>> from sklearn.metrics import mean_squared_error, mean_absolute_error,
   r2_score
>>> print(f'R^2: {r2_score(y, predictions):.3f}')
R^2: 0.897
```

We are able to obtain an R^2 of 0.9 on the test set. You may observe that this doesn't outperform our previous standard neural network. The reason is that we have a relatively small training dataset of only eight thousand samples. RNNs typically require a larger dataset to excel.

The two RNN models we've explored up to this point followed the many-to-one structure. In our upcoming project, we'll create an RNN using the many-to-many structure, and the objective is to generate a "novel."

Writing your own War and Peace with RNNs

In this project, we'll work on an interesting language modeling problem – text generation.

An RNN-based text generator can write anything, depending on what text we feed it. The training text can be from a novel such as *A Game of Thrones*, a poem from Shakespeare, or the movie scripts for *The Matrix*. The artificial text that's generated should read similarly (but not identically) to the original one if the model is well trained. In this section, we are going to write our own *War and Peace* with RNNs, a novel written by the Russian author Leo Tolstoy. Feel free to train your own RNNs on any of your favorite books.

We will start with data acquisition and analysis before constructing the training set. After that, we will build and train an RNN model for text generation.

Acquiring and analyzing the training data

I recommend downloading text data for training from books that are not currently protected by copyright. Project Gutenberg (www.gutenberg.org) is a great place for this. It provides over 60,000 free e-books whose copyright has expired.

The original work, *War and Peace*, can be downloaded from http://www.gutenberg.org/ebooks/2600, but note that there will be some cleanup, such as removing the extra beginning section "*The Project Gutenberg EBook*," the table of contents, and the extra appendix "*End of the Project Gutenberg EBook of War and Peace*" of the plain text UTF-8 file (http://www.gutenberg.org/files/2600/2600-0.txt), required. So, instead of doing this, we will download the cleaned text file directly from https://cs.stanford.edu/people/karpathy/char-rnn/warpeace_input.txt. Let's get started:

1. First, we read the file and convert the text into lowercase:

```
>>> with open('warpeace_input.txt', 'r', encoding="utf8") as fp:
        raw_text = fp.read()
>>> raw_text = raw_text.lower()
```

2. Then, we take a quick look at the training text data by printing out the first 200 characters:

```
>>> print(raw_text[:200])
"well, prince, so genoa and lucca are now just family estates of the
buonapartes. but i warn you, if you don't tell me that this means war,
if you still try to defend the infamies and horrors perpetr
```

3. Next, we count the number of unique words:

```
>>> all_words = raw_text.split()
>>> unique_words = list(set(all_words))
>>> print(f'Number of unique words: {len(unique_words)}')
Number of unique words: 39830
```

    Then, we count the total number of characters:

```
>>> n_chars = len(raw_text)
>>> print(f'Total characters: {n_chars}')
Total characters: 3196213
```

4. From these 3 million characters, we obtain the unique characters, as follows:

```
>>> chars = sorted(list(set(raw_text)))
>>> vocab_size = len(chars)
>>> print(f'Total vocabulary (unique characters): {vocab_size}')
Total vocabulary (unique characters): 57
>>> print(chars)
['\n', ' ', '!', '"', '"', '(', ')', '*', ',', '-', '.', '/', '0', '1',
'2', '3', '4', '5', '6', '7', '8', '9', ':', ';', '=', '?', 'a', 'b',
'c', 'd', 'e', 'f', 'g', 'h', 'i', 'j', 'k', 'l', 'm', 'n', 'o', 'p',
'q', 'r', 's', 't', 'u', 'v', 'w', 'x', 'y', 'z', 'à', 'ä', 'é', 'ê', '\
ufeff']
```

The raw training text is made up of 57 unique characters and close to 40,000 unique words. Generating words, which requires computing 40,000 probabilities at one step, is far more difficult than generating characters, which requires computing only 57 probabilities at one step. Hence, we treat a character as a token, and the vocabulary here is composed of 57 characters.

So, how can we feed the characters to the RNN model and generate output characters? Let's see in the next section.

Constructing the training set for the RNN text generator

Recall that in a synced "many-to-many" RNN, the network takes in a sequence and simultaneously produces a sequence; the model captures the relationships among the elements in a sequence and reproduces a new sequence based on the learned patterns. As for our text generator, we can feed in fixed-length sequences of characters and let it generate sequences of the same length, where each output sequence is one character shifted from its input sequence. The following example will help you understand this better.

Say that we have a raw text sample, "learning," and we want the sequence length to be 5. Here, we can have an input sequence, "learn," and an output sequence, "earni." We can put them into the network as follows:

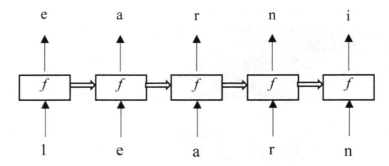

Input sequence: learn

Output sequence: earni

Figure 12.12: Feeding a training set ("learn," "earni") to the RNN

We've just constructed a training sample ("learn," "earni"). Similarly, to construct training samples from the entire original text, first, we need to split the original text into fixed-length sequences, X; then, we need to ignore the first character of the original text and split shift it into sequences of the same length, Y. A sequence from X is the input of a training sample, while the corresponding sequence from Y is the output of the sample. Let's say we have a raw text sample, "machine learning by example," and we set the sequence length to 5. We will construct the following training samples:

Input	Output
machi	achin
ne□le	e□lea
arnin	rning
g□by□	□by□e
examp	xampl

Figure 12.13: Training samples constructed from "machine learning by example"

Here, □ denotes space. Note that the remaining subsequence, "le," is not long enough, so we simply ditch it.

We also need to one-hot encode the input and output characters since neural network models only take in numerical data. We simply map the 57 unique characters to indices from 0 to 56, as follows:

```
>>> index_to_char = dict((i, c) for i, c in enumerate(chars))
>>> char_to_index = dict((c, i) for i, c in enumerate(chars))
>>> print(char_to_index)
{'\n': 0, ' ': 1, '!': 2, '"': 3, "'": 4, '(': 5, ')': 6, '*': 7, ',': 8, '-':
9, '.': 10, '/': 11, '0': 12, '1': 13, '2': 14, '3': 15, '4': 16, '5': 17, '6':
18, '7': 19, '8': 20, '9': 21, ':': 22, ';': 23, '=': 24, '?': 25, 'a': 26,
'b': 27, 'c': 28, 'd': 29, 'e': 30, 'f': 31, 'g': 32, 'h': 33, 'i': 34, 'j':
35, 'k': 36, 'l': 37, 'm': 38, 'n': 39, 'o': 40, 'p': 41, 'q': 42, 'r': 43,
's': 44, 't': 45, 'u': 46, 'v': 47, 'w': 48, 'x': 49, 'y': 50, 'z': 51, 'à':
52, 'ä': 53, 'é': 54, 'ê': 55, '\ufeff': 56}
```

For instance, the character c becomes a vector of length 57 with 1 in index 28 and 0s in all other indices; the character h becomes a vector of length 57 with 1 in index 33 and 0s in all other indices.

Now that the character lookup dictionary is ready, we can construct the entire training set, as follows:

```
>>> import numpy as np
>>> text_encoded = np.array(char_to_index[ch] for ch in raw_text],
                            dtype=np.int32)
>>> seq_length = 40
>>> chunk_size = seq_length + 1
>>> text_chunks = np.array([text_encoded[i:i+chunk_size]
                    for i in range(len(text_encoded)-chunk_size+1)])
```

Here, we set the sequence length to 40 and obtain training samples of a length of 41 where the first 40 elements represent the input, and the last 40 elements represent the target.

Next, we initialize the training dataset object and data loader, which will be used for model training:

```
>>> import torch
>>> from torch.utils.data import Dataset
>>> class SeqDataset(Dataset):
        def __init__(self, text_chunks):
            self.text_chunks = text_chunks
        def __len__(self):
            return len(self.text_chunks)
        def __getitem__(self, idx):
            text_chunk = self.text_chunks[idx]
            return text_chunk[:-1].long(), text_chunk[1:].long()

>>> seq_dataset = SeqDataset(torch. from_numpy (text_chunks))
```

```
>>> batch_size = 64
>>> seq_dl = DataLoader(seq_dataset, batch_size=batch_size, shuffle=True,
                        drop_last=True)
```

We just create a data loader in batches of 64 sequences, shuffle the data at the beginning of each epoch, and drop any remaining data points that don't fit into a complete batch.

We finally got the training set ready and it is time to build and fit the RNN model. Let's do this in the next section.

Building and training an RNN text generator

We first build the RNN model as follows:

```
>>> class RNN(nn.Module):
        def __init__(self, vocab_size, embed_dim, rnn_hidden_dim):
            super().__init__()
            self.embedding = nn.Embedding(vocab_size, embed_dim)
            self.rnn_hidden_dim = rnn_hidden_dim
            self.rnn = nn.LSTM(embed_dim, rnn_hidden_dim,
                            batch_first=True)
            self.fc = nn.Linear(rnn_hidden_dim, vocab_size)

        def forward(self, x, hidden, cell):
            out = self.embedding(x).unsqueeze(1)
            out, (hidden, cell) = self.rnn(out, (hidden, cell))
            out = self.fc(out).reshape(out.size(0), -1)
            return out, hidden, cell

        def init_hidden(self, batch_size):
            hidden = torch.zeros(1, batch_size, self.rnn_hidden_dim)
            cell = torch.zeros(1, batch_size, self.rnn_hidden_dim)
            return hidden, cell
```

This class defines a sequence-to-sequence model that takes tokenized input, converts token indices into dense vector representation with an embedding layer, processes the dense vectors through an LSTM layer, and generates logits for the next token in the sequence.

In this class, the init_hidden method initializes the hidden state and cell state of the LSTM. It takes batch_size as a parameter, which is used to determine the batch size for the initial states. Two tensors are created: hidden and cell, both initialized with zeros. The forward method receives two additional inputs, hidden and cell, which correspond to the many-to-many architecture of our RNN model.

One more thing to note, we use logits as outputs of the model here instead of probabilities, as we will sample from the predicted logits to generate new sequences of characters.

Now, let's train the RNN model we just defined as follows:

1. First, we specify the embedding dimension and the size of the LSTM hidden layer, and initiate the RNN model object:

```
>>> embed_dim = 256
>>> rnn_hidden_dim = 512
>>> model = RNN(vocab_size, embed_dim, rnn_hidden_dim)
>>> model = model.to(device)
>>> model
RNN(
  (embedding): Embedding(57, 256)
  (rnn): LSTM(256, 512, batch_first=True)
  (fc): Linear(in_features=512, out_features=57, bias=True)
)
```

A relatively high embedding dimension (like 256) allows for capturing richer semantic information about words. This can be beneficial for tasks like text generation. However, excessively high dimensions can increase computational cost and might lead to overfitting. 256 provides a good balance between these factors.

Text generation often requires the model to learn long-term dependencies between words in a sequence. A hidden layer size of 512 offers a good capacity to capture these complex relationships.

2. The next task is to define a loss function and an optimizer. In the case of multiclass classification, where there is a single logit output for each target character, we utilize CrossEntropyLoss as the appropriate loss function:

```
>>> loss_fn = nn.CrossEntropyLoss()
>>> optimizer = torch.optim.Adam(model.parameters(), lr=0.003)
```

3. Now, we train the model for 10,000 epochs. In each epoch, we train our many-to-many RNN on one training batch selected from the data loader, and we display the training loss for every 500 epochs:

```
>>> num_epochs = 10000
>>> for epoch in range(num_epochs):
        hidden, cell = model.init_hidden(batch_size)
        seq_batch, target_batch = next(iter(seq_dl))
        seq_batch = seq_batch.to(device)
        target_batch = target_batch.to(device)
        optimizer.zero_grad()
        loss = 0
        for c in range(seq_length):
```

```
                    pred, hidden, cell = model(seq_batch[:, c],
                                                hidden.to(device),
                                                cell.to(device))
                loss += loss_fn(pred, target_batch[:, c])
            loss.backward()
            optimizer.step()
            loss = loss.item()/seq_length
            if epoch % 500 == 0:
                print(f'Epoch {epoch} - loss: {loss:.4f}')
Epoch 0 - loss: 4.0255
Epoch 500 - loss: 1.4560
Epoch 1000 - loss: 1.2794
...
8500 loss: - 1.2557
Epoch 9000 - loss: 1.2014
Epoch 9500 - loss: 1.2442
```

For each element in a given sequence, we feed the recurrent layer with the previous hidden state along with the current input.

4. Model training is complete, and now it's time to assess its performance. We can generate text by providing a few starting words, for instance:

```
>>> from torch.distributions.categorical import Categorical
>>> def generate_text(model, starting_str, len_generated_text=500):
        encoded_input = torch.tensor([char_to_index[s] for s in starting_str])
        encoded_input = torch.reshape(encoded_input, (1, -1))
        generated_str = starting_str
        model.eval()
        hidden, cell = model.init_hidden(1)
        for c in range(len(starting_str)-1):
            _, hidden, cell = model(encoded_input[:, c].view(1), hidden, cell)
        last_char = encoded_input[:, -1]
            for _ in range(len_generated_text):
                logits, hidden, cell = model(last_char.view(1), hidden, cell)
                logits = torch.squeeze(logits, 0)
                last_char = Categorical(logits=logits).sample()
                generated_str += str(index_to_char[last_char.item()])
            return generated_str
>>> model.to('cpu')
>>> print(generate_text(model, 'the emperor', 500))
```

```
the emperor!" said he.
"finished! it's all with moscow, it's not get bald hills!" he added the
civer with whom and desire to change. they really asked the imperor's
field!" she said. alpaty. there happed the cause of the longle matestood
itself. "the mercy tiresist between paying so impressions, and till
the staff offsicilling petya, the chief dear body, returning quite
dispatchma--he turned and ecstatically. "ars doing her dome." said
rostov, and the general feelings of the bottom would be the pickled ha
```

We generate a 500-character text beginning with our given input "the emperor." Specifically, we first initialize the hidden and cell state for the RNN model. This is required to start generating text. Then, in the for loop, we iterate over the characters in the starting text except the last one. For each character in the input, we pass it through the model, updating the hidden and cell states. To generate the next character index, we predict the logits for all possible characters and sample it based on the logits utilizing a Categorical distribution. With that, we've successfully used a many-to-many type of RNN to generate text.

Feel free to tweak the model so that the RNN-based text generator can write a more realistic and interesting version of *War and Peace*.

An RNN with a many-to-many structure is a type of sequence-to-sequence (seq2seq) model that takes in a sequence and outputs another sequence. A typical example is machine translation, where a sequence of words from one language is transformed into a sequence in another language. The state-of-the-art seq2seq model is the **Transformer** model, and it was developed by Google Brain. We will discuss it in the next chapter.

Summary

In this chapter, we worked on three NLP projects: sentiment analysis, stock price prediction, and text generation using RNNs. We started with a detailed explanation of the recurrent mechanism and different RNN structures for different forms of input and output sequences. You also learned how LSTM improves vanilla RNNs.

In the next chapter, we will focus on the Transformer, a recent state-of-the-art sequential learning model, and generative models.

Exercises

1. Use a bi-directional recurrent layer (it is easy enough to learn about it by yourself) and apply it to the sentiment analysis project. Can you beat what we achieved? Hint: set the bidirectional argument to True in the LSTM layer.
2. Feel free to fine-tune the hyperparameters in the text generator, and see whether you can generate a more realistic and interesting version of *War and Peace*.
3. Can you train an RNN model on any of your favorite books in order to write your own version?

Unlock this book's exclusive benefits now

This book comes with additional benefits designed to elevate your learning experience.

Note: Have your purchase invoice ready before you begin.

https://www.packtpub.com/unlock/9781835085622

13

Advancing Language Understanding and Generation with the Transformer Models

In the previous chapter, we focused on RNNs and used them to deal with sequence learning tasks. However, RNNs may easily suffer from the vanishing gradient problem. In this chapter, we will explore the Transformer neural network architecture, which is designed for sequence-to-sequence tasks and is particularly well suited for **Natural Language Processing (NLP)**. The key innovation is the self-attention mechanism, allowing the model to weigh different parts of the input sequence differently, and enabling it to capture long-range dependencies more effectively than RNNs.

We will learn two cutting-edge models utilizing the Transformer architecture and delve into their practical applications, such as sentiment analysis and text generation. Expect enhanced performance on tasks previously covered in the preceding chapter.

We will cover the following topics in this chapter:

- Understanding self-attention
- Exploring the Transformer's architecture
- Improving sentiment analysis with **Bidirectional Encoder Representations from Transformers (BERT)** and Transformers
- Generating text using **Generative Pre-trained Transformers (GPT)**

Understanding self-attention

The Transformer neural network architecture revolves around the self-attention mechanism. So, let's first kick off the chapter by looking at this. **Self-attention** is a mechanism used in machine learning, particularly in NLP and computer vision. It allows a model to weigh the importance of different parts of the input sequence.

Self-attention is a specific type of attention mechanism. In traditional attention mechanisms, the importance weights are between two different sets of input data. For example, an attention-based English-to-French translation model may focus on specific parts (e.g., nouns, verbs) of the English source sentences that are relevant to the current French target word being generated. However, in self-attention, the importance weighting operates between any two elements within the same input sequence. It focuses on how different parts in the same sequence relate to each other. Used for English-to-French translation, the self-attention model analyzes how each English word interacts with every other English word. By understanding these relationships, the model can generate a more nuanced and accurate French translation.

In the context of NLP, traditional RNNs process input sequences sequentially. Because of this sequential processing style, RNNs can only handle shorter sequences well, and capture shorter-range dependencies among tokens. On the contrary, a self-attention-powered model can simultaneously process all input tokens in a sequence. For a given token, the model assigns different attention weights to different tokens based on their relevance to the given token, regardless of their positions. As a result, it can capture the relationships between different tokens in the input, and their long-range dependencies. Self-attention-based models outperform RNNs in several sequence-to-sequence tasks such as machine translation, text summarization, and query answering.

Let's discuss how **self-attention** plays a key role in sequence learning tasks in the following examples:

"I read *Python Machine Learning by Example* by Hayden Liu and it is indeed a great book." Apparently, *it* here refers to *Python Machine Learning by Example*. When the Transformer model processes this sentence, self-attention will associate *it* with *Python Machine Learning by Example*.

We can use a self-attention-based model to summarize a document (e.g., this chapter). Self-attention isn't limited by the order of sentences, unlike sequential learning RNNs, and it can identify relationships between sentences (even distant ones), which ensures the summary reflects the overall information.

Given a token in an input sequence, self-attention allows the model to look at the other tokens in the sequence at different attention levels. In the next section, we'll look at a more detailed explanation of how the self-attention mechanism works.

Key, value, and query representations

The self-attention mechanism is applied to each token in a sequence. Its goal is to represent every token by an embedding vector that captures long-range context. The embedding vector of an input token is composed of three vectors: key, value, and query. For a given token, a self-attention-based model learns these three vectors in order to compute the embedding vector.

The following describes the meaning of each of the three vectors. To aid comprehension, we use an analogy that likens a model's understanding of a sequence to a detective investigating a crime scene with a bunch of clues. To solve the case (understand the meaning of the sequence), the detective needs to figure out which clues (tokens) are most important and how clues are connected (how tokens relate to each other):

- The key vector, K, represents the core information of a token. It captures the key but not the detailed information a token holds. In our detective analogy, the key vector of a clue might contain information about a witness who saw the crime, but not the details they saw.

- The value vector, V, holds the full information of a token. In our detective example, the value vector of a clue could be the detailed statement from the witness.

- Finally, the query vector, Q, represents the importance of understanding a given token in the context of the whole sequence. It is a question about a token's relevance to the current task. During the detective's investigation, their focus can change depending on what they are currently looking for. It can be the weapon used, the victim, the motivation, or something else. The query vector represents the detective's current focus in the investigation.

These three vectors are derived from the input token's embeddings. Let's discuss how they work together using the detective example again:

- First, we calculate the attention scores based on the key and query vector. Based on a query vector, Q, the model analyses each token and computes the relevance score between its key vector, K, and the query vector, Q. A high score indicates a high importance of the token to the context. In the detective example, they try to figure out how relevant a clue is to the current investigation focus. For instance, the detective would think clue A about the building is highly relevant when they are looking into the crime scene location. Note that the detective doesn't look at the details of clue A yet, just like the attention scores are computed based on the key vector, not the value vector. The model (the detective) will use the value vector (details of a clue) in the next stage – embedding vector generation.

- Next, we generate the embedding vector for a token using the value vectors, V, and the attention weights. The detective has decided how much weight (attention scores) is assigned to each clue, and now they will combine the details (value vector) of each clue to create a comprehensive understanding (embedding vector) of the crime scene.

The terminology, "query," "key," and "value," is inspired by the information retrieval systems.

Let's illustrate using a search engine example. Given a query, the search engine undergoes a matching process against the key of each document candidate and comes up with a ranking score individually. Based on the detailed information and the ranking score of a document, the search engine creates a search result page with the retrieval of specific associated values.

We've discussed what key, value, and query vectors are in self-attention mechanisms, and how they work together to allow the model to capture important information from an input sequence. The generated context-aware embedding vector encapsulates the relationships between tokens in the sequence. I hope the detective's crime scene analogy helped you gain a better understanding. In the next section, we will see how the context-aware embedding vector is generated mathematically.

Attention score calculation and embedding vector generation

We have a sequence of tokens $(x_1, x_2, ...x_i, ...x_n)$. Here, n is the length of the sequence.

For a given token, the calculation of attention score begins by computing the similarity score between each token in the sequence and the token in question. The similarity score is computed by taking the dot product of the query vector of the current token and the key vector of the other tokens:

$$s_{ij} = q_i \cdot k_j$$

Here, s_{ij} is the similarity score between token x_i and x_j, q_i is the query vector of x_i, and k_j is the query vector of x_j. The similarity score measures how relevant a token x_j is to the current token x_i.

You may notice that the raw similarity scores do not directly reflect relative relevance (they can be negative or greater than 1). Recall that the softmax function normalizes raw scores, converting them into probabilities that sum up to 1. Hence, we need to normalize them using a softmax function. The normalized scores are the attention scores we are looking for:

$$a_{ij} = softmax\left(\frac{q_i \cdot k_j}{\sqrt{d}}\right) = \frac{\exp(a_{ij})}{\sum_{l=1}^{n} \exp(a_{il})}$$

Here, a_{ij} is the similarity score between token x_i and x_j, d is the dimension of the key vector, and the division of $\sqrt{d}$ is for scaling. The attention weights $(a_{i1}, a_{i2}, ..., a_{in})$ convey the probability distribution over all other tokens in the sequence with respect to the current token.

With the normalized attention weights available, we can now compute the embedding vector for the current token. The embedding vector z_i is a weighted sum of the value vectors v of all tokens in the sequence, where each weight is the attention score a_{ij} between the current token x_i and the respective token x_j:

$$z_i = \sum_{j=1}^{n} a_{ij} \cdot v_j$$

This weighted sum vector is considered the context-aware representation of the current token, taking into account its relationship with all other tokens.

Take the sequence *python machine learning by example* as an example; we take the following steps to calculate the self-attention embedding vector for the first word, *python*:

1. We calculate the dot products between each word in the sequence and the word *python*. They are $q_1 \cdot k_1$, $q_1 \cdot k_2$, $q_1 \cdot k_3$, $q_1 \cdot k_4$, and $q_1 \cdot k_5$. Here, q_1 is the query vector for the first word, and k_1 to k_5 are the key vectors for the five words, respectively.

2. We normalize the resulting dot products with division and `softmax` activation to find the attention weights:

$$a_{11} = softmax\left(\frac{q_1 \cdot k_1}{\sqrt{d}}\right)$$

$$a_{12} = softmax\left(\frac{q_1 \cdot k_2}{\sqrt{d}}\right)$$

$$a_{13} = softmax\left(\frac{q_1 \cdot k_3}{\sqrt{d}}\right)$$

$$a_{14} = softmax\left(\frac{q_1 \cdot k_4}{\sqrt{d}}\right)$$

$$a_{15} = softmax\left(\frac{q_1 \cdot k_5}{\sqrt{d}}\right)$$

3. Then, we multiply the resulting attention weights by the value vectors, v_1, v_2, v_3, v_4, v_5, and add up the results:

$$z_1 = a_{11}.v_1 + a_{12}.v_2 + a_{13}.v_3 + a_{14}.v_4 + a_{15}.v_5$$

z_1 is the context-aware embedding vector for the first word, *python*, in the sequence. We repeat this process for each remaining word in the sequence to obtain its context-aware embedding.

During training in a self-attention mechanism, the key, query, and value vectors are created by three weight matrices, W_k, W_q, and W_v, using a linear transformation:

$$k_i = x_i W_k$$

$$q_i = x_i W_q$$

$$v_i = x_i W_v$$

Here, W_k is the weight matrix for the key transformation, W_q is the weight matrix for the query transformation, and W_v is the weight matrix for the value transformation. These three weight matrices are learnable parameters. During the model training process, they get updated typically using gradient-based optimization algorithms.

Now, let's see how we can simulate the calculation of z_1 in PyTorch:

1. First, assume we have the following integer representation mapping for the tokens in the input `python machine learning by example`:

```
>>> import torch
>>> sentence = torch.tensor(
        [0, # python
         8, # machine
         1, # learning
```

```
        6, # by
        2] # example
    )
```

Each integer corresponds to the index of the respective token in the vocabulary.

2. We also assume we have embeddings ready to use for our simulated vocabulary:

```
>>> torch.manual_seed(0)
>>> embed = torch.nn.Embedding(10, 16)
>>> sentence_embed = embed(sentence).detach()
>>> sentence_embed
tensor([[-1.1258, -1.1524, -0.2506, -0.4339,  0.8487,  0.6920, -0.3160,
         -2.1152,
          0.3223, -1.2633,  0.3500,  0.3081,  0.1198,  1.2377,  1.1168,
         -0.2473],
        [-0.8834, -0.4189, -0.8048,  0.5656,  0.6104,  0.4669,  1.9507,
         -1.0631,
         -0.0773,  0.1164, -0.5940, -1.2439, -0.1021, -1.0335, -0.3126,
          0.2458],
        [-1.3527, -1.6959,  0.5667,  0.7935,  0.5988, -1.5551, -0.3414,
          1.8530,
          0.7502, -0.5855, -0.1734,  0.1835,  1.3894,  1.5863,  0.9463,
         -0.8437],
        [ 1.6459, -1.3602,  0.3446,  0.5199, -2.6133, -1.6965, -0.2282,
          0.2800,
          0.2469,  0.0769,  0.3380,  0.4544,  0.4569, -0.8654,  0.7813,
         -0.9268],
        [-0.6136,  0.0316, -0.4927,  0.2484,  0.4397,  0.1124,  0.6408,
          0.4412,
         -0.1023,  0.7924, -0.2897,  0.0525,  0.5229,  2.3022, -1.4689,
         -1.5867]])
```

Here, our simulated vocabulary has 10 tokens, and the embedding size is 16. detach() is used to create a new tensor that shares the same underlying data as the original tensor but is detached from the computation graph. Also, note that you may get different embedding results due to some non-deterministic operations.

3. Next, we assume we have the following three weight matrices, W_k, W_q, and W_v:

```
>>> d = sentence_embed.shape[1]
>>> w_key = torch.rand(d, d)
>>> w_query = torch.rand(d, d)
>>> w_value = torch.rand(d, d)
```

For matrix operations, it is essential that vectors Q and K share the same dimensions to ensure they operate within a consistent feature space. However, vector V is allowed to have different dimensions. In this example, we will maintain uniform dimensions for all three vectors for the sake of simplicity. So, we choose 16 as the common dimension.

4. Now, we can compute the key vector k_1, query vector q_1, and value vector v_1 for the *python* token accordingly:

```
>>> token1_embed = sentence_embed[0]
>>> key_1 = w_key.matmul(token1_embed)
>>> query_1 = w_query.matmul(token1_embed)
>>> value_1 = w_value.matmul(token1_embed)
```

Take a look at k_1:

```
>>> key_1
tensor([-1.1371, -0.5677, -0.9324, -0.3195, -2.8886, -1.2679, -1.1153,
         0.2904,  0.3825,  0.3179, -0.4977, -3.8230,  0.3699, -0.3932,
        -1.8788, -3.3556])
```

5. We can also directly compute the key matrix (composed of key vectors for individual tokens) as follows:

```
>>> keys = sentence_embed.matmul(w_key.T)
>>> keys[0]
tensor([-1.1371, -0.5677, -0.9324, -0.3195, -2.8886, -1.2679, -1.1153,
         0.2904,  0.3825,  0.3179, -0.4977, -3.8230,  0.3699, -0.3932,
        -1.8788, -3.3556])
```

Similarly, the value matrix can be directly computed as follows:

```
>>> values = sentence_embed.matmul(w_value.T)
```

6. With the key matrix and the query vector q_1, we obtain the attention weight vector $a_1 = (a_{11}, a_{12}, a_{13}, a_{14}, a_{15})$:

```
>>> import torch.nn.functional as F
>>> a1 = F.softmax(query_1.matmul(keys.T) / d ** 0.5, dim=0)
>>> a1
tensor([3.2481e-01, 4.2515e-01, 6.8915e-06, 2.5002e-01, 1.5529e-05])
```

7. Finally, we multiply the resulting attention weights by the value vectors to obtain the context-aware embedding vector for the first token:

```
>>> z1 = a1.matmul(values)
>>> z1
tensor([-0.7136, -1.1795, -0.5726, -0.4959, -0.6838, -1.6460, -0.3782,
-1.0066,
```

```
        -0.4798, -0.8996, -1.2138, -0.3955, -1.3302, -0.3832, -0.8446,
    -0.8470])
```

This is the self-attention version of the embedding vector for the *python* token based on the three toy weight matrices, W_k, W_q, and W_v.

In practice, we typically employ more than one set of trainable weight matrices, W_k, W_q, and W_v. That is why self-attention is often called **multi-head self-attention**. Each attention head has its own set of learnable parameters for the key, query, and value transformations. Using multiple attention heads can capture different aspects of relationships within a sequence. Let's dig into this next.

Multi-head attention

The single-head attention mechanism is effective but may not capture diverse relationships within the sequence. Multi-head attention extends this by employing multiple sets of query, key, and value matrices (multiple "heads"). Each head operates **independently** and can attend to different parts of the input sequence **in parallel**. This allows the model to capture diverse relationships simultaneously.

Using the previous example sequence, *python machine learning by example*, one attention head might focus on local dependencies, identifying "machine learning" as a noun phrase; another attention head might emphasize semantic relationships, inferring that the "examples" are about "machine learning." It's like having multiple analysts examining the same sentence. Each analyst focuses on a different aspect (for instance, one on grammar, one on word order, and another on sentiment). By combining their insights, you get a more comprehensive understanding of the sentence.

Finally, the outputs from all attention heads are concatenated and linearly transformed to produce the final attention output.

In this section, we presented a self-attention mechanism featuring trainable parameters. In the upcoming section, we will delve into the Transformer architecture, which centers around the self-attention mechanism.

Exploring the Transformer's architecture

The Transformer architecture was proposed as an alternative to RNNs for sequence-to-sequence tasks. It heavily relies on the self-attention mechanism to process both input and output sequences.

We'll start by looking at the high-level architecture of the Transformer model (image based on that in the paper *Attention Is All You Need*, by Vaswani et al.):

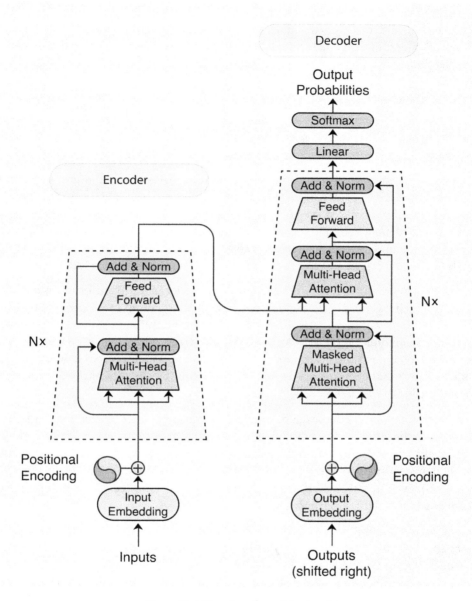

Figure 13.1: Transformer architecture

As you can see, the Transformer consists of two parts: the **encoder** (the big rectangle on the left-hand side) and the **decoder** (the big rectangle on the right-hand side). The encoder encrypts the input sequence. It has a **multi-head attention layer** and a regular feedforward layer. On the other hand, the decoder generates the output sequence. It has a masked multi-head attention (we will talk about this in detail later) layer, along with a multi-head attention layer and a regular feedforward layer.

At step t, the Transformer model takes in input steps $x_1, x_2, ..., x_t$ and output steps $y_1, y_2, ..., y_{t-1}$. It then predicts y_t. This is no different from the many-to-many RNN model. In the next section, we will explore the important elements of the Transformer that set it apart from RNNs, including the encoder-decoder structure, positional encoding, and layer normalization.

The encoder-decoder structure

The encoder-decoder structure is the key element in the Transformer architecture. It leverages the model's ability to handle sequence-to-sequence tasks. Below is a breakdown of the encoder component and the decoder component with an analogy to help you understand.

The **encoder** component processes the input sequence and creates a context representation. Typically, the encoder component is a stack of encoders. Each encoder consists of a self-attention layer and a feedforward neural network. We've covered that the self-attention allows each token to attend to other tokens in the sequence. Unlike sequential models like RNNs, relations between tokens (even the distant ones) are captured in the Transformer, and the feedforward neural network adds non-linearity to the model's learning capacity. We've seen this in deep neural networks before.

Imagine you want to order a meal at a restaurant. The encoder in the Transformer works in a similar way as you reading the menu and generating your own understanding. The encoder takes in the menu (input sequence of words). It analyzes the words using self-attention, just like you read the descriptions of each dish and their ingredients (relationships between words). After the encoder (it could be a stack of encoders) digests the information from the menu, it creates a condensed representation, the encoded context that captures the essence of the menu. This output of the encoder is like your own comprehension of the menu.

For the encoder component composed of multiple identical encoder blocks, the output from each encoder serves as the input for the subsequent block in the stacked structure. This stacked approach offers a powerful way to capture more complex relationships and create a richer understanding of the input sequence. We've seen a similar approach in RNNs. In case you are wondering, six encoder blocks were employed in the original design in *Attention Is All You Need*. The number of encoders is not magical and is subject to experimentation.

The **decoder** component utilizes the context representation provided by the encoder to generate the output sequence. Similar to the encoder, the decoder component also consists of multiple stacked decoder blocks. Similarly, each decoder block contains a self-attention layer and a feedforward neural network. However, the self-attention in the decoder is slightly different from the encoder one. It attends to the output sequence but it only considers the context it has already built. This means for a given token, the decoder self-attention only considers the relationships between previously processed tokens and the current tokens. Recall that the self-attention in the encoder can attend to the whole input sequence at once. Hence, we call the self-attention in the decoder **masked self-attention**.

Besides a masked self-attention layer and a feedforward neural network, the decoder block has an additional attention layer called **encoder-decoder attention**. It attends to the context representation provided by the encoder so that the generated output sequence is relevant to the encoded context.

Going back to the restaurant ordering analogy. Now, you (the decoder) want to place an order (generate the output sequence). The decoder uses encoder-decoder self-attention to consider the encoded context (your understanding of the menu). Encoder-decoder self-attention ensures the output generation is based on your comprehension of the menu, not someone else's. The decoder uses **masked self-attention** to generate the output word by word (your order of dishes). Masked self-attention ensures you don't "peek" at future dishes (words) you haven't "ordered" (generated) yet. With each generated word (the dish you order), the decoder can refine its understanding of the desired output sequence (meal) based on the encoded context.

Similar to the encoder, in the stacked decoder component, the output from each decoder block serves as the input for the subsequent block. Due to the encoder-decoder self-attention, the number of decoder blocks is usually the same as the encoder blocks.

During training, the model is provided with both the input and the target output sequences. It learns to generate the target output sequence by minimizing the difference between its predictions and the actual target sequence.

In this section, we've delved into the Transformer's encoder-decoder structure. The encoder stack extracts a context representation of the input sequence. The decoder generates the output sequence one token at a time, attending to both the encoded context and previously generated tokens.

Positional encoding

While powerful, self-attention struggles to differentiate between the importance of elements based solely on their content. For instance, given the sentence, "The white fox jumps over the brown dog," self-attention might assign similar importance to "fox" and "dog" simply because they share similar grammatical roles (nouns). To address this limitation, **positional encoding** is introduced to inject positional information into self-attention.

The positional encoding is a fixed-size vector that contains positional information of the token in the sequence. It is typically calculated based on mathematical functions. One common approach is to use a combination of sine and cosine functions as follows:

$$PE(pos, 2i) = sin\left(\frac{pos}{10000^{2i/d}}\right)$$

$$PE(pos, 2i + 1) = cos\left(\frac{pos}{10000^{2i/d}}\right)$$

Here, i is the dimension index, pos is the position of the token, and d is the dimension of the embedding vector. $PE(pos, 2i)$ denotes the $2i^{th}$ dimension of the positional encoding for position pos; $PE(pos, 2i+1)$ represents the $2i+1^{th}$ dimension of the positional encoding for position pos. $\frac{pos}{10000^{2i/d}}$ introduces different frequencies for different dimensions.

Let's try to encode the positions of the words in a simple sentence, "Python machine learning" using a four-dimensional vector. For the first word, "Python," we have the following:

$$PE(pos = 0, 0) = sin(0) = 0$$

$$PE(pos = 0, 1) = \cos(0) = 1$$

$$PE(pos = 0, 2) = \sin(0) = 0$$

$$PE(pos = 0, 3) = \cos(0) = 1$$

For the second word "machine," we have the following:

$$PE(pos = 1, 0) = \sin\left(\frac{1}{1}\right) = 0.8$$

$$PE(pos = 1, 1) = \cos\left(\frac{1}{1}\right) = 0.5$$

$$PE(pos = 1, 2) = \sin\left(\frac{1}{100}\right) = 0$$

$$PE(pos = 1, 3) = \cos\left(\frac{1}{100}\right) = 1$$

So, we have positional encoding [0, 1, 0, 1] for "Python," and [0.8, 0.5, 0, 1] for "machine." We will leave the third word as an exercise for you.

After the positional encoding vectors are computed for each position, they are then added to the embeddings of the corresponding tokens. As you can see in the Transformer architecture in *Figure 13.1*, the positional encoding is added to the input embedding before the input embedding is fed into the encoder component. Similarly, the positional encoding is added to the output embedding before the output embedding is fed into the decoder component. Now, when self-attention learns the relationships between tokens, it considers both the content (token themselves) and their positional information.

Positional encoding is an important complement to self-attention in Transformers. Since the Transformer doesn't inherently understand the sequential order of tokens like RNNs do, the additional positional encoding allows it to capture sequential dependencies effectively. In the next section, we will look at another crucial component in Transformers, layer normalization.

Layer normalization

A Transformer has many layers (blocks of encoders and decoders that are composed of multi-head self-attention), and it can suffer from exploding or vanishing gradients. This makes it difficult for the network to learn effectively during training. **Layer normalization** helps address this by normalizing the outputs of each layer.

Normalization is applied independently to each layer's activations, including the self-attention layer and the feedforward network. This means that each layer's outputs are kept within a specific range to prevent gradients from becoming too large or too small. As a result, layer normalization can stabilize the training process and improve the Transformer's learning efficiency.

We've gained a deep understanding of the Transformer architecture and its components, including the encoder, decoder, multi-head self-attention, masked self-attention, positional encoding, and layer normalization. Next, we will learn about models, BERT and GPT, that are based on the Transformer architecture, and will work on their applications, including sentiment analysis and text generation.

Improving sentiment analysis with **BERT** and **Transformers**

BERT (`https://arxiv.org/abs/1810.04805v2`) is a model based on the Transformer architecture. It has achieved significant success in various language understanding tasks in recent years.

As its name implies, bidirectional is one significant difference between BERT and earlier Transformer models. Traditional models often process sequence in a unidirectional manner, but BERT processes the entire context bidirectionally. This bidirectional context understanding makes the model more effective in capturing nuanced relationships in a sequence.

BERT is basically a stack of trained Transformer's encoders. It is pre-trained on large amounts of unlabeled text data in a self-supervised manner. During pre-training, it focuses on understanding the meaning of text in context. After pre-training, BERT can be fine-tuned for specific downstream tasks.

Let's first talk about the pre-training works.

Pre-training BERT

The goal of pre-training BERT is to capture rich contextualized representations of words. It involves training the model on **a large corpus** of unlabeled text data in a self-supervised manner. The pre-training process consists of two main tasks: the **Masked Language Model (MLM)** task and the **Next Sentence Prediction (NSP)** task. Here is the breakdown.

MLM

In MLM, random words in a sentence are replaced with a special token [MASK]. BERT takes the modified sentence as input and tries to predict the original masked word based on the surrounding words. In this fill-in-the-blank game, BERT is trained to understand the meaning and context of words.

It is worth noting that during the MLM task, the model is trained using the **bidirectional** context—both the left and right context of each masked word. This improves the masked word prediction accuracy. As a result, by going through a large number of training examples, BERT gets better at understanding word meanings and capturing relationships in different contexts.

NSP

The NSP task helps the model understand relationships between sentences and discourse-level information. During training, pairs of sentences are randomly sampled from the training corpus. For each pair, there is a 50% chance that the second sentence follows the first in the original text and a 50% chance that it doesn't. Two sentences concatenated together to form a training sample. Special [CLS] and [SEP] tokens are used to format the training sample:

- The [CLS] (classification) token is added at the beginning of the training sample. The output corresponding to the [CLS] token is used to represent the entire input sequence for classification tasks.
- The [SEP] (separator) token separates two concatenated input sentences.

BERT's pre-training process is depicted in the following diagram (note that "C" in the diagram is short for "Class"):

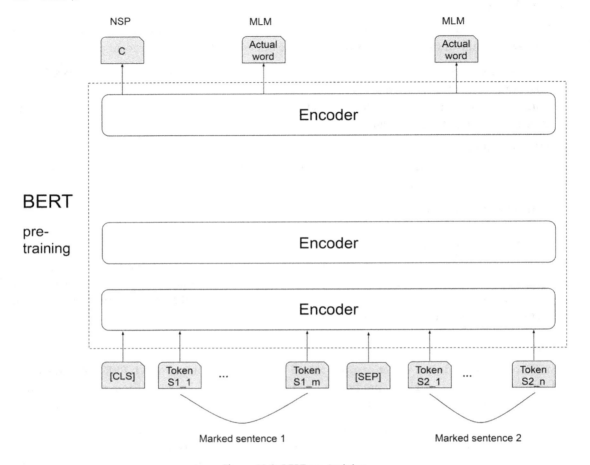

Figure 13.2: BERT pre-training

As you can see in the pre-training diagram, the model is trained on concatenated sentences to predict whether the second sentence follows the first one. It receives the correct label (is next sentence or not) as feedback and adjusts its parameters to improve its prediction accuracy. Through this sentence matchmaking task, BERT learns to understand how sentences relate to each other and how ideas flow coherently within a text.

Interestingly, the diagram shows that BERT's pre-training combines MLM and NSP tasks simultaneously. The model predicts masked tokens within a sentence while also determining whether a sentence pair follows sequentially.

While the MLM task focuses on word-level contextualization, the NSP task contributes to BERT's broader understanding of sentence relationships. After pre-training, BERT can be **fine-tuned** for specific downstream NLP tasks. Let's see how we do this next.

Fine-tuning of BERT

BERT is usually fine-tuned for targeted tasks such as sentiment analysis or named entity recognition. Task-specific layers are added on top of the pre-trained model, and the model is trained on labeled data relevant to the task. Here's the step-by-step process of fine-tuning:

1. We first gather labeled data specific to the downstream task.

2. Next, we need to select a pre-trained BERT model. Various options are available, such as BERT-base and BERT-large. You should use the one suitable for the downstream task and computational capabilities.

3. Based on the selected BERT model, we use the corresponding tokenizer to tokenize the input. Various tokenizers are available in the Hugging Face tokenizers package (https://huggingface.co/docs/tokenizers/python/latest/index.html). But you should use the one that matches the model.

4. Here comes the fun part – architecture modification. We can add task-specific layers on top of the pre-trained BERT model. For instance, you can add a single neuron with a sigmoid activation for sentiment analysis.

5. We then define the task-specific loss function and objective for training. Use the sentiment analysis example again; you can use the binary cross-entropy loss function.

6. Finally, we train the modified BERT model on the labeled data.

7. We usually perform hyperparameter tuning to find the optimal model configurations including learning rate, batch size, and regularization.

Fine-tuning BERT leverages the knowledge acquired during pre-training and adapts it to a specific task. With this transfer learning strategy, BERT doesn't need to start from scratch, so it learns new things faster and needs less data. In the next section, we'll utilize BERT to improve the sentiment prediction on movie reviews.

Fine-tuning a pre-trained BERT model for sentiment analysis

In *Chapter 12, Making Predictions with Sequences Using Recurrent Neural Networks,* we developed an LSTM model for movie review sentiment prediction. We will fine-tune a pre-trained BERT model for the same task in the following steps:

1. First, we read the IMDb review data from the PyTorch built-in datasets:

```
>>> from torchtext.datasets import IMDB
>>> train_dataset = list(IMDB(split='train'))
>>> test_dataset = list(IMDB(split='test'))
>>> print(len(train_dataset), len(test_dataset))
25000 25000
```

We just load 25,000 training samples and 25,000 test samples.

2. Then, we separate the raw data into text and label data, as we will need to tokenize the text data:

```
>>> train_texts = [train_sample[1] for train_sample in train_dataset]
>>> train_labels = [train_sample[0] for train_sample in train_dataset]
>>> test_texts = [test_sample[1] for test_sample in test_dataset]
>>> test_labels = [test_sample[0] for test_sample in test_dataset]
```

We have finished data preparation and move on to the tokenization step.

3. Now, we need to pick a suitable pre-trained model and the corresponding tokenizer. We choose the `distilbert-base-uncased` model given that we have limited computational resources. Think of it as a smaller ("distilled"), faster version of BERT. It keeps most of the power of BERT but with fewer parameters. The "uncased" part just means that the model was trained on lowercase text. To make things work smoothly, we'll use the `distilbert-base-uncased` tokenizer that matches the model.

4. If you have not installed Hugging Face's transformers package yet, you can do so in the command line as follows:

```
pip install transformers==4.32.1
```

Or, with the following:

```
conda install -c huggingface transformers=4.32.1
```

As of the current writing, we are using version `4.32.1`. Feel free to replicate the process using this version, as the transformer package undergoes frequent updates.

5. Pre-trained models from Hugging Face can be loaded and then downloaded and cached locally. We load the `distilbert-base-uncased` tokenizer as follows:

```
>>> import transformers
>>> from transformers import DistilBertTokenizerFast
>>> tokenizer = DistilBertTokenizerFast.from_pretrained('distilbert-base-
uncased')
```

We can also manually download a `transformers` model ahead of time, and read it from the specified local path. To fetch the `distilbert-base-uncased` pre-trained model and tokenizer, you can search `distilbert-base-uncased` at https://huggingface.co/models, and go to the clickable link of `distilbert-base-uncased`. Model and tokenizer files can be found in the **Files** tab or at https://huggingface.co/distilbert-base-uncased/tree/main directly. Download all files **except** `flax_model.msgpack`, `model.safetensors`, `rust_model.ot`, and `tf_model.h5` (as we only need the PyTorch model), and put them in the folder called `distilbert-base-uncased`. Finally, we will be able to load the tokenizer from the `distilbert-base-uncased` path as follows:

```
>>> tokenizer = DistilBertTokenizerFast.from_pretrained(
                                        'distilbert-base-uncased',
                                        local_files_only=True)
```

6. Now, we tokenize the input text from the train and test datasets:

```
>>> train_encodings = tokenizer(train_texts, truncation=True,
                                padding=True)
>>> test_encodings = tokenizer(test_texts, truncation=True, padding=True)
```

Take a look at the encoding result of the first train sample:

```
>>> train_encodings[0]
Encoding(num_tokens=512, attributes=[ids, type_ids, tokens, offsets,
attention_mask, special_tokens_mask, overflowing])
```

Here, the resulting encoding object can only hold up to 512 tokens. If the original text is longer, it will be truncated. Attributes are a list of information associated with the encoding process, including ids (representing the token IDs) and attention_mask (specifying which tokens should be attended to and which should be ignored).

7. Next, we encapsulate all data fields, including the labels within a Dataset class:

```
>>> import torch
>>> class IMDbDataset(torch.utils.data.Dataset):
    def __init__(self, encodings, labels):
        self.encodings = encodings
        self.labels = labels

    def __getitem__(self, idx):
        item = {key: torch.tensor(val[idx])
            for key, val in self.encodings.items()}
        item['labels'] = torch.tensor([0., 1.]
                                if self.labels[idx] == 2
                                else [1., 0.])
        return item

    def __len__(self):
        return len(self.labels)
```

Note that we transform the positive label (denoted by "2") and the negative label (denoted by "1") into the formats [0, 1] and [1, 0] respectively. We make this adjustment to align with the labeling format required by DistilBERT.

8. We then generate the custom Dataset objects for the train and test data:

```
>>> train_encoded_dataset = IMDbDataset(train_encodings, train_labels)
>>> test_encoded_dataset = IMDbDataset(test_encodings, test_labels)
```

9. Based on the resulting datasets, we create batch data loaders and get ready for model fine-tuning and evaluation:

```
>>> batch_size = 32
>>> train_dl = torch.utils.data.DataLoader(train_encoded_dataset,
                                            batch_size=batch_size,
                                            shuffle=True)
>>> test_dl = torch.utils.data.DataLoader(test_encoded_dataset,
                                           batch_size=batch_size,
                                           shuffle=False)
```

10. After completing the data preparation and tokenization, the next step is loading the pre-trained model and fine-tuning it with the dataset we've just prepared. The code for loading the pre-trained model is provided here:

```
>>> from transformers import DistilBertForSequenceClassification
>>> device = torch.device("cuda" if torch.cuda.is_available() else "cpu")
>>> model = DistilBertForSequenceClassification.from_pretrained(
                                            'distilbert-base-uncased',
                                            local_files_only=True)
>>> model.to(device)
```

We load the pre-trained `distilbert-base-uncased` model as mentioned earlier. We also ensure the model is placed on the specified computing device (GPU highly recommended if available) for training and inference.

The `transformers` package offers a variety of pre-trained models. You can explore them at https://huggingface.co/docs/transformers/index#supported-models-and-frameworks.

11. We set the corresponding `Adam` optimizer with a learning rate of 0.00005 as follows:

```
>>> optimizer = torch.optim.Adam(model.parameters(), lr=5e-5)
```

12. Now, we define a training function responsible for training (fine-tuning) the model for one iteration:

```
>>> def train(model, dataloader, optimizer):
    model.train()
    total_loss = 0
    for batch in dataloader:
        optimizer.zero_grad()

        input_ids = batch['input_ids'].to(device)
        attention_mask = batch['attention_mask'].to(device)
        labels = batch['labels'].to(device)
        outputs = model(input_ids, attention_mask=attention_mask,
```

```
                            labels=labels)
            loss = outputs['loss']

            optimizer.zero_grad()
            loss.backward()
            optimizer.step()

            total_loss += loss.item()*len(batch)

    return total_loss/len(dataloader.dataset)
```

This is similar to the train function we employed, except BERT needs both token IDs and attention_mask as inputs.

13. Similarly, we define the evaluation function responsible for evaluating the model accuracy:

```
>>> def evaluate(model, dataloader):
    model.eval()
    total_acc = 0
    with torch.no_grad():
        for batch in dataloader:

            input_ids = batch['input_ids'].to(device)
            attention_mask = batch['attention_mask'].to(device)
            labels = batch['labels'].to(device)
            outputs = model(input_ids, attention_mask=attention_mask)
            logits = outputs['logits']

            pred = torch.argmax(logits, 1)
            total_acc += (pred == torch.argmax(labels, 1)).float().sum().item()

    return  total_acc/len(dataloader.dataset)
```

14. We then train the model for one iteration and display the train loss and accuracy at the end of it:

```
>>> torch.manual_seed(0)
>>> num_epochs = 1
>>> for epoch in range(num_epochs):
>>>     train_loss = train(model, train_dl, optimizer)
>>>     train_acc = evaluate(model, train_dl)
>>>     print(f'Epoch {epoch+1} - loss: {train_loss:.4f} -
                accuracy: {train_acc:.4f}')
Epoch 1 - loss: 0.0242 - accuracy: 0.9642
```

The training process takes a while and the training accuracy is 96%. You may train with more iterations if resources and time allow.

15. Finally, we evaluate the performance on the test set:

```
>>> test_acc = evaluate(model, test_dl)
>>> print(f'Accuracy on test set: {100 * test_acc:.2f} %')
Accuracy on test set: 92.75 %
```

We obtained a test accuracy of 93% with just one epoch in fine-tuning the pre-trained DistilBERT model. This marks a significant enhancement compared to the 86% test accuracy attained with LSTM in *Chapter 12*.

Best practice

If you are dealing with large models or datasets on GPU, it is recommended to monitor GPU memory usage to avoid running out of GPU memory.

In PyTorch, you can use `torch.cuda.mem_get_info()` to check the GPU memory usage. It tells you information about the available and allocated GPU memory. Another trick is `torch.cuda.empty_cache()`. It attempts to release all unused cached memory held by the GPU memory allocator back to the system. Finally, if you're done with a model, you can run `del model` to free up the memory it was using.

Using the Trainer API to train Transformer models

The Trainer API included in Hugging Face is a shortcut for training Transformer-based models. It lets you fine-tune those pre-trained models on your own tasks with an easy high-level interface. No more wrestling with tons of training code like we did in the previous section.

We will fine-tune the BERT model more conveniently using the `Trainer` API in the following steps:

1. Load the pre-trained model again and create the corresponding optimizer:

```
>>> model = DistilBertForSequenceClassification.from_pretrained(
                                       'distilbert-base-uncased',
                                       local_files_only=True)
>>> model.to(device)
>>> optim = torch.optim.Adam(model.parameters(), lr=5e-5)
```

2. To execute the `Trainer` scripts, it is necessary to have the accelerate package installed. Use the following command to install `accelerate`:

```
conda install -c conda-forge accelerate
```

or

```
pip install accelerate
```

3. Next, we prepare the necessary configurations and initialize a `Trainer` object for training the model:

```
>>> from transformers import Trainer, TrainingArguments
>>> training_args = TrainingArguments(
        output_dir='./results',
        num_train_epochs=1,
        per_device_train_batch_size=32,
        logging_dir='./logs',
        logging_steps=50,
    )
>>> trainer = Trainer(
    model=model,
    args=training_args,
    train_dataset=train_encoded_dataset,
    optimizers=(optim, None)
)
```

Here, the `TrainingArguments` configuration defines the number of training epochs, the batch size for training, and the number of steps between each logging of training metrics. We also tell the Trainer what model to use, which data to train on, and what optimizer to use – the second element (`None`) means there's no learning rate scheduler this time.

4. You may notice that the `Trainer` initialization in the previous step does not involve evaluation metrics and test datasets. Let's add them and rewrite the initialization:

```
>>> from datasets import load_metric
>>> import numpy as np
>>> metric = load_metric("accuracy")
>>> def compute_metrics(eval_pred):
        logits, labels = eval_pred
        pred = np.argmax(logits, axis=-1)
        return metric.compute(predictions=pred,
                              references=np.argmax(labels, 1))

>>> trainer = Trainer(
        model=model,
        compute_metrics=compute_metrics,
        args=training_args,
        train_dataset=train_encoded_dataset,
        eval_dataset=test_encoded_dataset,
        optimizers=(optim, None)
    )
```

The `compute_metrics` function calculates the accuracy based on the predicted and true labels. The `Trainer` will use this metric to measure how well the model performs on the specified test set (`test_encoded_dataset`).

5. Now, we train the model with just one line of code:

```
>>> trainer.train()
Step       Training Loss
50         0.452500
100        0.321200
150        0.325800
200        0.258700
250        0.244300
300        0.239700
350        0.256700
400        0.234100
450        0.214500
500        0.240600
550        0.209900
600        0.228900
650        0.187800
700        0.194800
750        0.189500

TrainOutput(global_step=782, training_loss=0.25071206544061453,
metrics={'train_runtime': 374.6696, 'train_samples_per_second': 66.725,
'train_steps_per_second': 2.087, 'total_flos': 3311684966400000.0,
'train_loss': 0.25071206544061453, 'epoch': 1.0})
```

The model was just trained for one epoch as we specified, and training loss was displayed for every 50 steps.

6. With another line of code, we can evaluate the trained model on the test dataset:

```
>>> print(trainer.evaluate())
{'eval_loss': 0.18415148556232452, 'eval_accuracy': 0.929, 'eval_
runtime': 122.457, 'eval_samples_per_second': 204.153, 'eval_steps_per_
second': 25.519, 'epoch': 1.0}
```

We obtained the same test accuracy of 93% utilizing the Trainer API, and it required significantly less code.

Best practice

Here are some best practices for fine-tuning BERT:

- **Data is king:** You should prioritize high-quality and well-labeled data.
- **Start small:** You can begin with smaller pre-trained models like BERT-base or DistilBERT. They're less demanding on your computational power compared to larger models like BERT-large.
- **Automate hyperparameter tuning:** You may utilize automated hyperparameter tuning libraries (e.g., Hyperopt, Optuna) to search for optimal hyperparameters. This can save you time and let your computer do the heavy lifting.
- **Implement early stopping:** You should monitor validation loss during training. If it stops getting better after a while, hit the brakes. This early stopping strategy can prevent unnecessary training iterations. Remember, fine-tuning BERT can take some time and resources.

In this section, we discussed BERT, a model based on a Transformer encoder, and leveraged it to enhance sentiment analysis. In the following section, we will explore another Transformer-based mode, **GPT**.

Generating text using GPT

BERT and GPT are both state-of-the-art NLP models based on the Transformer architecture. However, they differ in their architectures, training objectives, and use cases. We will first learn more about GPT and then generate our own version of *War and Peace* with a fine-tuned GPT model.

Pre-training of GPT and autoregressive generation

GPT (*Improving Language Understanding by Generative Pre-training* by Alec Radford et al. 2018) is a **decoder-only** Transformer architecture, while BERT is encoder only. This means GPT utilizes masked self-attention in the decoders and emphasizes predicting the next token in a sequence.

Think of BERT like a super detective. It gets a sentence with some words hidden (masked) and has to guess what they are based on the clues (surrounding words) in both directions, like looking at a crime scene from all angles. GPT, on the other hand, is more like a creative storyteller. It is pre-trained using an **autoregressive** language model objective. It starts with a beginning word and keeps adding words one by one, using the previous words as inspiration, similar to how we write a story. This process repeats until the desired sequence length is reached.

The word "autoregressive" means it generates text one token at a time in a sequential manner.

Both BERT and GPT are pre-trained on large-scale datasets. However, due to their training methods, they have different strengths and use cases for fine-tuning. BERT is a master at grasping how words and sentences connect. It can be fine-tuned for tasks like sentiment analysis and text classification. On the other hand, GPT is better at creating grammatically correct and smooth-flowing text. This makes it ideal for tasks like text generation, machine translation, and summarization.

In the next section, we will write our own version of *War and Peace* as we did in *Chapter 12, Making Predictions with Sequences Using Recurrent Neural Networks*, but by fine-tuning a GPT model this time.

Writing your own version of *War and Peace* with GPT

For illustrative purposes, we'll employ GPT-2, a model that is more potent than GPT-1 yet smaller in size than GPT-3, and open source, to generate our own version of *War and Peace* in the following steps:

1. Before we start, let's quickly look at how to generate text using the GPT-2 model:

```
>>> from transformers import pipeline, set_seed
>>> generator = pipeline('text-generation', model='gpt2')
>>> set_seed(0)
>>> generator("I love machine learning",
              max_length=20,
              num_return_sequences=3)
[{'generated_text': 'I love machine learning, so you should use machine
learning as your tool for data production.\n\n'},
{'generated_text': 'I love machine learning. I love learning and I love
algorithms. I love learning to control systems.'},
{'generated_text': 'I love machine learning, but it would be pretty
difficult for it to keep up with the demands and'}]
```

This code snippet uses the GPT-2 model to generate text based on the prompt "I love machine learning," and it produces three alternative sequences with a maximum length of 20 tokens each.

2. To generate our own version of *War and Peace*, we need to fine-tune the GPT-2 model based on the original *War and Peace* text. We first need to load the GPT-2 based tokenizer:

```
>>> from transformers import TextDataset, GPT2Tokenizer
>>> tokenizer = GPT2Tokenizer.from_pretrained('gpt2', local_files_only=True)
```

3. Next, we create a `Dataset` object for the tokenized *War and Peace* text:

```
>>> text_dataset = TextDataset(tokenizer=tokenizer,
                               file_path='warpeace_input.txt',
                               block_size=128)
```

We prepare a dataset (`text_dataset`) by tokenizing the text from the `'warpeace_input.txt'` file (which we used in *Chapter 12*). We set the `tokenizer` parameter to the previously created GPT-2 tokenizer, and `block_size` to `128`, specifying the maximum length of each sequence of tokens.

```
>>> len(text_dataset)
>>> 6176
```

We generated 6176 training samples based on the original text.

4. Recall in *Chapter 12*, we had to create the training data manually. Thankfully, this time we utilize the `DataCollatorForLanguageModeling` class from Hugging Face to automatically generate the input sequence and output token. This data collator is specifically designed for language modeling tasks, where we're trying to predict masked tokens. Let's see how to create a data collator as follows:

```
>>> from transformers import DataCollatorForLanguageModeling
>>> data_collator = DataCollatorForLanguageModeling(tokenizer=tokenizer,
                                                    mlm=False)
```

The data collator helps organize and batch the input data for training the language model. We set the `mlm` parameter to False. If set to `True`, it would enable masked language modeling, where tokens are randomly masked for the model to predict as we did in BERT. In our case here, it's turned off, meaning that the model is trained using an autoregressive approach. This approach is ideal for tasks like text generation, where the model learns to create new text one word at a time.

5. After completing the tokenization and data collation, the next step is loading the pre-trained GPT-2 model and creating the corresponding optimizer:

```
>>> import torch
>>> from transformers import GPT2LMHeadModel
>>> model = GPT2LMHeadModel.from_pretrained('gpt2')
>>> model.to(device)
>>> optim = torch.optim.Adam(model.parameters(), lr=5e-5)
```

We use `GPT2LMHeadModel` for text generation. It predicts the probability distribution of the next token in a sequence.

6. Next, we prepare the necessary configurations and initialize a `Trainer` object for training the model:

```
>>> from transformers import Trainer, TrainingArguments
>>> training_args = TrainingArguments(
        output_dir='./gpt_results',
        num_train_epochs=20,
        per_device_train_batch_size=16,
        logging_dir='./gpt_logs',
        save_total_limit=1,
        logging_steps=500,
    )
```

```
>>> trainer = Trainer(
    model=model,
    args=training_args,
    data_collator=data_collator,
    train_dataset=train_dataset,
    optimizers=(optim, None)
)
```

The model will be trained on the `text_dataset` dataset divided into 16 sample batches for 20 epochs. We also set the limit on the total number of checkpoints to save. In this case, only the latest checkpoint will be saved. This is to reduce space consumption. In the `trainer`, we provide the `DataCollatorForLanguageModeling` instance to organize the data for training.

7. Now, we are ready to train the model:

```
>>> trainer.train()
Step      Training Loss
500          3.414100
1000         3.149500
1500         3.007500
2000         2.882600
2500         2.779100
3000         2.699200
3500         2.621700
4000         2.548800
4500         2.495400
5000         2.447600
5500         2.401400
6000         2.367600
6500         2.335500
7000         2.315100
7500         2.300400

TrainOutput(global_step=7720, training_loss=2.640813370936893,
metrics={'train_runtime': 1408.7655, 'train_samples_per_second': 87.68,
'train_steps_per_second': 5.48, 'total_flos': 8068697948160000.0, 'train_
loss': 2.640813370936893, 'epoch': 20.0})
```

8. After the GPT-2 model is trained based on *War and Peace*, we finally use it to generate our own version. We first develop the following function to generate text based on a given model and prompt text:

```
>>> def generate_text(prompt_text, model, tokenizer, max_length):
        input_ids = tokenizer.encode(prompt_text,
                                return_tensors="pt").to(device)
```

```
        # Generate response
        output_sequences = model.generate(
            input_ids=input_ids,
            max_length=max_length,
            num_return_sequences=1,
            no_repeat_ngram_size=2,
            top_p=0.9,
        )

        # Decode the generated responses
        responses = []
        for response_id in output_sequences:
            response = tokenizer.decode(response_id,
                                        skip_special_okens=True)
            responses.append(response)

        return responses
```

In the generation process, we first tokenize the input prompt text using the tokenizer and convert the tokenized sequence to a PyTorch tensor. Then, we generate text based on the tokenized input using the given model. Here, no_repeat_ngram_size prevents repeating the same n-gram phrases to keep things fresh. Another interesting setting, top_p, controls the diversity of the generated text. It considers the tokens that are most probable instead of only the most probable one. Finally, we decode the generated response using the tokenizer, translating it back to human language.

9. We use the same prompt, "the emperor," and here is our version of *War and Peace* in 100 words:

```
>>> prompt_text = "the emperor"
>>> responses = generate_text(prompt_text, model, tokenizer, 100)
>>> for response in responses:
        print(response)
the emperor's, and the Emperor Francis, who was in attendance on him, was
present.

The Emperor was present because he had received the news that the French
troops were advancing on Moscow, that Kutuzov had been wounded, the
Emperor's wife had died, a letter from Prince Andrew had come from Prince
Vasili, Prince Bolkonski had seen at the palace, news of the death of the
Emperor, but the most important news was that …
```

We've successfully generated our own version of *War and Peace* using the fine-tuned GPT-2 model. It reads better than the LSTM version in *Chapter 12*.

GPT is a decoder-only Transformer architecture. It's all about making things up on the fly, one token at a time. Unlike some other Transformer-based models, it doesn't need a separate step (encoding) to understand what you feed it. This lets it focus on generating text that flows naturally, like a creative writer. Once GPT is trained on a massive pile of text, we can fine-tune it for specific tasks with smaller datasets. Think of it like teaching a master storyteller to write about a specific topic, like history or science fiction. In our example, we generated our own version of *War and Peace* by fine-tuning a GPT-2 model.

While BERT focuses on bidirectional pre-training, GPT is autoregressive and predicts the next word in a sequence. You may wonder whether there is a model that combines aspects of both bidirectional understanding and auto-regressive generation. The answer is yes – **Bidirectional and Auto-Regressive Transformers (BART)**. It was introduced by Facebook AI (*BART: Bidirectional and Autoregressive Transformers for Sequence-to-Sequence Learning* by Lewis et al., 2019), and designed to combine both strengths.

Summary

This chapter was all about Transformer, a powerful neural network architecture designed for sequence-to-sequence tasks. Its key ingredient, self-attention, lets the model focus on the most important parts of the information it's looking at in a sequence.

We worked on two NLP projects: sentiment analysis and text generation using two state-of-the-art Transformer models, BERT and GPT. We observed an elevated performance compared to what we did in the last chapter. We also learned how to fine-tune these Transformers with the Hugging Face library, a one-stop shop for loading pre-trained models, performing different NLP tasks, and fine-tuning models on your own data. Plus, it throws in some bonus tools for chopping up text, checking how well the model did, and even generating some text of its own.

In the next chapter, we will focus on another OpenAI cutting-edge model, CLIP, and will implement natural language-based image search.

Exercises

1. Can you compute the positional encoding for the third word "learning" in the example sentence "python machine learning" using a four-dimensional vector?

2. Can you fine-tune a BERT model for topic classification? You can take the newsgroups dataset as an example.

3. Can you fine-tune a BART model (https://huggingface.co/facebook/bart-base) to write your own version of *War and Peace*?

Join our book's Discord space

Join our community's Discord space for discussions with the authors and other readers:

`https://packt.link/yuxi`

14

Building an Image Search Engine Using CLIP: a Multimodal Approach

In the previous chapter, we focused on Transformer models such as BERT and GPT, leveraging their capabilities for sequence learning tasks. In this chapter, we'll explore a multimodal model, which seamlessly connects visual and textual data. With its dual encoder architecture, this model learns the relationships between visual and textual concepts, enabling it to excel in tasks involving image and text. We will delve into its architecture, key components, and learning mechanisms, leading to a practical implementation of the model. We will then build a multimodal image search engine with text-to-image and image-to-image capabilities. To top it all off, we will tackle an awesome zero-shot image classification project!

We will cover the following topics in this chapter:

- Introducing the CLIP model
- Getting started with the dataset
- Architecting the CLIP model
- Finding images with words

Introducing the CLIP model

We have explored computer vision in *Chapter 11, Categorizing Images of Clothing with Convolutional Neural Networks*, and NLP in *Chapter 12, Making Predictions with Sequences Using Recurrent Neural Networks*, and *Chapter 13, Advancing Language Understanding and Generation with the Transformer Models*. In this chapter, we will delve into a model that bridges the realms of computer vision and NLP, the **Contrastive Language–Image Pre-Training** (CLIP) model developed by OpenAI. Unlike traditional models that are specialized for either computer vision or natural language processing, CLIP is trained to understand both **modalities** (image and text) in a unified manner. Hence, CLIP excels at understanding and generating relationships between images and natural language.

A modality in ML/AI is a specific way of representing information. Common modalities include text, images, audio, video, and even sensor data.

Excited to delve into the workings of CLIP? Let's explore and discover more about how it works!

Understanding the mechanism of the CLIP model

CLIP is designed to learn representations of images and corresponding textual descriptions simultaneously. The model learns to associate similar pairs and disassociate dissimilar pairs of images and text. Its unique architecture (see *Figure 14.1* below) enables it to develop semantic connections between images and their textual descriptions:

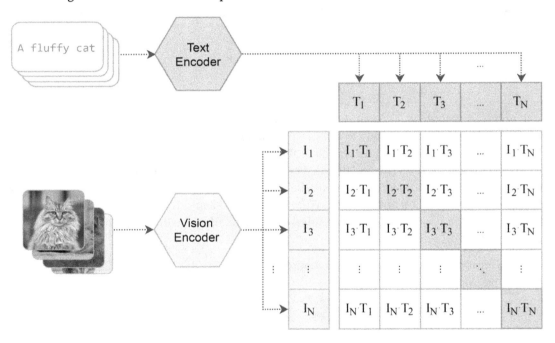

Figure 14.1: CLIP architecture (image based on Figure 1 in "Learning Transferable Visual Models From Natural Language Supervision": https://arxiv.org/pdf/2103.00020.pdf)

As you can see, it utilizes a dual-encoder architecture that integrates both vision and text encoder. The output from the vision encoder and the output from the text encoder are projected into a shared space. It then evaluates the placement of these image-text pairs based on their similarity. This shared semantic space allows CLIP to perform various vision-language tasks, such as image classification, object detection, and image retrieval.

Here are the key components of the CLIP model.

Vision encoder

The vision encoder (also called image encoder) in CLIP is responsible for processing and encoding image inputs. It is typically implemented as a **CNN**. Recall that CNNs are well suited for image-related tasks, as they can effectively capture hierarchical features in images. The output of the vision encoder for an input image is a fixed-size vector representation. The embedding vector captures the semantic content of the image.

There are two main architectures used for the vision encoder. The first version is a modified **ResNet mode** based on the ResNet-50 model. Additionally, the average pooling layer is substituted with an attention pooling mechanism. This attention pooling is realized as a single layer of multi-head attention, with the query conditioned on the recently introduced **Vision Transformer** (https://huggingface. co/google/vit-base-patch16-224). It includes an additional normalization layer just before the Transformer as the only adjustment.

It is important to note that the generated visual embeddings exist in **a shared space** with embeddings from the text encoder. This shared space projection enables direct comparisons between visual and textual representations. If an image and a textual description are semantically related, they would be mapped to closer points in this space. For example, an image of a cat and the corresponding text "a fluffy cat" would be close together in this space, indicating their semantic similarity.

The vision encoder is pre-trained on a large and diverse dataset containing images and their associated textual descriptions. For example, OpenAI mentioned in the CLIP paper (https://openai.com/index/ clip) that their model was trained on a collection of 400 million image-text pairs obtained from crawling the internet. The pre-training process allows the vision encoder to learn rich and generalized visual representations. Furthermore, the learned representations are task-agnostic. Hence, we can fine-tune a CLIP model for a wide range of text-image applications.

Text encoder

Similarly, the text encoder is responsible for processing and encoding textual inputs. The process begins with tokenization. The tokenized text is then passed through an embedding layer and converted into a fixed-size high-dimensional vector. Additionally, to preserve important sequential information in the text, we add **positional encoding** to the embeddings. The resulting embeddings can capture the semantic content of the text.

The text encoder is implemented as a Transformer with a particular architecture. For instance, the OpenAI team utilized a 12-level, 512-wide model with 8 attention heads and 63 million parameters in total, and its maximum sequence length was restricted to 76.

As mentioned earlier, the text embeddings are in a shared space with the embeddings from the vision encoder. This allows direct comparisons between visual and textual inputs and cross-modal understanding. Similarly, pre-training on a diverse dataset enables the model to learn generalizable contextual understanding from various linguistic contexts. The text encoder can be fine-tuned for various downstream tasks in collaboration with the vision encoder.

Contrastive learning

Contrastive learning is the training strategy in CLIP. It teaches the model to differentiate between similar and dissimilar image-text pairs. During training, CLIP is presented with positive and negative image-text pairs. A positive pair consists of an image and description that are semantically related. On the other hand, a negative pair is formed by combining an image with a randomly chosen description, creating mismatches.

Contrastive learning focuses on bringing embeddings of positive pairs closer together in the shared embedding space, while pushing embeddings of negative pairs further apart. This separation is achieved through a **contrastive loss function**. Let's break down the contrastive loss calculation:

1. Embedding generation:

   Given N images I and the corresponding text descriptions T, the CLIP model first creates image embeddings I_{emb} and text embeddings T_{emb} using its dual encoder (vision encoder and text encoder) architecture.

2. Similarity matrix calculation:

   Since embeddings I_{emb} and T_{emb} are in the same space, we can calculate pair-wise similarities S. For image i and text j, their similarity s_{ij} is the cosine similarity image embedding $I_{emb\,i}$ and text embedding $T_{emb\,j}$:

   $$s_{ij} = cos\left(I_{emb\,i}, T_{emb\,j}\right)$$

   Here, $1 \leq i \leq N, 1 \leq j \leq N$. The goal is to maximize the similarity of image and text embeddings for N positive pairs, while minimizing the similarity for the embeddings of $N^2 - N$ negative pairings.

3. Target matrix creation:

   Next, we construct the target ("ideal") matrix Y for learning. Here, $y_{ij} = 1$ if image i and text j are a positive pair (diagonal elements); $y_{ij} = 0$ for all other pairs (off-diagonal elements).

4. Cross-entropy loss computation:

   With the similarity matrix S and target matrix Y, we then compute the cross-entropy loss for both image and text modalities. Here is the loss for image alignment:

   $$loss_I = CrossEntropyLoss\left(S, Y, axis = 0\right)$$

   It measures how well the model predicts the correct image given a text description.

   The loss for text alignment, measuring how well the model predicts the correct description given an image, is:

   $$loss_T = CrossEntropyLoss\left(S, Y, axis = 1\right)$$

5. Final loss computation:

The contrastive loss is the average of the image-based loss and the text-based loss:

$$loss = (loss_I + loss_T)/2$$

During training, the model's parameters are updated to minimize the contrastive loss function. This drives the model to learn embeddings that align correctly paired images and text, while pushing apart mismatched pairs.

The contrastive learning objective contributes to effective cross-modal retrieval and understanding. Along with the dual-encoder architecture, a pre-trained CLIP model can perform various downstream image-text tasks without task-specific retraining. So what are those typical applications and scenarios? Let's see next.

Exploring applications of the CLIP model

In this section, we will explain some common applications and use cases for the CLIP model.

Zero-shot image classification

In a zero-shot learning setup, CLIP is presented with a task it has not been explicitly trained on. For instance, it might be asked to classify images into unseen categories, or to generate descriptions for images without having seen similar examples during pre-training.

The first zero-shot application is image classification based on textual descriptions. We don't need to perform any task-specific training thanks to the model's pre-trained knowledge. The model can categorize the images based on their alignment with the descriptions.

For example, given three unseen images (*Image 1*: a photo of a red vintage car on a city street; *Image 2*: a painting depicting a red car in an urban setting; *Image 3*: a cartoon illustration of a city with a red car), we pass the query text "A vintage red car parked on a city street" to the CLIP model. It may correctly rank *Image 1* as the most relevant to the query, with a red vintage car on a city street. *Image 3* may be ranked the least relevant due to its cartoon style, which is least aligned with the query.

The zero-shot learning capability makes CLIP useful for tasks where labeled examples are scarce or even unavailable. We've seen it can categorize images even for categories never seen during pre-training. In the next section, we will use it for zero-shot text classification.

Zero-shot text classification

Similarly, CLIP can classify new textual descriptions based on images. In zero-shot setting, we don't need to provide labeled examples for fine-tuning. The model can categorize text inputs based on their alignment with the images.

For example, given a query image of a mountain landscape and three potential descriptions (*Text 1*: "A view of a serene mountain range," *Text 2*: "The majestic peaks and valleys of the mountains," and *Text 3*: "Hiking trails in mountainous regions."), the CLIP model may score *Text 1* the most relevant due to its highest alignment with the query image.

We've talked about CLIP for zero-shot image and text classification. In fact, we can extend it to content retrieval. Let's see the next section.

Image and text retrieval

CLIP can be used to retrieve images relevant to a given text query, and vice versa,

For example, in image retrieval, we pass the text query "playful puppies" to an image search engine. The CLIP model retrieves images that best match the description of playful puppies. It also ranks them based on their alignment with the text query. Similarly, CLIP can also be used to retrieve and rank captions for images that accurately describe their content.

We've demonstrated CLIP's cross-modal retrieval abilities. In the next section, let's look at its adoption in cross-modal generation.

Image and text generation

Beyond retrieving from the existing content pool, we can use CLIP to generate images based on textual prompts or to provide textual descriptions for images.

For instance, we can generate artistic images by giving the CLIP model a prompt like "a surreal painting of a robot riding a bicycle." We can also ask the CLIP model to describe a picture of a modern kitchen. It may answer with "A contemporary kitchen design."

In fact, CLIP can answer many questions about the given images, beyond just providing descriptions.

Visual question answering (VQA)

CLIP can be adapted for visual question-answering tasks. It can be used to answer questions about images based on its knowledge of both visual and textual modalities. For example, we can use the model to answer questions like "What kind of animal is this?" or "How many people are in the photo?"

Transfer learning

Finally, we can fine-tune CLIP on specific downstream tasks, such as object detection, sentiment analysis, and custom classification tasks.

During pre-training, the CLIP model gains a generalized understanding across modalities from a diverse range of images and text. Leveraging transfer learning, we don't need to perform extensive task-specific training. It can be adopted for a wide range of vision and NLP applications.

Excited to start implementing CLIP? Let's begin by delving into the dataset containing images and captions that we'll use for the training process.

Getting started with the dataset

We are going to use the `Flickr8k` dataset (`https://hockenmaier.cs.illinois.edu/8k-pictures.html`), created by M. Hodosh, P. Young, and J. Hockenmaier, described in *Framing Image Description as a Ranking Task: Data, Models and Evaluation Metrics, Journal of Artificial Intelligence Research*, Volume 47, pages 853–899 (`https://www.jair.org/index.php/jair/article/view/10833/25855`). It is commonly employed in various computer vision tasks, particularly image captioning.

The `Flickr8k` dataset contains 8,000 images collected from the Flickr photo-sharing website. These images cover a diverse range of scenes, objects, and activities. Each image in the dataset is associated with five English sentences. These sentences serve as captions and provide textual descriptions of the image content.

One common use of the `Flickr8k` dataset is image captioning, where the goal is to train models to generate human-like captions for images. The `Flickr8k` dataset is often used by researchers and practitioners as a benchmark for image captioning models. It allows us to evaluate the ability of models to understand and describe visual content in natural language.

There is also an extended version called `Flickr30k`, which contains 30,000 images with corresponding captions. The larger dataset provides a more extensive and diverse set of images for training and evaluation, but it consumes more computational resources. So we will focus on the `Flickr8k` dataset in this chapter.

Obtaining the Flickr8k dataset

To obtain the `Flickr8k` dataset, simply submit a request at `https://illinois.edu/fb/sec/1713398`. Upon request, dataset links will be emailed to you. One of the links will lead you to a downloadable file, `Flickr8k_Dataset.zip`, from which you can extract 8,091 image files. Another link will direct you to a downloadable file called `Flickr8k_text.zip`. We will use the extracted file `Flickr8k.token.txt`, which contains the raw captions of the `Flickr8k` dataset. The first column is in the format of "image path # caption number," and the second column is the corresponding caption.

The dataset is also available on Kaggle, such as `https://www.kaggle.com/datasets/adityajn105/flickr8k/data`. The `captions.txt` file contains information mirroring that of the `Flickr8k.token.txt` file, but it is easier to use, as the first column contains only the image paths. For simplicity, we will use the `captions.txt` file instead of the original `Flickr8k.token.txt` file.

Loading the Flickr8k dataset

After extracting all the images from `Flickr8k_Dataset.zip` and getting the caption text file ready, we can now load the `Flickr8k` dataset into a custom PyTorch Dataset object. Follow the steps below:

1. First, we import the necessary packages:

```
>>> import os
>>> from PIL import Image
>>> import torch
>>> from torch.utils.data import Dataset, DataLoader
>>> import torchvision.transforms as transforms
```

Here, the `Image` package will be used to load the image files.

2. We then set the image directory and the caption file path as follows:

```
>>> image_dir = "flickr8k/Flicker8k_Dataset"
>>> caption_file = "flickr8k/captions.txt"
```

Here, we put all the extracted images in the `flickr8k/Flicker8k_Dataset` folder and the `captions.txt` file in the same root directory, `flickr8k`.

3. Next, we load the `DistilBRET` tokenizer as we did in the previous chapter, *Advancing Language Understanding and Generation with the Transformer Models*:

```
>>> from transformers import DistilBertTokenizer
>>> tokenizer = DistilBertTokenizer.from_pretrained(
                                    'distilbert-base-uncased')
```

4. Now, we create a custom PyTorch Dataset class for the `Flickr8k` dataset, as follows:

```
>>> class Flickr8kDataset(Dataset):
        def __init__(self, image_dir, caption_file):
            self.image_dir = image_dir
            self.transform = transforms.Compose([
                            transforms.Resize((224, 224)),
                            transforms.ToTensor(),
                        ])
            self.image_paths, self.captions =
            self.read_caption_file(caption_file)

        def read_caption_file(self, caption_file):
            image_paths = []
            captions = []

            with open(caption_file, "r") as file:
                lines = file.readlines()
                for line in lines[1:]:
                    parts = line.strip().split(",")
                    image_paths.append(os.path.join(self.image_dir,
                                            parts[0]))
                    captions.append(parts[1])

            self.caption_encodings = tokenizer(captions, truncation=True,
                                        padding=True,
                                        max_length=200)

            return image_paths, captions

        def __len__(self):
            return len(self.image_paths)
```

```
        def __getitem__(self, idx):
            item = {key: torch.tensor(val[idx]) for key, val in
                                      self.caption_encodings.items()}

            caption = self.captions[idx]
            item["caption"] = caption

            img_path = self.image_paths[idx]
            img = Image.open(img_path).convert("RGB")
            img = self.transform(img)
            item['image'] = img

            return item
```

Upon initialization, we define an image transformation function using the transforms module from torchvision, including resizing the images to (224, 224) pixels and converting them to tensors; we read the caption file line by line, extract image paths and captions, and store them in the image_paths and captions lists. The captions are tokenized and encoded using the given tokenizer, with options for truncation, padding, and a maximum length of 200 tokens. Results are stored in caption_encodings.

Upon retrieving an item from the dataset, the tokenized and encoded captions are stored in the item object along with the original caption. The image at the corresponding index is also loaded, transformed, and added to the item.

5. We initiate an instance of the custom Dataset class, as follows:

```
>>> flickr8k_dataset = Flickr8kDataset(image_dir=image_dir,
                                       caption_file=caption_file)
```

Take a look at one data sample:

```
>>> item_sample = next(iter(flickr8k_dataset))
{'input_ids': tensor([ 101, 1037, 2775, 1999, 1037, 5061, 4377, 2003,
        8218, 2039, 1037, 2275, 1997, 5108, 1999, 2019, 4443, 2126, 1012,
         102,    0,    0,    0,    0,    0,    0,    0,    0,    0,    0,
           0,    0,    0,    0,    0,    0,    0,    0,    0]),
 'attention_mask': tensor([1, 1, 1, 1, 1, 1, 1, 1, 1, 1, 1, 1, 1, 1, 1,
           1, 1, 1, 1, 1, 0, 0, 0, 0, 0, 0, 0, 0, 0, 0, 0, 0, 0, 0, 0, 0,
           0, 0, 0]),
 'caption': 'A child in a pink dress is climbing up a set of stairs in an
entry way.',
 'image': tensor([[[0.3216, 0.4353, 0.4549,  ..., 0.0157, 0.0235, 0.0235],
         [0.3098, 0.4431, 0.4667,  ..., 0.0314, 0.0275, 0.0471],
         [0.3020, 0.4588, 0.4745,  ..., 0.0314, 0.0275, 0.0392],
```

```
        ...,
       [0.7294, 0.5882, 0.6706,  ..., 0.8314, 0.6471, 0.6471],
       [0.6902, 0.6941, 0.8627,  ..., 0.8235, 0.6588, 0.6588],
       [0.8118, 0.8196, 0.7333,  ..., 0.8039, 0.6549, 0.6627]],


      [[0.3412, 0.5020, 0.5255,  ..., 0.0118, 0.0235, 0.0314],
       [0.3294, 0.5059, 0.5412,  ..., 0.0353, 0.0392, 0.0824],
       [0.3098, 0.5176, 0.5529,  ..., 0.0353, 0.0510, 0.0863],
       ...,
       [0.4235, 0.3137, 0.4784,  ..., 0.8667, 0.7255, 0.7216],
       [0.3765, 0.5059, 0.6627,  ..., 0.8549, 0.7216, 0.7216],
       [0.4941, 0.5804, 0.4784,  ..., 0.8392, 0.7216, 0.7216]],


      [[0.3804, 0.4902, 0.4980,  ..., 0.0118, 0.0157, 0.0196],
       [0.3608, 0.5059, 0.5176,  ..., 0.0275, 0.0235, 0.0235],
       [0.3647, 0.5255, 0.5333,  ..., 0.0196, 0.0235, 0.0275],
       ...,
       [0.1216, 0.1098, 0.2549,  ..., 0.9176, 0.8235, 0.7961],
       [0.0784, 0.1804, 0.2902,  ..., 0.9137, 0.8118, 0.7843],
       [0.1843, 0.2588, 0.2824,  ..., 0.9176, 0.8039, 0.7686]]])}
```

The caption is `A child in a pink dress is climbing up a set of stairs in an entryway`.
Let's display the image itself using the following script:

```
>>> import matplotlib.pyplot as plt
>>> import numpy as np
>>> npimg = item_sample['image'].numpy()
>>> plt.imshow(np.transpose(npimg, (1, 2, 0)))
```

Figure 14.2: Image of the Flickr8k data sample (photo by Rick & Brenda Beerhorst, Flickr:
https://www.flickr.com/photos/studiobeerhorst/1000268201/)

🔍 **Quick tip:** Need to see a high-resolution version of this image? Open this book in the next-gen Packt Reader or view it in the PDF/ePub copy.

🔒 **The next-gen Packt Reader** and a **free PDF/ePub copy** of this book are included with your purchase. Unlock them by scanning the QR code below or visiting https://www.packtpub.com/unlock/9781835085622.

6. The last step of data preparation is to create a `DataLoader` object to handle batching and shuffling. We set the batch size to 32 and initiate a `DataLoader`, based on the previously created dataset:

```
>>> batch_size = 32
>>> data_loader = DataLoader(flickr8k_dataset, batch_size=batch_size,
shuffle=True)
```

Now that the dataset is prepared, let's proceed to develop the CLIP model in the following section.

Architecting the CLIP model

The vision encoder and text encoder are the two main components of the CLIP model. We will start with the vision encoder.

Vision encoder

Implementing the vision encoder is quite straightforward. We leverage the PyTorch `vision` library, which provides access to various pre-trained image models, including `ResNets` and `VisionTransformer`. Here, we opt for ResNet50 as our vision encoder as an example.

The vision encoder ensures each image is encoded into a fixed-size vector, with the dimensionality matching the model's output channels (in the case of ResNet50, the vector size is 2048):

```
>>> import torch.nn as nn
>>> from torchvision.models import resnet50
>>> class VisionEncoder(nn.Module):
        def __init__(self):
            super().__init__()
            pretrained_resnet50 = resnet50(pretrained=True)
            self.model = nn.Sequential(*list(
                                     pretrained_resnet50.children())[:-1])

            for param in self.model.parameters():
                param.requires_grad = False

        def forward(self, x):
            x= self.model(x)
            x = x.view(x.size(0), -1)
            return x
```

Upon initialization, we load the pre-trained ResNet50 model. Then, we remove the final classification layer, as we are using the ResNet50 model as a feature extractor instead of a classifier. Here, we freeze the model's parameters by setting their `requires_grad` trainable attribute to false. You can also fine-tune the pre-trained ResNet50 model component by making the parameters trainable. The `forward` method is used to extract image embeddings from input images.

We just implemented the `VisionEncoder` module based on the pre-trained ResNet50 model. We use the model's hidden layer output as the fixed-size vector representation for each image. Since we ignore its final classification layer, the ResNet50 model in this case is used as an image feature extractor.

We will continue with the text encoder module next.

Text encoder

For simplicity, we will employ DistilBERT as the text encoder. We extract the complete representation of a sentence by utilizing the final representations of the [CLS] token. The expectation is that this representation captures the overall meaning of the sentence (the image caption in this case). Conceptually, this is similar to the process applied to images, where they are transformed into fixed-size vectors. For DistilBERT (and BERT as well), each token's output representation is a vector with a size of 768.

We implement the text encoder using the following code:

```
>>> from transformers import DistilBertModel
>>> class TextEncoder(nn.Module):
        def __init__(self):
            super().__init__()
            self.model = DistilBertModel.from_pretrained(
                                         'distilbert-base-uncased')
            for param in self.model.parameters():
                param.requires_grad = False

        def forward(self, input_ids, attention_mask=None):
            outputs = self.model(input_ids=input_ids,
                                 attention_mask=attention_mask)
            return outputs.last_hidden_state[:, 0, :]
```

Upon initialization, we first load a pre-trained DistilBERT model from the Hugging Face Transformers library. Then, we freeze the parameters of the DistilBERT model by setting requires_grad to False for all parameters. Again, you can also fine-tune the pre-trained DistilBERT model by making the parameters trainable. In the forward pass, we feed the input into the DistilBERT model and extract the last hidden state from the model's outputs. Finally, we return the vector corresponding to the [CLS] token, as the embedding representation of the input caption.

We just implemented the TextEncoder module to encode textual input using the DistilBERT model. It uses the [CLS] token representation as the fixed-size vector representation of the input text sequence. Similar to what we did in the vision encoder for simplicity, we freeze the parameters in the DistilBERT model and use it as a text feature extractor without further training.

Projection head for contrastive learning

Having encoded both images and texts into fixed-size vectors (2,048 for images and 768 for text), the next step is to project them into a shared space. This process enables the comparison of image and text embedding vectors. We can later train the CLIP model to distinguish between relevant and non-relevant image-text pairs.

We develop the following head projection module to transform the initial 2,048-dimensional image vectors or 768-dimensional text vectors into a shared 256-dimensional space:

```python
>>> class ProjectionHead(nn.Module):
        def __init__(self, embedding_dim, projection_dim=256, dropout=0.1):
            super().__init__()
            self.projection = nn.Linear(embedding_dim, projection_dim)
            self.gelu = nn.GELU()
            self.fc = nn.Linear(projection_dim, projection_dim)
            self.dropout = nn.Dropout(dropout)
            self.layer_norm = nn.LayerNorm(projection_dim)

        def forward(self, x):
            projection = self.projection(x)
            x = self.gelu(projection)
            x = self.fc(x)
            x = self.dropout(x)
            x = projection + x
            x = self.layer_norm(x)
            return x
```

Here, we first create a linear projection layer to transform input vectors from the size of `embedding_dim` to `projection_dim`. We then apply the **Gaussian Error Linear Unit** (GELU) activation function to introduce non-linearity. We add another fully connected layer and incorporate a dropout layer for regularization. Finally, we apply layer normalization for more efficient training.

GELU is an activation function that introduces non-linearity into neural networks. It is defined as:

$$GELU(x) = 0.5x\left(1 + tanh\left(\sqrt{\frac{2}{\pi}}(x + 0.044715x^3)\right)\right)$$

Compared to ReLU, GELU is much more complex (as you can see) and, hence, has smoother gradients. Also, GELU tends to perform better than ReLU in deeper or more complex networks. However, ReLU remains popular due to its simplicity and effectiveness in many scenarios.

Best Practice

Layer normalization is used in deep neural networks to normalize the inputs of a layer. It aims to improve training stability and model generalization. Unlike batch normalization, which normalizes across the entire batch of data, layer normalization normalizes across the features for each individual data sample.

For each data point, layer normalization is applied independently across the features. Hence, layer normalization is beneficial for training with small mini-batches or online training, while batch normalization is more suitable for large batches or large datasets. They are both valuable techniques for stabilizing training in DL. You can choose one based on factors like dataset size and batch size.

In this context, `embedding_dim` represents the size of the input vectors (2,048 for images and 768 for text), while `projection_dim` denotes the size of the output vectors, 256 in our case.

In summary, this projection head module is designed to transform both input image and text representation vectors into the same lower-dimensional space. As well as using linear projection, we add non-linearity and employ regularization techniques, such as dropout and layer normalization. The resulting projected vectors will become the building blocks for contrastive learning. Let's figure out how they are used to learn semantical relationships between images and text in the next section.

CLIP model

This section is where the real excitement unfolds! We utilize the previously constructed modules to implement the primary CLIP model, as follows:

```
>>> import torch.nn.functional as F
>>> class CLIPModel(nn.Module):
        def __init__(self, image_embedding=2048, text_embedding=768):
            super().__init__()
            self.vision_encoder = VisionEncoder()
            self.text_encoder = TextEncoder()
            self.image_projection = ProjectionHead(embedding_dim=image_embedding)
            self.text_projection = ProjectionHead(embedding_dim=text_embedding)

        def forward(self, batch):
            image_features = self.vision_encoder(batch["image"])
            text_features = self.text_encoder(
                input_ids=batch["input_ids"],
                attention_mask=batch["attention_mask"]
            )
```

```
        image_embeddings = self.image_projection(image_features)
        text_embeddings = self.text_projection(text_features)

        logits = text_embeddings @ image_embeddings.T
        images_similarity = image_embeddings @ image_embeddings.T
        texts_similarity = text_embeddings @ text_embeddings.T
        targets = F.softmax((images_similarity + texts_similarity)/2 , dim=-1)
        texts_loss = F.cross_entropy(logits, targets)
        images_loss = F.cross_entropy(logits.T, targets.T)
        loss = (images_loss + texts_loss) / 2
        return loss.mean()
```

The initialization is self-explanatory, where we create instances of the `VisionEncoder` and `TextEncoder` for images and text, respectively, and their corresponding head projection `ProjectionHead` instances. In the forward pass, we encode the input images into fixed-size vectors using the vision encoder, and we encode the input texts using the text encoder. Recall that the output size for encoded image and text vectors is 2,048 and 768, respectively. Subsequently, we employ separate projection modules to project the encoded vectors into a shared space, as previously mentioned. In this shared space, both encodings assume a similar shape (256 in our case). Following this, we compute the contrastive loss. Here are the details:

1. First, we calculate the similarity between text and image embeddings using matrix multiplication (`text_embeddings @ image_embeddings.T`). Here, the `@` operator in PyTorch performs matrix multiplication, or dot product in this context, and `.T` is the transpose operation that we discussed previously. Recall that in linear algebra, calculating the dot product is a common method to gauge the similarity between two vectors. A higher result suggests greater similarity.

2. Next, we compute the similarities between image embeddings themselves, and the similarities between text respectively.

3. We then combine the image and text similarities to create target distributions.

4. We calculate the cross-entropy loss between the predicted logits and target distributions for both images and texts.

5. Finally, we compute the final contrastive loss as the average of the image and text losses.

We just developed a module to train the CLIP model using a contrastive loss. The model takes a batch containing both image and text data, encodes them, and then projects them into a shared space. It then calculates the similarities and computes the contrastive loss. The goal is to bring similar pairs of image and text representations closer together and push dissimilar pairs further apart.

Now that all the modules are prepared, it's time to commence training the CLIP model.

Finding images with words

In this section, we will first train a CLIP model that we implemented in the previous sections. We will then use the trained model to retrieve images given a query. Finally, we will use a pre-trained CLIP model to perform image searches and zero-shot predictions.

Training a CLIP model

Let's train a CLIP model in the following steps:

1. First, we create a CLIP model and move it to system device (either a GPU or CPU):

    ```
    >>> device = torch.device("cuda" if torch.cuda.is_available() else "cpu")
    >>> model = CLIPModel().to(device)
    ```

2. Next, we initialize an Adam optimizer to train the model and set the learning rate:

    ```
    >>> optimizer = torch.optim.Adam(model.parameters(), lr=0.001)
    ```

3. As we did in previous chapters, we define the following training function to update the model:

    ```
    >>> def train(model, dataloader, optimizer):
            model.train()
            total_loss = 0
            b = 0
            for batch in dataloader:
                optimizer.zero_grad()
                batch = {k: v.to(device) for k, v in batch.items()
                                          if k != "caption"}
                loss = model(batch)
                optimizer.zero_grad()
                loss.backward()
                optimizer.step()

                total_loss += loss.item()*len(batch)

            return total_loss/len(dataloader.dataset)
    ```

4. We train the model for three epochs:

    ```
    >>> num_epochs = 3
    >>> for epoch in range(num_epochs):
            train_loss = train(model, data_loader, optimizer)
            print(f'Epoch {epoch+1} - loss: {train_loss:.4f}')
    Epoch 1 - loss: 0.2551
    Epoch 2 - loss: 0.1504
    Epoch 3 - loss: 0.1274
    ```

The training completes after three epochs. Now, let's proceed to conduct an image search using the trained CLIP model.

Obtaining embeddings for images and text to identify matches

To find matching images for a text query (or vice versa), the key process involves obtaining the projected embeddings for both the image candidates and the text query. The goal is to fetch the image that achieves the highest similarity score between its embedding and the text embedding.

For illustrative purposes, we'll utilize a single batch of image data as the pool of image candidates. Let's explore the steps involved in searching for the pertinent image within this sample pool:

1. First, we sample a batch of 32 data points from the `data_loader`:

```
>>> torch.manual_seed(0)
>>> data_loader = DataLoader(flickr8k_dataset, batch_size=batch_size,
                             shuffle=True)
>>> sample_batch = next(iter(data_loader))
```

2. Next, we compute the projected embeddings for the sampled images using the previously trained CLIP model:

```
>>> batch_image_features = model.vision_encoder(sample_batch["image"].
to(device))
>>> batch_image_embeddings = model.image_projection(batch_image_features)
```

3. We now define the image search function, as follows:

```
>>> def search_top_images(model, image_embeddings, query, n=1):
        encoded_query = tokenizer([query])
        batch = {
            key: torch.tensor(values).to(device)
            for key, values in encoded_query.items()
        }
        model.eval()
        with torch.no_grad():
            text_features = model.text_encoder(
                input_ids=batch["input_ids"],
                attention_mask=batch["attention_mask"])
            text_embeddings = model.text_projection(text_features)

        dot_similarity = text_embeddings @ image_embeddings.T
        values, indices = torch.topk(dot_similarity.squeeze(0), n)
        return indices
```

Here, we first compute the projected text embeddings for a given text query. Next, we compute the dot product similarity between the text embedding and the precomputed image embeddings for each image candidate. We retrieve the top-n indices corresponding to the highest similarity scores. Don't forget to set the trained model to evaluation mode, indicating that no gradients should be computed during inference.

4. Let's observe its performance now! First, we search for "a running dog" using the image search function we just defined and display the search results:

```
>>> query = "a running dog"
>>> top_image_ids = search_top_images(model, batch_image_embeddings,
query, 2)
>>> print("Query:", query)
>>> for id in top_image_ids:
        image = sample_batch["image"][id]
        npimg = image.numpy()
        plt.imshow(np.transpose(npimg, (1, 2, 0)))
        plt.title(f"Query: {query}")
        plt.show()
```

The following screenshot shows the result:

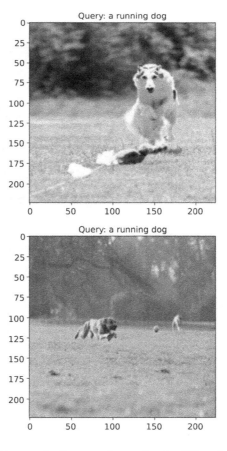

Figure 14.3: Retrieved images for the query "a running dog" (Top photo by Ron Mandsager, Flickr: https://www.flickr.com/photos/remandsager/3540416981/; bottom photo by Rob Burns-Sweeney, Flickr: https://www.flickr.com/photos/mulberryphotographic/3368207495/)

The two retrieved images are highly pertinent to the query.

5. Let's try another query, "kids jumping into a pool", before we end this section:

```
>>> query = " kids jumping into a pool "
>>> top_image_ids = search_top_images(model, batch_image_embeddings,
query)
>>> print("Query:", query)
>>> for id in top_image_ids:
        image = sample_batch["image"][id]
        npimg = image.numpy()
        plt.imshow(np.transpose(npimg, (1, 2, 0)))
        plt.title(f"Query: {query}")
        plt.show()
```

The following screenshot shows the result:

Figure 14.4: Retrieved image for the query "kids jumping into a pool" (photo by Alecia, Flickr:
https://www.flickr.com/photos/jnjsmom2007/2602415701/)

The retrieved image is exactly what we are looking for.

The CLIP model we implemented employs the pre-trained ResNet50 model as the vision encoder and the pre-trained DistilBERT model as the text encoder. Remember that we kept the parameters frozen for both ResNet50 and DistilBERT, utilizing them as image and text feature extractors. If desired, you can fine-tune these models by allowing their parameters to be trainable. This will be the exercise for this chapter. We trained our CLIP model using the Flickr8k dataset and performed image searches as a performance evaluation. Starting from the next section, we will use the pre-trained CLIP model, which learns from a much larger and more diverse dataset to perform image search, image-to-image search, and zero-shot prediction.

Image search using the pre-trained CLIP model

A popular library SentenceTransformers (https://www.sbert.net/index.html) offers a wrapper for the OpenAI CLIP model. The SentenceTransformer package is developed for sentence and text embeddings. It provides pre-trained models to encode sentences into high-dimensional vectors in a semantic space.

Let's perform the following tasks to search images, using a pre-trained CLIP model from SentenceTransformer:

1. First things first, install the SentenceTransformers library using the following command:

    ```
    pip install -U sentence-transformers
    ```

    or

    ```
    conda install -c conda-forge sentence-transformers
    ```

2. Import the SentenceTransformers library and load the pre-trained CLIP model:

    ```
    >>> from sentence_transformers import SentenceTransformer, util
    >>> model = SentenceTransformer('clip-ViT-B-32')
    ```

    Here, we use the **Vision Transformer** (ViT)-based CLIP model. The "B-32" designation refers to the size of the ViT model, which means it has 32 times more parameters than the base ViT model.

3. Next, we need to compute the image embeddings for all the `Flickr8k` image candidates using the CLIP model we just loaded:

```
>>> import glob
>>> image_paths = list(glob.glob('flickr8k/Flicker8k_Dataset/*.jpg'))

>>> all_image_embeddings = []
>>> for img_path in image_paths:
        img = Image.open(img_path)
        all_image_embeddings.append(model.encode(img, convert_to_
tensor=True))
```

The `model.encode()` method can take in text or images and generate corresponding embeddings. Here, we store all the resulting image embeddings in `all_image_embeddings`.

4. Similar to what we did in the previous section, we define the image search function as follows:

```
>>> def search_top_images(model, image_embeddings, query, top_k=1):
        query_embeddings = model.encode([query], convert_to_tensor=True,
                                         show_progress_bar=False)
        hits = util.semantic_search(query_embeddings,  image_embeddings,
                                     top_k=top_k)[0]
        return hits
```

Here, we use the `model.encode()` method again to obtain the embeddings for the given query. We employ the `util.semantic_search` utility function to fetch the top k images for the given text query, based on the similarities of their embeddings.

5. Now, let's search for "`a swimming dog`", using the image search function we just defined, and display the search results:

```
>>> query = "a swimming dog"
>>> hits = search_top_images(model, all_image_embeddings, query)
>>> for hit in hits:
        img_path = image_paths[hit['corpus_id']]
        image = Image.open(img_path)
        plt.imshow(image)
        plt.title(f"Query: {query}")
        plt.show()
```

The following screenshot shows the result:

Figure 14.5: Retrieved image for the query "a swimming dog" (photo by Julia, Flickr: https://www.flickr.com/photos/drakegsd/408233586/)

This is very accurate.

6. We can go beyond a text-to-image search and perform an **image-to-image** search:

```
>>> image_query =
        Image.open("flickr8k/Flicker8k_Dataset/240696675_7d05193aa0.jpg")
```

We will take a random image, 240696675_7d05193aa0.jpg, as the query image, feed it to the image search function, and display the retrieved images that follow the query image:

```
>>> hits = search_top_images(model, all_image_embeddings, image_query, 3)
[1:]
>>> plt.imshow(image_query)
>>> plt.title(f"Query image")
>>> plt.show()

>>> for hit in hits:
        img_path = image_paths[hit['corpus_id']]
        image = Image.open(img_path)
        plt.imshow(image)
        plt.title(f"Similar image")
        plt.show()
```

Note that we skip the first retrieved image because it is the query image.

The following screenshot shows the result:

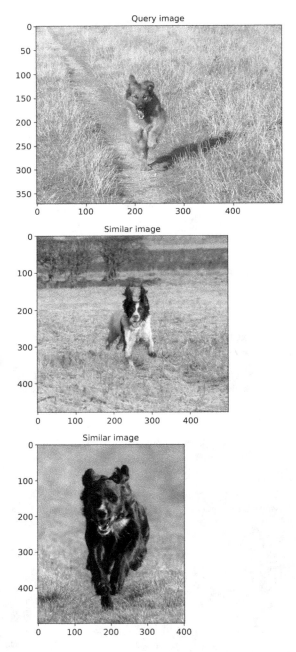

Figure 14.6: Query image and similar images (top photo by Rose, Flickr: https://www.flickr. com/photos/rosespics/240696675/; middle photo by Mark Dowling, Flickr: https://www. flickr.com/photos/markdowlrods/421932359/; bottom photo by Rob, Flickr: https://www. flickr.com/photos/mind_the_goat/3419634480/)

We can see that the retrieved images are highly similar to the query image.

The pre-trained CLIP model excels in both text-to-image and image-to-image search tasks. Notably, the model may not have been specifically trained on the `Flickr8k` dataset, but it performs well in zero-shot learning, as you saw in this section. Finally, let's take a look at another example of zero-shot prediction – classifying the CIFAR-100 dataset.

Zero-shot classification

In the final part of this chapter, we will utilize the CLIP model to classify the `CIFAR-100` dataset. Both `CIFAR-10` and `CIFAR-100` (https://www.cs.toronto.edu/~kriz/cifar.html) are labeled subsets derived from the 80 Million Tiny Images dataset, a collection curated by Alex Krizhevsky, Vinod Nair, and Geoffrey Hinton. The `CIFAR-100` dataset comprises 60,000 32x32 color images. These images are categorized into 100 classes, with each class containing exactly 600 images. We can load the dataset directly from PyTorch, which includes 50,000 images for training and 10,000 images for testing.

Let's perform the following tasks to classify the `CIFAR-100` dataset:

1. Firstly, we load the `CIFAR-100` dataset from PyTorch:

```
>>> from torchvision.datasets import CIFAR100
>>> cifar100 = CIFAR100(root="CIFAR100", download=True, train=False)
```

Here, we only load the testing subset with 10,000 samples.

2. Examine the classes of the dataset:

```
>>> print(cifar100.classes)
>>> print("Number of classes in CIFAR100 dataset:", len(cifar100.
classes))
['apple', 'aquarium_fish', 'baby', 'bear', 'beaver', 'bed', 'bee',
'beetle', 'bicycle', 'bottle', 'bowl', 'boy', 'bridge', 'bus',
'butterfly', 'camel', 'can', 'castle', 'caterpillar', 'cattle',
'chair', 'chimpanzee', 'clock', 'cloud', 'cockroach', 'couch', 'crab',
'crocodile', 'cup', 'dinosaur', 'dolphin', 'elephant', 'flatfish',
'forest', 'fox', 'girl', 'hamster', 'house', 'kangaroo', 'keyboard',
'lamp', 'lawn_mower', 'leopard', 'lion', 'lizard', 'lobster', 'man',
'maple_tree', 'motorcycle', 'mountain', 'mouse', 'mushroom', 'oak_tree',
'orange', 'orchid', 'otter', 'palm_tree', 'pear', 'pickup_truck',
'pine_tree', 'plain', 'plate', 'poppy', 'porcupine', 'possum', 'rabbit',
'raccoon', 'ray', 'road', 'rocket', 'rose', 'sea', 'seal', 'shark',
'shrew', 'skunk', 'skyscraper', 'snail', 'snake', 'spider', 'squirrel',
'streetcar', 'sunflower', 'sweet_pepper', 'table', 'tank', 'telephone',
'television', 'tiger', 'tractor', 'train', 'trout', 'tulip', 'turtle',
'wardrobe', 'whale', 'willow_tree', 'wolf', 'woman', 'worm']

Number of classes in CIFAR100 dataset: 100
```

There are 100 classes.

3. We start with one sample:

```
>>> sample_index = 0
>>> img, class_id = cifar100[index]
>>> print(f"Class of the sample image: {class_id} - {cifar100.
classes[class_id]}")
Class of the sample image: 49 - mountain
```

We will see if we can classify it correctly as a "mountain."

4. We then generate the image embeddings for the selected data sample using the pre-trained CLIP model:

```
>>> sample_image_embeddings = model.encode(img, convert_to_tensor=True)
```

5. Now, here's the clever approach. We consider each of the 100 classes as a textual description, and our goal is to find the most appropriate description for a given image, in order to classify it. Therefore, we must generate text embeddings for each of the 100 classes:

```
>>> class_text = model.encode(cifar100.classes, convert_to_tensor=True)
```

6. Let's search for the best class text description for the given image, as follows:

```
>>> hits = util.semantic_search(sample_image_embeddings, class_text,
top_k=1)[0]
>>> pred = hits[0]['corpus_id']
>>> print(f"Predicted class of the sample image: {pred}")
Predicted class of the sample image: 49
```

We can correctly predict the right class for the sample image. What about the whole dataset? Let's evaluate its performance in the next steps.

7. Similarly, we compute the image embeddings for all images in the dataset:

```
>>> all_image_embeddings = []
>>> class_true = []
>>> for img, class_id in cifar100:
        class_true.append(class_id)
        all_image_embeddings.append(model.encode(img, convert_to_
tensor=True))
```

We also record the true class information.

8. Now, we search for the best class text description for each of the images:

```
>>> class_pred = []
>>> for hit in util.semantic_search(all_image_embeddings, class_text,
top_k=1):
        class_pred.append(hit[0]['corpus_id'])
```

Finally, we evaluate the classification accuracy:

```
>>> from sklearn.metrics import accuracy_score
>>> acc = accuracy_score(class_true, class_pred)
>>> print(f"Accuracy of zero-shot classification: {acc * 100}%")
Accuracy of zero-shot classification: 55.15%
```

We attain a classification accuracy of 55% on the 100-class CIFAR dataset.

This project demonstrates how CLIP can be used for zero-shot classification. The model predicts the textual labels for images without having seen specific image-label pairs during training. Feel free to adjust the textual descriptions and images based on your specific use case. For example, you may group several similar fine classes into one coarse class, such as "boy," "girl," "man," and "woman" into "people."

Zero-shot classification using CLIP is powerful, but its performance is limited by training data used in a pre-trained model. Intuitively, this can be improved by leveraging pre-trained models on larger datasets. Another approach is knowledge distillation, which transfers knowledge from a complex and high-performance model to a smaller and faster model. You can read more about knowledge distillation in *Distilling the Knowledge in a Neural Network* (2015) by Geoffrey Hinton, Oriol Vinyals, and Jeff Dean (https://arxiv.org/abs/2006.05525).

Summary

This chapter introduced CLIP, a powerful DL model designed for cross-modal tasks, such as finding relevant images based on textual queries or vice versa. We learned that the model's dual encoder architecture and contrastive learning mechanism enable it to understand both images and text in a shared space.

We implemented our customized versions of CLIP models, using the DistilBERT and ResNet50 models. Following an exploration of the Flickr8k dataset, we built a CLIP model and explored its capabilities in text-to-image and image-to-image searches. CLIP excels at zero-shot transfer learning. We showcased this by using a pre-trained CLIP model for image search and CIFAR-100 classification.

In the next chapter, we will focus on the third type of machine learning problem: reinforcement learning. You will learn how the reinforcement learning model learns by interacting with the environment to reach its learning goal.

Exercises

1. Fine-tune the pre-trained ResNet50 and DistilBERT models employed in our self-implemented CLIP model.

2. Can you perform zero-shot classification on the 10-class `CIFAR-10` dataset?

3. Fine-tune the CLIP model using the training set of the `CIFAR-100` dataset, and see if you can get better performance on the test set.

References

- *Learning Transferable Visual Models From Natural Language Supervision*, by Alec Radford et al.

- `Flickr8k` dataset: *Framing Image Description as a Ranking Task: Data, Models and Evaluation Metrics*, *Journal of Artificial Intelligence Research*, Volume 47, pages 853–899

- `CIFAR-100` dataset: *Learning Multiple Layers of Features from Tiny Images*, Alex Krizhevsky, 2009.

Unlock this book's exclusive benefits now

This book comes with additional benefits designed to elevate your learning experience.

Note: Have your purchase invoice ready before you begin.   `https://www.packtpub.com/unlock/9781835085622`

15

Making Decisions in Complex Environments with Reinforcement Learning

In the previous chapter, we focused on multimodal models for image and text co-learning. The last chapter of this book will be about reinforcement learning, which is the third type of machine learning task mentioned at the beginning of the book. You will see how learning from experience and learning by interacting with the environment differs from previously covered supervised and unsupervised learning.

We will cover the following topics in this chapter:

- Setting up the working environment
- Introducing reinforcement learning with examples
- Solving the FrozenLake environment with dynamic programming
- Performing Monte Carlo learning
- Solving the Taxi problem with the Q-learning algorithm

Setting up the working environment

Let's get started with setting up the working environment needed for this chapter, including Gymnasium (which builds upon OpenAI Gym), the toolkit that gives you a variety of environments to develop your learning algorithms on.

Introducing OpenAI Gym and Gymnasium

OpenAI Gym was a toolkit for developing and comparing reinforcement learning algorithms. It provided a collection of environments, or "tasks," in which reinforcement learning agents can interact and learn. These environments range from simple grid-world games to complex simulations of real-world scenarios, allowing researchers and developers to experiment with a wide variety of reinforcement learning algorithms. It was developed by OpenAI, focused on building safe and beneficial **Artificial General Intelligence (AGI)**.

Some key features of OpenAI Gym included:

- **Environment interface:** Gym provided a consistent interface for interacting with environments, allowing agents to observe states, take actions, and receive rewards (we will learn about these terms).

- **Extensive collection of environments:** Gym offered a diverse set of environments, including classic control tasks, Atari games, robotics simulations, and more. This allowed researchers and developers to evaluate algorithms across various domains.

- **Easy-to-use API:** Gym's API was straightforward and easy to use, making it accessible to both beginners and experienced researchers. Developers could quickly prototype and test reinforcement learning algorithms using Gym's environments.

- **Benchmarking:** Gym facilitated benchmarking by providing standardized environments and evaluation metrics. This enabled researchers to compare the performance of different algorithms on common tasks.

- **Community contributions:** Gym was an open-source project, and the community actively contributed new environments, algorithms, and extensions to the toolkit. This collaborative effort helped to continuously expand and improve Gym's capabilities.

Overall, OpenAI Gym served as a valuable resource for the reinforcement learning community, providing a standardized platform for research, experimentation, and benchmarking.

Gym was a pioneering library and set the standard for simplicity for many years. However, it is no longer actively maintained by the OpenAI team. Recognizing this, some developers took the initiative to create **Gymnasium** (https://gymnasium.farama.org/index.html), with approval from OpenAI. Gymnasium emerged as a successor to Gym, and the original developers from OpenAI occasionally contribute to its development, ensuring its reliability and continuity. In this chapter, we will be using Gymnasium, which is a **maintained fork** of Gym.

Installing Gymnasium

One way to install the Gymnasium library is via `pip`, as follows:

```
pip install gymnasium
```

It is recommended to install the `toy-text` extension using the following command:

```
pip install gymnasium [toy-text]
```

The toy-text extension provides additional toy text-based environments, such as the FrozenLake environment (discussed later), for reinforcement learning experimentation.

After the installation, you can check the available Gymnasium environments by running the following code:

```
>>> import gymnasium as gym
>>> print(gym.envs.registry.keys())
dict_keys(['CartPole-v0', 'CartPole-v1', 'MountainCar-v0',
'MountainCarContinuous-v0', 'Pendulum-v1', 'Acrobot-v1', 'phys2d/
CartPole-v0', 'phys2d/CartPole-v1', 'phys2d/Pendulum-v0', 'LunarLander-v2',
'LunarLanderContinuous-v2', 'BipedalWalker-v3', 'BipedalWalkerHardcore-v3',
'CarRacing-v2', 'Blackjack-v1', 'FrozenLake-v1', 'FrozenLake8x8-v1',
'CliffWalking-v0', 'Taxi-v3', 'tabular/Blackjack-v0', 'tabular/
CliffWalking-v0', 'Reacher-v2', 'Reacher-v4', 'Pusher-v2', 'Pusher-v4',
'InvertedPendulum-v2', 'InvertedPendulum-v4', 'InvertedDoublePendulum-v2',
'InvertedDoublePendulum-v4', 'HalfCheetah-v2', 'HalfCheetah-v3',
'HalfCheetah-v4', 'Hopper-v2', 'Hopper-v3', 'Hopper-v4', 'Swimmer-v2',
'Swimmer-v3', 'Swimmer-v4', 'Walker2d-v2', 'Walker2d-v3', 'Walker2d-v4',
'Ant-v2', 'Ant-v3', 'Ant-v4', 'Humanoid-v2', 'Humanoid-v3', 'Humanoid-v4',
'HumanoidStandup-v2', 'HumanoidStandup-v4', 'GymV21Environment-v0',
'GymV26Environment-v0'])
```

💡 **Quick tip:** Enhance your coding experience with the **AI Code Explainer** and **Quick Copy** features. Open this book in the next-gen Packt Reader. Click the **Copy** button (**1**) to quickly copy code into your coding environment, or click the **Explain** button (**2**) to get the AI assistant to explain a block of code to you.

📖 **The next-gen Packt Reader** is included for free with the purchase of this book. Unlock it by scanning the QR code below or visiting
https://www.packtpub.com/unlock/9781835085622.

You can see lists of the environments at `https://gymnasium.farama.org/environments/toy_text/` and `https://gymnasium.farama.org/environments/atari/`, including walking, moon landing, car racing, and Atari games. Feel free to play around with Gymnasium by going through its introduction at `https://gymnasium.farama.org/content/basic_usage/`.

To benchmark different reinforcement learning algorithms, we need to apply them in a standardized environment. Gymnasium is the perfect place for this, with a number of versatile environments. This is similar to using datasets such as MNIST, ImageNet, and Thomson Reuters News as benchmarks in supervised and unsupervised learning.

Gymnasium has an easy-to-use interface for the reinforcement learning environments that we can write **agents** to interact with. So what's reinforcement learning? What's an agent? Let's see in the next section.

Introducing reinforcement learning with examples

In this chapter, we will first introduce the elements of reinforcement learning along with an interesting example, then will move on to how we measure feedback from the environment, and follow with the fundamental approaches to solve reinforcement learning problems.

Elements of reinforcement learning

You may have played Super Mario (or Sonic) when you were young. During the video game, you control Mario to collect coins and avoid obstacles at the same time. The game ends if Mario hits an obstacle or falls in a gap, and you try to get as many coins as possible before the game ends.

Reinforcement learning is very similar to the Super Mario game. Reinforcement learning is about learning what to do. It involves observing situations in the environment and determining the right actions in order to maximize a numerical reward.

Here is the list of elements in a reinforcement learning task (we also link each element to Super Mario and other examples so it's easier to understand):

- **Environment:** The environment is a task or simulation. In the Super Mario game, the game itself is the environment. In self-driving, the road and traffic are the environment. In the context of Go playing chess, the board is the environment. The inputs to the environment are the actions sent from the **agent** and the outputs are **states** and **rewards** sent to the agent.

- **Agent:** The agent is the component that takes **actions** according to the reinforcement learning model. It interacts with the environment and observes the states to feed into the model. The goal of the agent is to solve the environment—that is, finding the best set of actions to maximize the rewards. The agent in the Super Mario game is Mario, and the autonomous vehicle is the agent for self-driving.

- **Action:** This is the possible movement of the agent. It is usually random in a reinforcement learning task at the beginning when the model starts to learn about the environment. Possible actions for Mario include moving left and right, jumping, and crouching.

- **States:** The states are the observations from the environment. They describe the situation in a numerical way at every time step. For a chess game, the state is the positions of all the pieces on the board. For Super Mario, the state includes the coordinates of Mario and other elements in the time frame. For a robot learning to walk, the position of its two legs is the state.

- **Rewards:** Every time the agent takes an action, it receives numerical feedback from the environment. The feedback is called the **reward**. It can be positive, negative, or zero. The reward in the Super Mario game can be, for example, +1 if Mario collects a coin, +2 if he avoids an obstacle, -10 if he hits an obstacle, or 0 for other cases.

The following diagram summarizes the process of reinforcement learning:

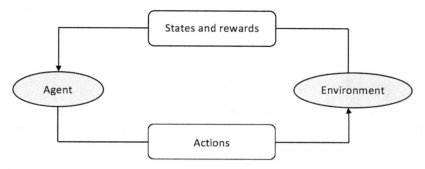

Figure 15.1: Reinforcement learning process

The reinforcement learning process is an iterative loop. At the beginning, the agent observes the initial state, s_0, from the environment. Then the agent takes an action, a_0, according to the model. After the agent moves, the environment is now in a new state, s_1, and it gives a feedback reward, R_1. The agent then takes an action, a_1, as computed by the model with inputs s_1 and R_1. This process continues until termination, completion, or for forever.

The goal of the reinforcement learning model is to maximize the total reward. So how can we calculate the total reward? Is it simply by summing up rewards at all the time steps? Let's see in the next section.

Cumulative rewards

At time step t, the **cumulative rewards** (also called **returns**) G_1 can be written as:

$$G_t = \sum_{k=0}^{T} R_{t+k+1}$$

Here, T is the termination time step or infinity. G_t means the total future reward after taking an action a_t at time t. At each time step t, the reinforcement learning model attempts to learn the best possible action in order to maximize G_t.

However, in many real-world cases, things don't work this way where we simply sum up all future rewards. Take a look at the following example:

Stock A rises 6 dollars at the end of day 1 and falls 5 dollars at the end of day 2. Stock B falls 5 dollars on day 1 and rises 6 dollars on day 2. After two days, both stocks rise 1 dollar. So, if we knew that, which one would we buy at the beginning of day 1? Obviously, stock A, because we won't lose money and can even profit 6 dollars if we sell it at the beginning of day 2.

Both stocks have the same total reward but we favor stock A as we care more about immediate return than distant return. Similarly in reinforcement learning, we discount rewards in the distant future and the discount factor is associated with the time horizon. Longer time horizons should have less impact on the cumulative rewards. This is because longer time horizons include more irrelevant information and consequently are of higher variance.

We define a discount factor γ with a value between 0 and 1. We rewrite the cumulative rewards incorporating the discount factor:

$$G_t = \sum_{k=0}^{T} \gamma^k R_{t+k+1}$$

As you can see, the larger the γ, the smaller the discount and vice versa. If $\gamma = 1$, there is literally no discount and the model evaluates an action based on the sum total of all future rewards. If $\gamma = 0$, the model only focuses on the immediate reward R_{t+1}.

Now that we know how to calculate the cumulative reward, the next thing to talk about is how to maximize it.

Approaches to reinforcement learning

There are two main approaches to solving reinforcement learning problems, which are about finding the optimal actions to maximize the cumulative rewards. One is a policy-based approach and the other is value-based.

Policy-based approach

A **policy** is a function π that maps each input state to an action:

$$a = \pi(s)$$

It can be either deterministic or stochastic:

- **Deterministic:** There is one-to-one mapping from the input state to the output action
- **Stochastic:** This gives a probability distribution over all possible actions $P(A = a|s)$

In the **policy-based** approach, the model learns the optimal policy that maps each input state to the best action. The agent directly learns the best course of action (policy) for any situation (state) it encounters.

In a policy-based algorithm, the model starts with a random policy. It then computes the value function of that policy. This step is called the **policy evaluation step**. After this, it finds a new and better policy based on the value function. This is the **policy improvement step**. These two steps repeat until the optimal policy is found.

Imagine you are training a race car driver. In a policy-based approach, you directly teach the driver the best manoeuvres (policy) to take on different parts of the track (states) to achieve the fastest lap time (reward). You don't tell them the estimated result (reward) of each turn, but rather guide them towards the optimal racing line through feedback and practice.

Value-based approach

The **value** V of a state is defined as the expected future cumulative reward to collect from the state:

$$V(s) = E\left[\sum_{k=0}^{T} \gamma^k R_{t+k+1} | s_t = S\right]$$

In the **value-based** approach, the model learns the optimal value function that maximizes the value of the input state. In other words, the agent takes an action to reach the state that achieves the largest value.

In a value-based algorithm, the model starts with a random value function. It then finds a new and improved value function in an iterative manner, until it reaches the optimal value function.

Now imagine you are a treasure hunter. In a value-based approach, you learn the estimated treasure (value) of different locations (states) in a maze. This helps you choose paths that lead to areas with higher potential treasure (rewards) without needing a pre-defined course of action (policy).

We've learned there are two main approaches to solving reinforcement learning problems. In the next section, let's see how to solve a concrete reinforcement learning example (FrozenLake) using a concrete algorithm, the dynamic programming method, in a policy-based and value-based way respectively.

Solving the FrozenLake environment with dynamic programming

We will focus on the policy-based and value-based dynamic programming algorithms in this section. But let's start by simulating the FrozenLake environment. It simulates a simple grid-world scenario where an agent navigates through a grid of icy terrain, represented as a frozen lake, to reach a goal tile.

Simulating the FrozenLake environment

FrozenLake is a typical OpenAI Gym (now Gymnasium) environment with **discrete** states. It is about moving the agent from the starting tile to the destination tile in a grid, and at the same time avoiding traps. The grid is either 4 * 4 (FrozenLake-v1), or 8 * 8 (FrozenLake8x8-v1). There are four types of tiles in the grid:

- **The starting tile:** This is state 0, and it comes with 0 reward.
- **The goal tile:** It is state 15 in the 4 * 4 grid. It gives +1 reward and terminates an episode.
- **The frozen tile:** In the 4 * 4 grid, states 1, 2, 3, 4, 6, 8, 9, 10, 13, and 14 are walkable tiles. It gives 0 reward.
- **The hole tile:** In the 4 * 4 grid, states 5, 7, 11, and 12 are hole tiles. It gives 0 reward and terminates an episode.

Here, an **episode** means a simulation of a reinforcement learning environment. It contains a list of states from the initial state to the terminal state, a list of actions and rewards. In the 4 * 4 FrozenLake environment, there are 16 possible states as the agent can move to any of the 16 tiles. And there are four possible actions: moving left (0), down (1), right (2), and up (3).

The tricky part of this environment is that, as the ice surface is slippery, the agent won't always move in the direction it intends and can move in any other walkable direction or stay unmoved with certain probabilities. For example, it may move to the right even though it intends to move up.

Now let's simulate the 4 * 4 FrozenLake environment by following these steps:

1. To simulate any OpenAI Gym environment, we need to first look up its name in the documentation at `https://gymnasium.farama.org/environments/toy_text/frozen_lake/`. We get `FrozenLake-v1` in our case.

2. We import the gym library and create a `FrozenLake` instance:

```
>>> env = gym.make("FrozenLake-v1" , render_mode="rgb_array")
>>> n_state = env.observation_space.n
>>> print(n_state)
16
>>> n_action = env.action_space.n
>>> print(n_action)
4
```

The environment is initialized with the `FrozenLake-v1` identifier. Additionally, the `render_mode` parameter is set to `rgb_array`, indicating that the environment should render its state as an RGB array, suitable for visualization purposes.

We also obtain the dimensions of the environment.

3. Every time we run a new episode, we need to reset the environment:

```
>>> env.reset()
(0, {'prob': 1})
```

It means that the agent starts with state 0. Again, there are 16 possible states, 0, 1, ..., 15.

4. We render the environment to display it:

```
>>> import matplotlib.pyplot as plt
>>> plt.imshow(env.render())
```

 If you encounter any error, you may install the `pyglet` library, which embeds a `Matplotlib` figure within a window using canvas rendering, using the following command:

```
pip install pyglet
```

You will see a 4 * 4 matrix representing the FrozenLake grid and the tile (state 0) where the agent is located:

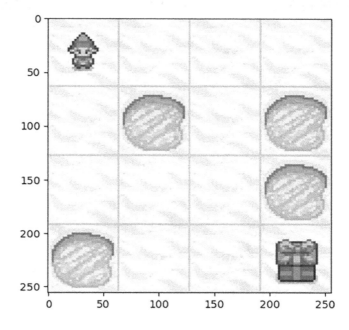

Figure 15.2: Initial state of FrozenLake

5. Let's now start moving the agent around. Let's take a right action since it is walkable:

```
>>> new_state, reward, terminated, truncated, info = env.step(2)
>>> is_done = terminated or truncated
>>> env.render()
>>> print(new_state)
4
>>> print(reward)
0.0
>>> print(is_done)
False
>>> print(info)
{'prob': 0.3333333333333333}
```

We take a "right" (2) action, but the agent moves down to state 4, at a probability of 33.33%, and gets 0 reward since the episode is not done yet.

Let's see the rendered result:

```
>>> plt.imshow(env.render())
```

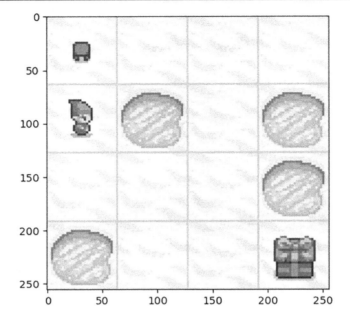

Figure 15.3: Result of the agent moving right

You may get a completely different result as the agent can move right to state 1 at a probability of 33.33%, or stay at state 0 at a probability of 33.33% due to the **slippery** nature of the frozen lake.

 In Gymnasium, **"terminated"** and **"truncated"** refer to different ways in which an episode can end in a reinforcement learning environment. When an episode is terminated, it means that the episode has ended naturally according to the rules of the environment. When an episode is truncated, it means that the episode is artificially terminated before it can end naturally.

6. Next, we define a function that simulates a FrozenLake episode under a given policy and returns the total reward (as an easy start, let's just assume discount factor $\gamma = 1$):

```
>>> def run_episode(env, policy):
...     state, _= env.reset()
...     total_reward = 0
...     is_done = False
...     while not is_done:
...         action = policy[state].item()
...         state, reward, terminated, truncated, info = env.step(action)
...         is_done = terminated or truncated
```

```
...             total_reward += reward
...         if is_done:
...             break
...     return total_reward
```

Here, `policy` is a PyTorch tensor, and `.item()` extracts the value of an element on the tensor.

7. Now let's play around with the environment using a random policy. We will implement a random policy (where random actions are taken) and calculate the average total reward over 1,000 episodes:

```
>>> import torch
>>> n_episode = 1000
>>> total_rewards = []
>>> for episode in range(n_episode):
...     random_policy = torch.randint(high=n_action, size=(n_state,))
...     total_reward = run_episode(env, random_policy)
...     total_rewards.append(total_reward)
...
>>> print(f'Average total reward under random policy:
            {sum(total_rewards)/n_episode}')
Average total reward under random policy: 0.016
```

On average, there is a 1.6% chance that the agent can reach the goal if we take random actions. This tells us it is not as easy to solve the FrozenLake environment as you might think.

8. As a bonus step, you can look into the transition matrix. The **transition matrix** $T(s, a, s')$ contains probabilities of taking action a from state s then reaching s'. Take state 6 as an example:

```
>>> print(env.env.P[6])
{0: [(0.3333333333333333, 2, 0.0, False), (0.3333333333333333, 5, 0.0,
True), (0.3333333333333333, 10, 0.0, False)], 1: [(0.3333333333333333, 5,
0.0, True), (0.3333333333333333, 10, 0.0, False), (0.3333333333333333,
7, 0.0, True)], 2: [(0.3333333333333333, 10, 0.0, False),
(0.3333333333333333, 7, 0.0, True), (0.3333333333333333, 2, 0.0, False)],
3: [(0.3333333333333333, 7, 0.0, True), (0.3333333333333333, 2, 0.0,
False), (0.3333333333333333, 5, 0.0, True)]}
```

The keys of the returning dictionary 0, 1, 2, 3 represent four possible actions. The value of a key is a list of tuples associated with the action. The tuple is in the format of (transition probability, new state, reward, is terminal state or not). For example, if the agent intends to take action 1 (down) from state 6, it will move to state 5 (H) with 33.33% probability and receive 0 reward and the episode will end consequently; it will move to state 10 with 33.33% probability and receive 0 reward; it will move to state 7 (H) with 33.33% probability and receive 0 reward and terminate the episode.

We've experimented with the random policy in this section, and we only succeeded 1.6% of the time. But this gets you ready for the next section where we will find the optimal policy using the value-based dynamic programming algorithm, called the **value iteration algorithm**.

Solving FrozenLake with the value iteration algorithm

Value iteration is an iterative algorithm. It starts with random policy values V, and then iteratively updates the values based on the **Bellman optimality equation** (https://en.wikipedia.org/wiki/Bellman_equation) until the values converge.

It is usually difficult for the values to completely converge. Hence, there are two criteria of convergence. One is passing a fixed number of iterations, such as 1,000 or 10,000. Another one is specifying a threshold (such as 0.0001, or 0.00001) and we terminate the process if the changes of all values are less than the threshold.

Importantly, in each iteration, instead of taking the expectation (average) of values across all actions, it picks the action that maximizes the policy values. The iteration process can be expressed as follows:

$$V^*(s) := max_a \left[R(s, a, s') + \gamma \sum_s T(s, a, s') V^*(s') \right]$$

This is the representation of the Bellman equation for the state-value function V(s). Here, $V^*(s)$ is the optimal value function; $T(s, a, s')$ denotes the transition probability of moving to state s' from state s by taking action a; and $R(s, a, s')$ is the reward provided in state s' by taking action a.

Once we obtain the optimal values, we can easily compute the optimal policy accordingly:

$$\pi^*(s) := argmax_a \sum_s T(s, a, s')[R(s, a, s') + \gamma V^*(s')]$$

Let's solve the FrozenLake environment using the value iteration algorithm as follows:

1. First we set `0.99` as the discount factor, and `0.0001` as the convergence threshold:

```
>>> gamma = 0.99
>>> threshold = 0.0001
```

2. We develop the value iteration algorithm, which computes the optimal values:

```
>>> def value_iteration(env, gamma, threshold):
...     """
...     Solve a given environment with value iteration algorithm
...     @param env: Gymnasium environment
...     @param gamma: discount factor
...     @param threshold: the evaluation will stop once values for all
states are less than the threshold
...     @return: values of the optimal policy for the given environment
...     """
...     n_state = env.observation_space.n
```

```
...        n_action = env.action_space.n
...        V = torch.zeros(n_state)
...        while True:
...            V_temp = torch.empty(n_state)
...            for state in range(n_state):
...                v_actions = torch.zeros(n_action)
...                for action in range(n_action):
...                    for trans_prob, new_state, reward, _ in \
                                env.env.P[state][action]:
...                        v_actions[action] += trans_prob * (
                                reward + gamma * V[new_state])
...                V_temp[state] = torch.max(v_actions)
...            max_delta = torch.max(torch.abs(V - V_temp))
...            V = V_temp.clone()
...            if max_delta <= threshold:
...                break
...        return V
```

The value_iteration function does the following tasks:

- Starts with policy values as all 0s
- Updating the values based on the Bellman optimality equation:

$$V^*(s) := max_a \left[R(s, a, s') + \gamma \sum_{s} T(s, a, s') V^*(s') \right]$$

- Computing the maximal change of the values across all states
- Continuing to update the values if the maximal change is greater than the convergence threshold
- Otherwise, terminating the iteration process and returning the last values as the optimal values

3. We apply the algorithm to solve the FrozenLake environment along with the specified parameters:

```
>>> V_optimal = value_iteration(env, gamma, threshold)
```

Take a look at the resulting optimal values:

```
>>> print('Optimal values:\n', V_optimal)
Optimal values:
tensor([0.5404, 0.4966, 0.4681, 0.4541, 0.5569, 0.0000, 0.3572, 0.0000,
0.5905, 0.6421, 0.6144, 0.0000, 0.0000, 0.7410, 0.8625, 0.0000])
```

4. Since we have the optimal values, we can extract the optimal policy from the values. We develop the following function to do this:

```
>>> def extract_optimal_policy(env, V_optimal, gamma):
...     """
...     Obtain the optimal policy based on the optimal values
...     @param env: Gymnasium environment
...     @param V_optimal: optimal values
...     @param gamma: discount factor
...     @return: optimal policy
...     """
...     n_state = env.observation_space.n
...     n_action = env.action_space.n
...     optimal_policy = torch.zeros(n_state)
...     for state in range(n_state):
...         v_actions = torch.zeros(n_action)
...         for action in range(n_action):
...             for trans_prob, new_state, reward, _ in
                                env.env.P[state][action]:
...                 v_actions[action] += trans_prob * (
                        reward + gamma * V_optimal[new_state])
...         optimal_policy[state] = torch.argmax(v_actions)
...     return optimal_policy
```

5. Then we obtain the optimal policy based on the optimal values:

```
>>> optimal_policy = extract_optimal_policy(env, V_optimal, gamma)
```

Take a look at the resulting optimal policy:

```
>>> print('Optimal policy:\n', optimal_policy)
Optimal policy:
tensor([0., 3., 3., 3., 0., 0., 0., 0., 3., 1., 0., 0., 0., 2., 1., 0.])
```

This means the optimal action in state 0 is 0 (left), 3 (up) in state 1, etc. This doesn't seem very intuitive if you look at the grid. But remember that the grid is slippery and the agent can move in another direction than the desired one.

6. If you doubt that it is the optimal policy, you can run 1,000 episodes with the policy and gauge how good it is by checking the average reward, as follows:

```
>>> def run_episode(env, policy):
        state, _ = env.reset()
        total_reward = 0
        is_done = False
```

```
        while not is_done:
            action = policy[state].item()
            state, reward, terminated, truncated, info = env.step(action)
            is_done = terminated or truncated
            total_reward += reward
            if is_done:
                break
        return total_reward

>>> n_episode = 1000
>>> total_rewards = []
>>> for episode in range(n_episode):
...     total_reward = run_episode(env, optimal_policy)
...     total_rewards.append(total_reward)
```

Here, we define the `run_episode` function to simulate one episode. Then we print out the average reward over 1,000 episodes:

```
>>> print('Average total reward under the optimal policy:', sum(total_
rewards) / n_episode)
Average total reward under the optimal policy: 0.738
```

Value iteration is guaranteed to converge to the optimal value function for a finite environment with a finite state and action space. It provides a computationally efficient method for solving for the optimal policy in RL problems, especially when the dynamics of the environment are known. Under the optimal policy computed by the value iteration algorithm, the agent in FrozenLake reaches the goal tile 74% of the time.

Best practice

The discount factor is an important parameter in RL, especially for value-based models. A high factor (closer to 1) makes the agent prioritize long-term rewards, leading to more exploration, while a low factor (closer to 0) makes it focus on immediate rewards. Typical tuning strategies for the discount factor include grid search and random search. Both could be computationally expensive for large ranges. Adaptive tuning is another approach, where we dynamically adjust the factor during training. You can start with a medium value (such as 0.9). If the agent seems too focused on immediate rewards, converges fast, and ignores exploration, try increasing the discount factor. If the agent keeps exploring and never settles on a good policy, try decreasing the discount factor.

Can we do something similar with the policy-based approach? Let's see in the next section.

Solving FrozenLake with the policy iteration algorithm

The **policy iteration** algorithm has two components, policy evaluation and policy improvement. Similar to value iteration, it starts with an arbitrary policy and follows with a bunch of iterations.

In the policy evaluation step in each iteration, we first compute the values of the latest policy, based on the **Bellman expectation equation**:

$$V(s) := \sum_{s} T(s, a, s')[R(s, a, s') + \gamma V(s')]$$

In the policy improvement step, we derive an improved policy based on the latest policy values, again based on the Bellman optimality equation:

$$\pi(s) := argmax_a \sum_{s} T(s, a, s')[R(s, a, s') + \gamma V(s')]$$

These two steps repeat until the policy converges. At convergence, the latest policy and its value are the optimal policy and the optimal value.

Let's develop the policy iteration algorithm and use it to solve the FrozenLake environment as follows:

1. We start with the `policy_evaluation` function that computes the values of a given policy:

```
>>> def policy_evaluation(env, policy, gamma, threshold):
...     """
...         Perform policy evaluation
...         @param env: Gymnasium environment
...         @param policy: policy matrix containing actions and
...         their probability in each state
...         @param gamma: discount factor
...         @param threshold: the evaluation will stop once values
...         for all states are less than the threshold
...         @return: values of the given policy
...     """
...         n_state = policy.shape[0]
...         V = torch.zeros(n_state)
...         while True:
...             V_temp = torch.zeros(n_state)
...             for state in range(n_state):
...                 action = policy[state].item()
...                 for trans_prob, new_state, reward, _ in \
                                        env.env.P[state][action]:
...                     V_temp[state] += trans_prob * \
                                        (reward + gamma * V[new_state])
...             max_delta = torch.max(torch.abs-V - V_temp))
...             V = V_temp.clone()
```

```
...            if max_delta <= threshold:
...                break
...        return V
```

The function does the following tasks:

- Initializing the policy values with all 0s
- Updating the values based on the Bellman expectation equation
- Computing the maximal change of the values across all states
- If the maximal change is greater than the threshold, it keeps updating the values
- Otherwise, it terminates the evaluation process and returns the latest values

2. Next, we develop the second component, the policy improvement, in the following function:

```
>>> def policy_improvement(env, V, gamma):
...     """
...        Obtain an improved policy based on the values
...        @param env: Gymnasium environment
...        @param V: policy values
...        @param gamma: discount factor
...        @return: the policy
...     """
...        n_state = env.observation_space.n
...        n_action = env.action_space.n
...        policy = torch.zeros(n_state)
...        for state in range(n_state):
...            v_actions = torch.zeros(n_action)
...            for action in range(n_action):
...                for trans_prob, new_state, reward, _ in
                                    env.env.P[state][action]:
...                    v_actions[action] += trans_prob * (
                                reward + gamma * V[new_state])
...            policy[state] = torch.argmax(v_actions)
...        return policy
```

It derives a new and better policy from the input policy values based on the Bellman optimality equation.

3. With both components ready, we now develop the whole policy iteration algorithm:

```
>>> def policy_iteration(env, gamma, threshold):
...     """
...        Solve a given environment with policy iteration algorithm
...        @param env: Gymnasium environment
...        @param gamma: discount factor
```

```
...        @param threshold: the evaluation will stop once values for all
states are less than the threshold
...        @return: optimal values and the optimal policy for the given
environment
...    """
...        n_state = env.observation_space.n
...        n_action = env.action_space.n
...        policy = torch.randint(high=n_action,
                                   size=(n_state,)).float()
...        while True:
...            V = policy_evaluation(env, policy, gamma, threshold)
...            policy_improved = policy_improvement(env, V, gamma)
...            if torch.equal(policy_improved, policy):
...                return V, policy_improved
...            policy = policy_improved
```

This function does the following tasks:

- Initializing a random policy
- Performing policy evaluation to update the policy values
- Performing policy improvement to generate a new policy
- If the new policy is different from the old one, it updates the policy and runs another iteration of policy evaluation and improvement
- Otherwise, it terminates the iteration process and returns the latest policy and its values

4. Next, we use policy iteration to solve the FrozenLake environment:

```
>>> V_optimal, optimal_policy = policy_iteration(env, gamma, threshold)
```

5. Finally, we display the optimal policy and its values:

```
>>> print('Optimal values'\n', V_optimal)
Optimal values:
tensor([0.5404, 0.4966, 0.4681, 0.4541, 0.5569, 0.0000, 0.3572, 0.0000,
0.5905, 0.6421, 0.6144, 0.0000, 0.0000, 0.7410, 0.8625, 0.0000])
>>> print('Optimal policy'\n', optimal_policy)
Optimal policy:
tensor([0., 3., 3., 3., 0., 0., 0., 0., 3., 1., 0., 0., 0., 2., 1., 0.])
```

We got the same results as the value iteration algorithm.

Best practice

Policy-based methods rely on estimating the gradient of the expected reward with respect to the policy parameters. In practice, we use techniques like REINFORCE, which uses simple Monte Carlo estimates, and **Proximal Policy Optimization** (**PPO**) employing stable gradient estimation. You can read more here: `https://professional.mit.edu/course-catalog/reinforcement-learning` (*Chapter 8, Policy Gradient Methods*).

We have just solved the FrozenLake environment with the policy iteration algorithm. You may wonder how to choose between the value iteration and policy iteration algorithms. Take a look at the following table:

Scenario	Preference	Reason
A large number of actions	Policy iteration	Policy iteration can converge faster
A small number of actions	Value iteration	Less computation in value iteration
A fair policy exists (obtained either by intuition or domain knowledge)	Policy iteration	Policy iteration from a fair policy can converge faster
Others	No preference	Policy iteration and value iteration are comparable

Table 15.4: Choosing between the policy iteration and value iteration algorithms

We solved a reinforcement learning problem using dynamic programming methods. They require a fully known transition matrix and reward matrix of an environment. And they have limited scalability for environments with many states. In the next section, we will continue our learning journey with the Monte Carlo method, which has no requirement of prior knowledge of the environment and is much more scalable.

Performing Monte Carlo learning

Monte Carlo (**MC**)-based reinforcement learning is a **model-free** approach, which means it doesn't need a known transition matrix and reward matrix. In this section, you will learn about MC policy evaluation on Gymnasium's Blackjack environment, and solve the environment with MC control algorithms. Blackjack is a typical environment with an unknown transition matrix. Let's first simulate the Blackjack environment.

Simulating the Blackjack environment

Blackjack is a popular card game. The game has the following rules:

- The player competes against a dealer and wins if the total value of their cards is higher than the dealer's and doesn't exceed 21.
- Cards from 2 to 10 have values from 2 to 10.
- Cards J, K, and Q have a value of 10.
- The value of an ace can be either 1 or 11 (called a "usable" ace).
- At the beginning, both parties are given two random cards, but only one of the dealer's cards is revealed to the player. The player can request additional cards (called **hit**) or stop having any more cards (called **stick**). Before the player calls stick, the player will lose if the sum of their cards exceeds 21 (called **bust**). After the player sticks, the dealer keeps drawing cards until the sum of cards reaches 17. If the sum of the dealer's cards exceeds 21, the player will win. If neither of the two parties busts, the one with higher points will win or it may be a draw.

The Blackjack environment (https://gymnasium.farama.org/environments/toy_text/blackjack/) in Gymnasium is formulated as follows:

- An episode of the environment starts with two cards for each party, and only one from the dealer's cards is observed.
- An episode ends if there is a win or draw.
- The final reward of an episode is +1 if the player wins, -1 if the player loses, or 0 if there is a draw.
- In each round, the player can take any of the two actions, hit (1) or stick (0).

Now let's simulate the Blackjack environment and explore its states and actions:

1. First, create a `Blackjack` instance:

```
>>> env = gym.make('Blackjack'v1')
```

2. Reset the environment:

```
>>> env.reset(seed=0)
((11, 10, False), {})
```

It returns the initial state (a 3-dimensional vector):

- Player's current points (11 in this example)
- The points of the dealer's revealed card (10 in this example)
- Having a usable ace or not (False in this example)

The usable ace variable is True only if the player has an ace that can be counted as 11 without causing a bust. If the player doesn't have an ace, or has an ace but it busts, this state variable will become False.

For another state example (18, 6, True), it means that the player has an ace counted as 11 and a 7, and that the dealer's revealed card is value 6.

3. Let's now take some actions to see how the environment works. First, we take a hit action since we only have 11 points:

```
>>> env.step(1)
((12, 10, False), 0.0, False, False, {})
```

It returns a state (12, 10, False), a 0 reward, and the episode not being done (meaning False).

4. Let's take another hit since we only have 12 points:

```
>>> env.step(1)
((13, 10, False), 0.0, False, False, {})
```

5. We have 13 points and think it is good enough. Then we stop drawing cards by taking the stick action (0):

```
>>> env.step(0)
((13, 10, False), -1.0, True, False, {})
```

The dealer gets some cards and beats the player. So the player loses and gets a -1 reward. The episode ends.

Feel free to play around with the Blackjack environment. Once you feel comfortable with the environment, you can move on to the next section, MC policy evaluation on a simple policy.

Performing Monte Carlo policy evaluation

In the previous section, we applied dynamic programming to perform policy evaluation, which is the value function of a policy. However, it won't work in most real-life situations where the transition matrix is not known beforehand. In this case, we can evaluate the value function using the MC method.

To estimate the value function, the MC method uses empirical mean return instead of expected return (as in dynamic programming). There are two approaches to compute the empirical mean return. One is first-visit, which averages returns **only** for the **first occurrence** of a state *s* among all episodes. Another one is every-visit, which averages returns for **every occurrence** of a state *s* among all episodes.

Obviously, the first-visit approach has a lot less computation and is therefore more commonly used. I will only cover the first-visit approach in this chapter.

In this section, we experiment with a simple policy where we keep adding new cards until the total value reaches 18 (or 19, or 20 if you like). We perform first-visit MC evaluation on the simple policy as follows:

1. We first need to define a function that simulates a Blackjack episode under the simple policy:

```
>>> def run_episode(env, hold_score):
...     state , _ = env.reset()
...     rewards = []
...     states = [state]
...     while True:
...         action = 1 if state[0] < hold_score else 0
...         state, reward, terminated, truncated, info = env.step(action)
...         is_done = terminated or truncated
...         states.append(state)
...         rewards.append(reward)
...         if is_done:
...             break
...     return states, rewards
```

In each round of an episode, the agent takes a hit if the current score is less than `hold_score` or a stick otherwise.

2. In the MC settings, we need to keep track of states and rewards over all steps. And in first-visit value evaluation, we average returns only for the first occurrence of a state among all episodes. We define a function that evaluates the simple Blackjack policy with first-visit MC:

```
>>> from collections import defaultdict
>>> def mc_prediction_first_visit(env, hold_score, gamma, n_episode):
...     V = defaultdict(float)
...     N = defaultdict(int)
...     for episode in range(n_episode):
...         states_t, rewards_t = run_episode(env, hold_score)
...         return_t = 0
...         G = {}
...         for state_t, reward_t in zip(
...                         states_t[1::-1], rewards_t[::-1]):
...             return_t = gamma * return_t + reward_t
...             G[state_t] = return_t
...         for state, return_t in G.items():
...             if state[0] <= 21:
...                 V[state] += return_t
...                 N[state] += 1
```

```
...        for state in V:
...            V[state] = V[state] / N[state]
...        return V
```

The function performs the following tasks:

- Running n_episode episodes under the simple Blackjack policy with the run_episode function
- For each episode, computing the G returns for the first visit of each state
- For each state, obtaining the value by averaging its first returns from all episodes
- Returning the resulting values

Note that here we ignore states where the player busts, since we know their values are -1.

3. We specify the hold_score as 18 and the discount rate as 1 as a Blackjack episode is short enough, and will simulate 500,000 episodes:

```
>>> hold_score = 18
>>> gamma = 1
>>> n_episode = 500000
```

4. Now we plug in all variables to perform MC first-visit evaluation:

```
>>> value = mc_prediction_first_visit(env, hold_score, gamma, n_episode)
```

We then print the resulting values:

```
>>> print(value)
defaultdict(<class 'float'>, {(13, 10, False): -0.2743235693191795, (5,
10, False): -0.3903118040089087, (19, 7, True): 0.6293800539083558, (17,
7, True): -0.1297709923664122, (18, 7, False): 0.4188926663428849, (13,
7, False): -0.04472843450479233, (19, 10, False): -0.016647081864473168,
(12, 10, False): -0.24741546832491254, (21, 10, True):
......
......
......
2, 2, True): 0.07981220657276995, (5, 5, False): -0.25877192982456143,
(4, 9, False): -0.24497991967871485, (15, 5, True):
-0.011363636363636364, (15, 2, True): -0.08379888268156424, (5, 3,
False): -0.19078947368421054, (4, 3, False): -0.2987012987012987})
```

We have just computed the values for all possible 280 states:

```
>>> print('Number of states:', len(value))
Number of states: 280
```

We have just experienced computing the values of 280 states under a simple policy in the Blackjack environment using the MC method. The transition matrix of the Blackjack environment is not known beforehand. Moreover, obtaining the transition matrix (size 280 * 280) will be extremely costly if we go with the dynamic programming approach. In the MC-based solution, we just need to simulate a bunch of episodes and compute the empirical average values. In a similar manner, we will search for the optimal policy in the next section.

Performing on-policy Monte Carlo control

MC control is used to find the optimal policy for environments with unknown transition matrices. There are two types of MC control, on-policy and off-policy. In the **on-policy approach**, we execute the policy and evaluate and improve it iteratively; whereas in the off-policy approach, we train the optimal policy using data generated by another policy.

In this section, we focus on the on-policy approach. The way it works is very similar to the policy iteration method. It iterates between the following two phases, evaluation and improvement, until convergence:

- In the evaluation phase, instead of evaluating the state value, we evaluate the **action-value**, which is commonly called the **Q-value**. Q-value $Q(s, a)$ is the value of a state-action pair (s, a) when taking the action a in state s under a given policy. The evaluation can be conducted in a first-visit or an every-visit manner.

- In the improvement phase, we update the policy by assigning the optimal action in each state:

$$\pi(s) = argmax_a Q(s, a)$$

Let's now search for the optimal Blackjack policy with on-policy MC control by following the steps below:

1. We start by developing a function that executes an episode by taking the best actions under the given Q-values:

```
>>> def run_episode(env, Q, n_action):
...       state, _ = env.reset()
...       rewards = []
...       actions = []
...       states = []
...       action = torch.randint(0, n_action, [1]).item()
...       while True:
...           actions.append(action)
...           states.append(state)
...           state, reward, terminated, truncated, info = env.step(action)
...           is_done = terminated or truncated
...           rewards.append(reward)
```

```
...            if is_done:
...                break
...            action = torch.argmax(Q[state]).item()
...        return states, actions, rewards
```

This serves as the improvement phase. Specifically, it does the following tasks:

- Initializing an episode
- Taking a random action as an exploring start
- After the first action, taking actions based on the given Q-value table, that is $a = argmax_a Q(s, a)$
- Storing the states, actions, and rewards for all steps in the episode, which will be used for evaluation

2. Next, we develop the on-policy MC control algorithm:

```
>>> def mc_control_on_policy(env, gamma, n_episode):
...     G_sum = defaultdict(float)
...     N = defaultdict(int)
...     Q = defaultdict(lambda: torch.empty(env.action_space.n))
...     for episode in range(n_episode):
...         states_t, actions_t, rewards_t =
                    run_episode(env, Q, env.action_space.n)
...         return_t = 0
...         G = {}
...         for state_t, action_t, reward_t in zip(state_t[::-1],
                                                   actions_t[::-1],
                                                   rewards_t[::-1]):
...             return_t = gamma * return_t + reward_t
...             G[(state_t, action_t)] = return_t
...         for state_action, return_t in G.items():
...             state, action = state_action
...             if state[0] <= 21:
...                 G_sum[state_action] += return_t
...                 N[state_action] += 1
...                 Q[state][action] =
                            G_sum[state_action] / N[state_action]
...     policy = {}
...     for state, actions in Q.items():
...         policy[state] = torch.argmax(actions).item()
...     return Q, policy
```

This function does the following tasks:

- Randomly initializing the Q-values
- Running `n_episode` episodes
- For each episode, performing policy improvement and obtaining the training data; performing first-visit policy evaluation on the resulting states, actions, and rewards; and updating the Q-values
- In the end, finalizing the optimal Q-values and the optimal policy

3. Now that the MC control function is ready, we compute the optimal policy:

```
>>> gamma = 1
>>> n_episode = 500000
>>> optimal_Q, optimal_policy = mc_control_on_policy(env, gamma, n_
episode)
```

Take a look at the optimal policy:

```
>>> print(optimal_policy)
{(16, 8, True): 1, (11, 2, False): 1, (15, 5, True): 1, (14, 9, False):
1, (11, 6, False): 1, (20, 3, False): 0, (9, 6, False):
0, (12, 9, False): 0, (21, 2, True): 0, (16, 10, False): 1, (17, 5,
False): 0, (13, 10, False): 1, (12, 10, False): 1, (14, 10, False): 0,
(10, 2, False): 1, (20, 4, False): 0, (11, 4, False): 1, (16, 9, False):
0, (10, 8,
......

......
1, (18, 6, True): 0, (12, 2, True): 1, (8, 3, False): 1, (13, 3, True):
0, (4, 7, False): 1, (18, 8, True): 0, (6, 5, False): 1, (17, 6, True):
0, (19, 9, True): 0, (4, 4, False): 0, (14, 5, True): 1, (12, 6, True):
0, (4, 9, False): 1, (13, 4, True): 1, (4, 8, False): 1, (14, 3, True):
1, (12, 4, True): 1, (4, 6, False): 0, (12, 5, True): 0, (4, 2, False):
1, (4, 3, False): 1, (5, 4, False): 1, (4, 1, False): 0}
```

You may wonder if this optimal policy is really optimal and better than the previous simple policy (hold at 18 points). Let's simulate 100,000 Blackjack episodes under the optimal policy and the simple policy respectively:

1. We start with the function that simulates an episode under the simple policy:

```
>>> def simulate_hold_episode(env, hold_score):
...     state, _ = env.reset()
...     while True:
...         action = 1 if state[0] < hold_score else 0
...         state, reward, terminated, truncated, info = env.step(action)
```

```
...        is_done = terminated or truncated
...        if is_done:
...            return reward
```

2. Next, we work on the simulation function under the optimal policy:

```
>>> def simulate_episode(env, policy):
...     state, _ = env.reset()
...     while True:
...         action = policy[state]
...         state, reward, terminated, truncated, info = env.step(action)
...         is_done = terminated or truncated
...         if is_done:
...             return reward
```

3. We then run 100,000 episodes for both policies and keep track of their winning times:

```
>>> n_episode = 100000
>>> hold_score = 18
>>> n_win_opt = 0
>>> n_win_hold = 0
>>> for _ in range(n_episode):
...     reward = simulate_episode(env, optimal_policy)
...     if reward == 1:
...         n_win_opt += 1
...     reward = simulate_hold_episode(env, hold_score)
...     if reward == 1:
...         n_win_hold += 1
```

We print out the results as follows:

```
>>> print(f'Winning probability under the simple policy: {n_win_hold/n_
episode}')
Winning probability under the simple policy: 0.40256
>>>print(f'Winning probability under the optimal policy: {n_win_opt/n_
episode}')
Winning probability under the optimal policy: 0.43148
```

Playing under the optimal policy has a 43% chance of winning, while playing under the simple policy has only a 40% chance.

In this section, we solved the Blackjack environment with a model-free algorithm, MC learning. In MC learning, the Q-values are updated until the end of an episode. This could be problematic for long processes. In the next section, we will talk about Q-learning, which updates the Q-values for every step in an episode. You will see how it increases learning efficiency.

Solving the Blackjack problem with the Q-learning algorithm

Q-learning is also a model-free learning algorithm. It updates the Q-function for every step in an episode. We will demonstrate how Q-learning is used to solve the Blackjack environment.

Introducing the Q-learning algorithm

Q-learning is an **off-policy** learning algorithm that optimizes the Q-values based on data generated by a behavior policy. The behavior policy is a greedy policy where it takes actions that achieve the highest returns for given states. The behavior policy generates learning data and the target policy (the policy we attempt to optimize) updates the Q-values based on the following equation:

$$Q(s, a) := Q(s, a) + \alpha(r + \gamma max_{a'} Q(s', a') - Q(s, a))$$

Here, s' is the resulting state after taking action a from state s and r is the associated reward. $max_{a'} Q(s', a')$ means that the behavior policy generates the highest Q-value given state s'. Hyperparameters α and γ are the learning rate and discount factor respectively. The Q-learning equation updates the Q-value (estimated future reward) of a state-action pair based on the current Q-value, the immediate reward, and the potential future rewards the agent can expect by taking the best possible action in the next state.

Learning from the experience generated by another policy enables Q-learning to optimize its Q-values in every single step in an episode. We gain the information from a greedy policy and use this information to update the target values right away.

One more thing to note is that the target policy is epsilon-greedy, meaning it takes a random action with a probability of ϵ (value from 0 to 1) and takes a greedy action with a probability of $1 - \epsilon$. The epsilon-greedy policy combines **exploitation** and **exploration**: it exploits the best action while exploring different actions.

Developing the Q-learning algorithm

Now it is time to develop the Q-learning algorithm to solve the Blackjack environment:

1. We start with defining the epsilon-greedy policy:

```
>>> def epsilon_greedy_policy(n_action, epsilon, state, Q):
...     probs = torch.ones(n_action) * epsilon / n_action
...     best_action = torch.argmax(Q[state]).item()
...     probs[best_action] += 1.0 - epsilon
...     action = torch.multinomial(probs, 1).item()
...     return action
```

Given $|A|$ possible actions, each action is taken with a probability $\epsilon/|A|$, and the action with the highest state-action value is chosen with an additional probability $1 - \epsilon$.

2. We will start with a large exploration factor $\epsilon = 1.0$ and reduce it over time until it reaches 0.1. We define the starting and final ϵ as follows:

```
>>> epsilon = 1.0
>>> final_epsilon = 0.1
```

3. Next, we implement the Q-learning algorithm:

```
>>> def q_learning(env, gamma, n_episode, alpha, epsilon, final_epsilon):
        n_action = env.action_space.n
        Q = defaultdict(lambda: torch.zeros(n_action))
        epsilon_decay = epsilon / (n_episode / 2)
        for episode in range(n_episode):
            state, _ = env.reset()
            is_done = False
            epsilon = max(final_epsilon, epsilon - epsilon_decay)

            while not is_done:
                action = epsilon_greedy_policy(n_action, epsilon, state,
Q)
                next_state, reward, terminated, truncated, info = env.
step(action)
                is_done = terminated or truncated
                delta = reward + gamma * torch.max(
                                      Q[next_state]) - Q[state]
[action]
                Q[state][action] += alpha * delta
                total_reward_episode[episode] += reward
                if is_done:
                    break
                state = next_state
        policy = {}
        for state, actions in Q.items():
            policy[state] = torch.argmax(actions).item()
        return Q, policy
```

We initialize the action-value function Q and calculate the epsilon decay rate for the epsilon-greedy exploration strategy. For each episode, we let the agent take actions following the epsilon-greedy policy, and update the Q function for each step based on the off-policy learning equation. The exploration factor also decreases over time. We run n_episode episodes and finally extract the optimal policy from the learned action-value function Q by selecting the action with the maximum value for each state.

4. We then initiate a variable to store the performance of each of 10,000 episodes, measured by the reward:

```
>>> n_episode = 10000
>>> total_reward_episode = [0] * n_episode
```

5. Finally, we perform Q-learning to obtain the optimal policy for the Blackjack problem with the following hyperparameters:

```
>>> gamma = 1
>>> alpha = 0.003
>>> optimal_Q, optimal_policy = q_learning(env, gamma, n_episode, alpha,
                                           epsilon, final_epsilon)
```

Here, discount rate $\gamma = 1$ and learning rate $\alpha = 0.003$ for more exploration.

6. After 10,000 episodes of learning, we plot the rolling average rewards over the episodes as follows:

```
>>> rolling_avg_reward = [total_reward_episode[0]]
>>> for i, reward in enumerate(total_reward_episode[1:], 1):
        rolling_avg_reward.append((rolling_avg_reward[-1]*i + reward)/
(i+1))
>>> plt.plot(rolling_avg_reward)
>>> plt.title('Average reward over time')
>>> plt.xlabel('Episode')
>>> plt.ylabel('Average reward')
>>> plt.ylim([-1, 1])
>>> plt.show()
```

Refer to the following screenshot for the end result:

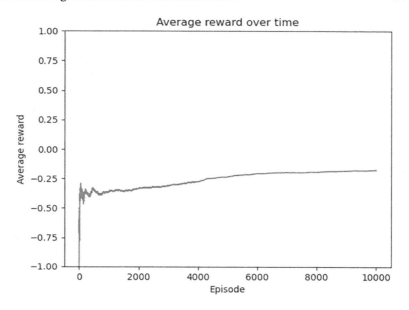

Figure 15.5: Average reward over episodes

The average reward steadily rises throughout the training process, eventually converging. This indicates that the model becomes proficient in solving the problem after training.

7. Finally, we simulate 100,000 episodes for the optimal policy we obtained using Q-learning and calculate the winning chance:

```
>>> n_episode = 100000
>>> n_win_opt = 0
>>> for _ in range(n_episode):
...     reward = simulate_episode(env, optimal_policy)
...     if reward == 1:
...         n_win_opt += 1

>>>print(f'Winning probability under the optimal policy: {n_win_opt/n_
episode}')
Winning probability under the optimal policy: 0.42398
```

Playing under the optimal policy has a 42% chance of winning.

In this section, we solved the Blackjack problem with off-policy Q-learning. The algorithm optimizes the Q-values in every single step by learning from the experience generated by a greedy policy.

Summary

This chapter commenced with configuring the working environment, followed by an examination of the core concepts of reinforcement learning, accompanied by practical examples. Subsequently, we delved into the FrozenLake environment, employing dynamic programming techniques such as value iteration and policy iteration to tackle it effectively. Monte Carlo learning was introduced for value estimation and control in the Blackjack environment. Finally, we implemented the Q-learning algorithm to address the same problem, providing a comprehensive overview of reinforcement learning techniques.

Exercises

1. Can you try to solve the 8 * 8 FrozenLake environment with the value iteration or policy iteration algorithm?
2. Can you implement the every-visit MC policy evaluation algorithm?

Join our book's Discord space

Join our community's Discord space for discussions with the authors and other readers:

`https://packt.link/yuxi`

packt.com

Subscribe to our online digital library for full access to over 7,000 books and videos, as well as industry leading tools to help you plan your personal development and advance your career. For more information, please visit our website.

Why subscribe?

- Spend less time learning and more time coding with practical eBooks and Videos from over 4,000 industry professionals
- Improve your learning with Skill Plans built especially for you
- Get a free eBook or video every month
- Fully searchable for easy access to vital information
- Copy and paste, print, and bookmark content

At www.packt.com, you can also read a collection of free technical articles, sign up for a range of free newsletters, and receive exclusive discounts and offers on Packt books and eBooks.

Other Books You May Enjoy

If you enjoyed this book, you may be interested in these other books by Packt:

Mastering PyTorch – Second Edition

Ashish Ranjan Jha

ISBN: 978-1-80107-430-8

- Implement text, vision, and music generation models using PyTorch
- Build a deep Q-network (DQN) model in PyTorch
- Deploy PyTorch models on mobile devices (Android and iOS)
- Become well versed in rapid prototyping using PyTorch with fastai
- Perform neural architecture search effectively using AutoML
- Easily interpret machine learning models using Captum
- Design ResNets, LSTMs, and graph neural networks (GNNs)
- Create language and vision transformer models using Hugging Face

Building LLM Powered Applications

Valentina Alto

ISBN: 978-1-83546-231-7

- Explore the core components of LLM architecture, including encoder-decoder blocks and embeddings
- Understand the unique features of LLMs like GPT-3.5/4, Llama 2, and Falcon LLM
- Use AI orchestrators like LangChain, with Streamlit for the frontend
- Get familiar with LLM components such as memory, prompts, and tools
- Learn how to use non-parametric knowledge and vector databases
- Understand the implications of LFMs for AI research and industry applications
- Customize your LLMs with fine tuning
- Learn about the ethical implications of LLM-powered applications

Packt is searching for authors like you

If you're interested in becoming an author for Packt, please visit authors.packtpub.com and apply today. We have worked with thousands of developers and tech professionals, just like you, to help them share their insight with the global tech community. You can make a general application, apply for a specific hot topic that we are recruiting an author for, or submit your own idea.

Share your thoughts

Now you've finished *Python Machine Learning By Example - Fourth Edition*, we'd love to hear your thoughts! Scan the QR code below to go straight to the Amazon review page for this book and share your feedback or leave a review on the site that you purchased it from.

https://packt.link/r/1835085628

Your review is important to us and the tech community and will help us make sure we're delivering excellent quality content.

Index

www.ingramcontent.com/pod-product-compliance
Lightning Source LLC
Chambersburg PA
CBHW060639060326
40690CB00020B/4454